AN INTRODUCTION TO SOLUTE TRANSPORT IN HETEROGENEOUS GEOLOGIC MEDIA

Over the past several decades, analyses of solute migration in aquifers have widely adopted the classical advection–dispersion equation. However, misunderstandings over advection–dispersion concepts, their relationship with the scales of heterogeneity, our observation and interest, and their ensemble mean nature have created furious debates about the concepts' validity. This book provides a unified and comprehensive overview and lucid explanations of the stochastic nature of solute transport processes at different scales. It also presents tools for analyzing solute transport and its uncertainty to meet our needs at different scales. Easy-to-understand physical explanations without complex mathematics make this book an invaluable resource for students, researchers, and professionals performing groundwater quality evaluations, management, and remediation.

TIAN-CHYI "JIM" YEH is a professor in the Department of Hydrology and Atmospheric Sciences at the University of Arizona. He is an internationally renowned leader in stochastic/numerical analysis and laboratory/field investigations of flow and solute transport in variably saturated heterogeneous geologic media. He was the lead author of *Flow through Heterogeneous Geologic Media* (Cambridge University Press, 2015) and pioneered the new generation of aquifer characterization technology: hydraulic tomography.

YANHUI DONG is an associate professor in the Institute of Geology and Geophysics, Chinese Academy of Sciences. He specializes in fluid flow, reactive transport, and porosity/permeability evolution in porous and fractured geological media with application to the management of groundwater resources, the disposal of high-level radioactive wastes, and the environmental impact assessment of hydrocarbon extraction activities.

SHUJUN YE is a professor in the School of Earth Sciences and Engineering at Nanjing University. She has more than 20 years of teaching and research experience in the numerical simulation of groundwater flow, multiphase flow, mass transport, land subsidence, and Earth fissures. She is the vice-chair of the UNESCO Land Subsidence International Initiative and a member of the Groundwater Technical Committee of the American Geophysical Union. She is the coauthor of *Hydrogeology: Strategy of the Disciplines Development in China* (2021).

AN INTRODUCTION TO SOLUTE TRANSPORT IN HETEROGENEOUS GEOLOGIC MEDIA

TIAN-CHYI "JIM" YEH
University of Arizona

YANHUI DONG
Chinese Academy of Sciences

SHUJUN YE
Nanjing University

Shaftesbury Road, Cambridge CB2 8EA, United Kingdom

One Liberty Plaza, 20th Floor, New York, NY 10006, USA

477 Williamstown Road, Port Melbourne, VIC 3207, Australia

314–321, 3rd Floor, Plot 3, Splendor Forum, Jasola District Centre, New Delhi – 110025, India

103 Penang Road, #05–06/07, Visioncrest Commercial, Singapore 238467

Cambridge University Press is part of Cambridge University Press & Assessment,
a department of the University of Cambridge.

We share the University's mission to contribute to society through the pursuit of
education, learning and research at the highest international levels of excellence.

www.cambridge.org
Information on this title: www.cambridge.org/9781316511183

DOI: 10.1017/9781009049511

First published 2023

A catalogue record for this publication is available from the British Library.

Library of Congress Cataloging-in-Publication Data
Names: Yeh, T.-C. (Tian-Chyi), author. | Dong, Yanhui, author. | Ye, Shujun, author.
Title: An introduction to solute transport in heterogeneous geologic media / Tian-Chyi Jim Yeh,
Yanhui Dong, Shujun Ye.
Description: Cambridge ; New York, NY : Cambridge University Press, 2023. | Includes bibliographical
references and index.
Identifiers: LCCN 2022029342 (print) | LCCN 2022029343 (ebook) | ISBN 9781316511183 (hardback) |
ISBN 9781009049511 (epub)
Subjects: LCSH: Water quality–Mathematical models. | Water–Analysis–Mathematical models. |
Soils–Solute movement–Mathematical models. | Transport theory–Mathematical models. |
Diffusion in hydrology. | Reaction-diffusion equations–Numerical solutions. | BISAC: SCIENCE /
Earth Sciences / Hydrology
Classification: LCC TD380 .Y44 2022 (print) | LCC TD380 (ebook) | DDC 628.1/61–dc23/eng/20220811
LC record available at https://lccn.loc.gov/2022029342
LC ebook record available at https://lccn.loc.gov/2022029343

ISBN 978-1-316-51118-3 Hardback

Contents

Preface

Environmental fluids migrating at multiple scales and affected by heterogeneity and factors of various scales is a well-known fact. However, there is widespread and profound confusion in the hydrology literature, textbooks, and community about advection–dispersion concepts; the relationship between heterogeneity and dispersion; the stochastic nature of velocity at different scales; and model, observation, and heterogeneity scale issues. Countless publications on anomalous dispersion phenomena and inventions of non-Fickian models, and textbooks' omission of the fundamental scale issues testify to this perplexity. Expressly, the hydrology community poorly understands the ensemble mean nature of Fick's law, Darcy's law, the advection–dispersion equation (ADE), and their validity relative to the observation, model and heterogeneity scales. Inadequate and insufficient explanations of theories and mathematical analyses in the literature and textbooks are likely to blame. The hydrology community and education deserve a unified and comprehensible articulation of these theories and concepts.

Scales of heterogeneity, models, observations, and our interests dictate the analysis of solute transport in geologic media. Textbooks and reference books have overlooked this fact and omitted the stochastic concepts (such as the volume average, ensemble mean, and ergodicity) embedded in all solute transport theories. This book first introduces stochastics, volume average, ensemble average, and well-mixed concepts. It then elucidates the solute transport theories and principles from molecular to fluid-dynamic scales and soil column to field scales based on the stochastic concepts. Below are summaries of each chapter of this book.

Chapter 1, "**Fundamental Concepts**," provides a review of basic differential equations and statistics. It then introduces the spatial and temporal stochastic process, volume average, and ensemble-average concepts behind the continuum, concentration, mixing concepts, and scale issues in our observations and theories.

Chapter 2, "**Well-Mixed Models for Surface Water Quality Analysis**," presents classical lumped or well-mixed water quality models in surface-water reservoirs to elucidate concepts in Chapter 1, inherent in the fundamental mass balance principle in all science and engineering fields. The mass and energy balance equations and their analytical solutions for various inputs are presented, refreshing readers' calculous skills and demonstrating the linkage between math and physics. We explain advection and ensemble mixing concepts in the later chapters using the solutions. More importantly, practical applications of the models as a first-cut analysis to environmental problems are visited.

Chapter 3, **"Well-Mixed Models for Subsurface Water Quality Analysis**," examines the lumped or well-mixed groundwater flow and water quality models. It starts with the water balance equation, demonstrates its application to estimate groundwater recharge, and then formulates the chemical mass balance for the groundwater system. Applications of the models to the management of strategy for road salt deicing operation follow. The chemical balance equations for the equilibrium and nonequilibrium reactive chemicals are developed afterward. We subsequently introduce the first-order analysis for sensitivity and uncertainty evaluations and their applications.

Chapter 4. Molecular Diffusion. Moving a step beyond the lumped models, Chapter 4 discusses diffusion theories of microscopic-scale motions of molecules, which are often beyond the scales of observation and our interests but have profound impacts on our understanding of large-scale solute transport phenomena. The stochastic analysis follows to examine the macroscopic-scale behavior of random motions of molecules and explains the ensemble mean nature of Fick's law and the diffusion equation. Analytical solutions to the diffusion equation are presented, and formulations of the advection–diffusion equations then come after. These equations quantify the combined effects of macroscopic-scale (i.e., fluid-dynamics-scale) movement of fluids and microscopic-scale random motions of fluid molecules on the solute behavior at the scale of our observations.

Chapter 5. Numerical Methods for Advection–Diffusion Equations. This chapter presents finite differences and methods of characteristics numerical solutions to one and two-dimensional ADE. Numerical stability and dispersion are explored, and finite element solution to ADE is then presented in this chapter.

Chapter 6. Shear Flow Dispersion. This chapter brings in the mechanic dispersion concept from the analysis of solute migration in a pipe under laminar shear flow situations. Taylor's shear flow dispersion analysis explains the rationales and the validity of Fick's law for fluid-dynamic scale velocity variations over the cross-section of the pipe due to friction and viscous effects. From the analysis, we investigate the roles of the molecule's random motion and the fluid-dynamic scale velocity variations in the ensemble-mean nature of the advection-dispersion concept. In the end, this chapter elucidates the scale issues associated with molecular diffusion and mechanic dispersion concepts.

Chapter 7. Solute Transport in Soil Columns. Upscaling the shear flow dispersion concept to the solute transport in porous media in soil columns is examined in Chapter 7. It illustrates the inherent averaging feature of Darcy's law for depicting water and solute migration through intricate and tortuous flow paths in soil columns. The need for Darcy-scale advection/dispersion conceptualization for solute transport in porous media is underscored, which overcomes the deficiency of the average aspect of Darcy-scale velocity. The analysis of the Peclet number follows behind to relate the dispersion coefficient to dispersivity and Darcy-scale velocity to avoid the scenario dependence of the dispersion coefficient. The linkage between the dispersivity of the soil column experiment and the grain size of the soil is articulated. Also, we explicate the Fickian and non-Fickian models for solution transport in porous media in a stochastic context. Chemical reaction models, including first-order decay, equilibrium, and nonequilibrium reactions, are integrated into the advection-dispersion equation to analyze reactive solute transport. It explores the effects of chemical reactions on solute transport.

Chapter 8. Parameter Estimation. Practical and intuitive graphical approaches and spatial and temporal moment methods with examples are presented in Chapter 7 to estimate the velocity, dispersivity, and chemical reaction parameters from soil-column tracer experiments.

Chapter 9. Solute Transport in Field-Scale Aquifers. This chapter deals with solute transport in field-scale aquifers. It first introduces and explains the need for the stochastic representation of hydrologic property heterogeneity in geologic media. Advection–dispersion models coupling with flow models of the equivalent homogeneous, geologic layer, and highly parameterized media at different scales of our interests are briefed. Also examined in this chapter are the macrodispersion concept developed over the past decades and its limitations.

Chapter 10. Field-Scale Solute Transport Experiments Under Natural Gradients. Reviews of several field experiments are given under natural-gradient conditions in aquifers with different types of heterogeneity. It discusses the pros and cons of microdispersion, geologic zonation, and dual-domain approaches.

Chapter 11. Field-Scale Tracer Experiments in Aquifers Under Forced Gradients. Forced-gradient tracer experiments, distinct from the natural gradient experiments, are reviewed. They demonstrate the role of large-scale velocity variation on the breakthroughs at the pumping well. Besides, the results of the experiments stress the importance of local-scale heterogeneity in the snapshots of the tracer.

Chapter 12. High-Resolution Characterization. At last, hydraulic and electrical resistivity tomography are introduced that enhance the characterization of aquifer heterogeneity and monitoring tracer plumes. Given examples stress the relevance of mapping domain heterogeneity and the effects of boundary conditions, controlling large-scale velocity fields on solute migration in aquifers. These examples further highlight the need for technologies rather than new upscaling theories. Specifically, this chapter promotes using high-resolution technology to characterize and predict Darcy-scale advective velocity at field sites as thoroughly as possible. The classical dispersion concept from lab-scale experiments becomes appropriate to overcome the effects of unresolved velocity variations.

This textbook is written for senior undergraduate and graduate students in the sciences and engineering fields. It also intends to serve as a reference volume for scientists, engineers, practitioners, and decision-makers to enhance their understanding of solute transport processes, scale issues, and uncertainty in groundwater quality evaluations, protections, and remediation.

Finally, the first author is grateful to his wife (Mei-Lin Yeh) for continuous support and Professors Lynn W. Gelhar and Allan L. Gutjahr for opening the stochastic realm's door for him. Thorough reviews of various chapters of the manuscript and assistance by Professors Ye Zhang (University of Wyoming, USA), Hongbin Zhan (Texas A&M University, USA), Walter Illman (University of Waterloo, Canada), Kuo-Chin Hsu (National Cheng Kung University, Taiwan, China), Xiuyu Liang (Southern University of Science and Technology, Mainland China), Bo Guo (The University of Arizona), Andrew Binley (University of Lancaster, UK), UK, Yu-li Wang (National Taiwan University, Taiwan, China) and Drs. Xiaoru Su (China University of Geosciences, Beijing, China) and Joseph Doetsch (Lufthansa Industry Solutions, Germany) are acknowledged.

1

Fundamental Concepts

1.1 Introduction

This chapter first reviews the linear first-order nonhomogeneous ordinary differential equation. An introduction to statistics and stochastic processes follows. Afterward, this chapter explains the stochastic fluid continuum concept and associated control volume, spatial- and ensemble-representative control volume concepts. It then uses the well-known solute concentration definition as an example to elucidate the volume- and spatial-, ensemble-average, and ergodicity concepts. This chapter provides the basic math and statistics knowledge necessary to comprehend the themes of this book. Besides, this chapter's homework exercises demonstrate the power of the widely available Microsoft Excel for scientific investigations.

1.2 Linear First-Order Nonhomogeneous Ordinary Differential Equation

Since many chapters and homework assignments use the first-order ordinary differential equation and its solution technique, we will briefly review a simple solution below. A revisit to calculus or an introduction to the differential equation would be helpful.

Consider an ordinary differential equation that has a form:

$$\frac{dC}{dt} + f(t)C = r(t) \tag{1.2.1}$$

where $r(t) \neq 0$. It has the general solution:

$$C(t) = e^{-h}\left[\int e^{h} r(t)dt + A\right] \tag{1.2.2}$$

where $h = \int f(t)dt$. Since t is time, the differential equation is called an initial value problem (also called the Cauchy problem). If t is a spatial coordinate, the mathematical problem is called a boundary value problem.

Consider the following ordinary differential equation,

$$\frac{dC}{dt} - C = e^{2t}. \tag{1.2.3}$$

Comparing Eq. (1.2.3) with Eq. (1.2.1), we see $f(t) = -1$, $r(t) = e^{2t}$, and $h = \int(-1)dt = -t$ in Eq. (1.2.2). Substituting these relationships into Eq. (1.2.2), we have the general solution for the ordinary differential equation,

$$\begin{aligned} C(t) &= e^t\left[\int e^{-t}e^{2t}dt + A\right] \\ &= e^t\left[\int e^t dt + A\right] \\ &= e^t[e^t + A] = e^{2t} + Ae^t \end{aligned} \tag{1.2.4}$$

in which A can be determined if an initial condition is specified. Otherwise, the problem is not well defined, and many possible solutions exist. As an example, given that C at $t = 0$ is C_0, we find that $A = C_0 - 1$. As a result, a particular solution (unique solution) for this initial condition is

$$C(t) = e^{2t} + (C_0 - 1)e^t. \tag{1.2.5}$$

1.2.1 Homework

1. Find the general solutions of the following differential equations

$$\text{a) } \frac{dy}{dt} - y = 3, \text{ b) } \frac{dy}{dt} + 2y = 6e^t, \text{ c) } \frac{dy}{dt} + y = \sin(t). \tag{1.2.6}$$

2. Derive the analytical solution to the initial value problem

$$\frac{dy}{dt} + y = (t+1)^2, y(0) = 0. \tag{1.2.7}$$

3. Evaluate the analytical solution of problem 2 and plot y as a function of t.

1.3 Random Variable

A random, aleatory, or stochastic variable is a quantity whose outcome is unpredictable or unknown, although anything can be viewed as random, as explained later. A mathematical treatment of random variables is the probability, an abstract statement about the likelihood of something happening or being the case.

Suppose we take a bottle of lake water to determine the chloride concentration (C_0). While the concentration may have a range of possible values (e.g., 0–1000 ppm), within this range, the exact value is unknown until measured. The concentration C_0 thus is conceptualized as a random variable to express our uncertainty about the concentration.

Similarly, repeated measurements of the concentration may yield a distribution of values due to the experimental error. If the measurement is precise, all the values are very close. If not, the values may be widely spread. Such a spread distribution of replicates often assesses the measurement method's precision or accuracy (an expression of our uncertainty about the measurement). In this case, the C_0 is a random variable, expressing our uncertainty about the instrument's measurement.

The above examples immediately lead to the notion that if the concentration is measured precisely (error-free), the concentration is a deterministic variable, not a random variable. However, this notion may not be necessary. Any precisely known, determined, or measured event is always an outcome of many possible events that could have occurred. For example, the head resulting from flipping a coin is undoubtedly a deterministic outcome since it has happened. On the other hand, it is the outcome of a random variable consisting of two possibilities (head or tail) – a flipping coin experiment. In this sense, everything that happened is the outcome of a random event. This thinking leads to probabilistic science (i.e., statistics).

In statistics, the set of all possible values of C_0 of the water sample is a **population.** This population could be a hypothetical and potentially infinite group of C_0 values conceived as a generalization from our knowledge. The likelihood of taking a value from the population is then viewed as chance or probability. Therefore, C_0 is a random variable with a **probability distribution** that describes all the possible values and likelihoods being sampled.

The random variable can be discrete or continuous. A discrete random variable has a specified finite or countable list of values (having a countable range). Typical examples of discrete random variables are the outcome of flipping a coin, throwing dice, or drawing lottery numbers. On the other hand, a continuous random variable has an uncountable list of values. For instance, the porosity of a porous medium, chloride concentration in the water, or any variables in science is a continuous random variable because it has a continuous spectrum of values rather than a countable list of values. Of course, we could approximate the continuous random variable using the discrete random variable by grouping them into a countable list.

In the discrete case, we can determine the probability of a random variable X equal to a given value x (i.e., $P(X = x)$) from all possible values of X. The set of all the possible values is the probability mass function (PMF). On the other hand, in the continuous random variable case, the probability that X is any particular value x is 0 since the random variable value varies indefinitely. In other words, finding

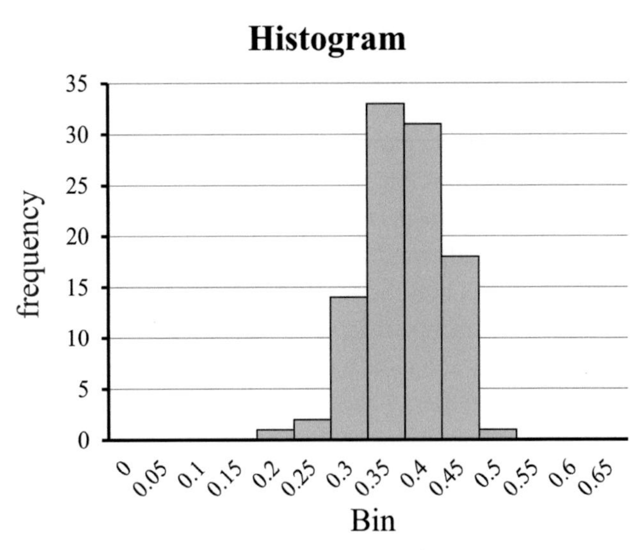

Figure 1.1 The histogram for the porosity value of a sandstone.

P(X = x) for a continuous random variable X is meaningless. Instead, we find the probability that X falls in some interval (a, b) – we find P(a < X < b) by using a probability density function (PDF).

The porosity of a sandstone formation is an excellent example to elucidate the PDF. Suppose the sandstone formation's porosity is reported to be 0.3. Intuition tells us that every sample taken from the formation will not be precisely 0.3. For example, randomly selecting a sandstone sample may sometimes find the porosity is 0.27 and others are not. We then ask the probability of a randomly selected sandstone having a porosity value between 0.25 and 0.35. That is to say, if we let X denote a randomly selected sandstone's porosity, we like to determine P(0.25 < X < 0.35).

Consider that we randomly select 100 core samples from the sandstone, determine their porosity, and create a histogram of the resulting porosity values (Fig. 1.1). This **histogram** describes the distribution of the number of samples in each range of porosity value (i.e., several bins or classes: 0–0.1, 0.1–0.2, and so on). As indicated in Fig. 1.1, most samples have a value close to 0.3; some are a bit more and some a bit less. This histogram illustrates that arbitrarily taking a core sample will likely have a sample with a porosity of 0.3. Alternatively, if we repeatedly take a sample from the 100 cores, we will get a core with a porosity of 0.3 most of the time. However, a probability of any particular value (e.g., 60%) does not guarantee that we will get 60 core samples with a porosity value of 0.3 after picking a sample over 100 times. This probability is merely an abstract concept that quantifies the chance that some event might happen.

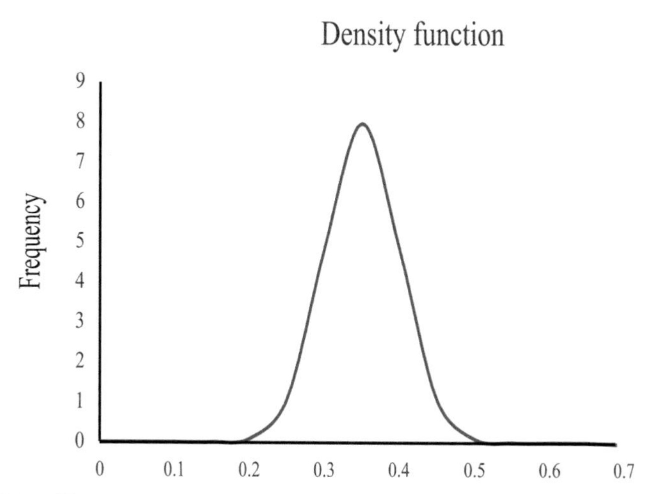

Figure 1.2 An illustration of a density function. The vertical axis is the density, and the horizontal axis is the porosity value.

Now, we decrease the interval to an infinitesimally small point. X's probability distribution becomes a curve like in Fig. 1.2, denoted as $f(x)$ and called a probability density function. It shows that the porosity value varies continuously from 0 to 0.7. Notice that the density function could have a value greater than 1, as the histogram could have.

A density histogram is defined as the area of each rectangle equals the corresponding class's relative frequency, and the entire histogram's area equals 1. That suggests that finding the probability that a continuous random variable X falls in some interval of values is finding the area under the curve $f(x)$ bounded by the endpoints of the interval. In this example, the probability of a randomly selected core sample having a porosity value between 0.20 and 0.30 is the area between the two values. A formal definition of the probability density function of a random variable is given below.

Definition. The **probability density function** (PDF) of a continuous random variable X is an integrable function $f(x)$ satisfying the following:

(1) $f(x) > 0$, for all x.

(2) The area under the curve $f(x)$ is 1, that is:

$$\int_{-\infty}^{\infty} f(x)dx = 1 \tag{1.3.1}$$

(3) The probability that x in a bin or class, say, $0.3 < x < 0.4$, is given by the integral of $f(x)$ over that interval:

$$P(0.3 < x < 0.4) = \int_{0.3}^{0.4} f(x)dx. \tag{1.3.2}$$

Again, notice that a probability density function, the number of occurrences of a value x and can be greater than 1, is not a probability.

A widely used PDF is

$$f(x) = \frac{1}{\sigma_x \sqrt{2\pi}} \exp\left(-\frac{(x-\mu_x)^2}{2\sigma_x^2}\right). \tag{1.3.3}$$

This PDF describes a normally distributed random variable X. In the equation, x is the random variable value, μ_x is the mean (i.e., the most likely) value of all x values, and σ_x^2 is the variance, indicating the likely deviation of x from the mean. Because the random variable has a normal (Gaussian) distribution, it has symmetric and bell-shaped distribution (Fig. 1.3). Thus, the mean and variance fully characterize a random variable with a normal distribution, as indicated by the equation. For this reason, most statistical analyses assume that random variables have a normal distribution, attributing to the mathematical simplicity of Eq. (1.3.3). As PDF is an abstract concept, we never have an infinite number of samples to disapprove of it.

Frequency distribution: A frequency distribution is a table or a graph (e.g., a histogram) that displays a summarized grouping of data divided into mutually exclusive classes and the number of occurrences in a class. It becomes a relative frequency distribution if the total number of samples in all classes normalizes the frequency.

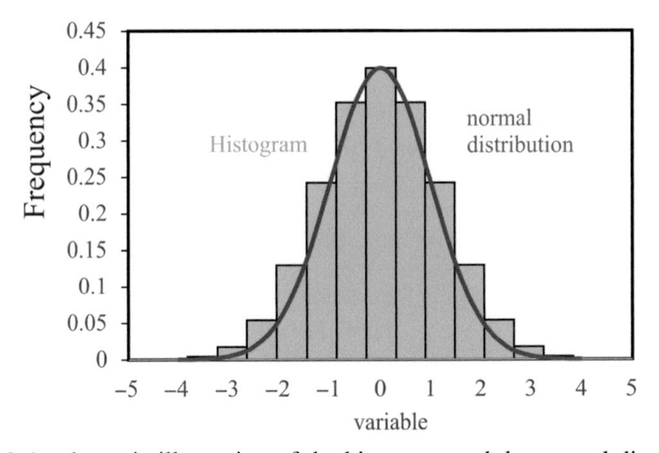

Figure 1.3 A schematic illustration of the histogram and the normal distribution.

Probability distribution: A probability distribution is a mathematical function that produces probabilities of different possible outcomes for an experiment and is an alias for relative frequency distribution.

Cumulative distribution function (CDF): A CDF is a mathematical functional form describing a probability distribution. For example, the cumulative distribution function of a real-valued random variable X, or just distribution function of X, evaluated at x, is the probability that X will take a value less than or equal to x.

Distribution Parameters: The following definitions are theoretical, based on an infinitely large population.

$$\text{Mean } E[\mathrm{X}] = \mu_x = \int_{-\infty}^{\infty} x f_{\mathrm{X}}(x) dx \tag{1.3.4}$$

where X is the random variable name and x represents the values of the random variable. $f_{\mathrm{X}}(x)$ denotes the probability density function (PDF) of the random variable, X. E represents the expected value (an average over the infinitely large population). The mean represents the most likely value of the random variable, for example, the average value of the porosity of a sandstone formation.

The variance, a measure of the spread of the distribution, is defined as

$$\text{Variance } \sigma_x^2 = E\left[(\mathrm{X} - \mu_x)^2\right] = \int_{-\infty}^{\infty} (x - \mu_x)^2 f_{\mathrm{X}}(x) dx \tag{1.3.5}$$

The standard deviation is the square root of the variance, the most likely deviation of the random variable value from its mean value. For example, it is often used to construct the upper or lower bound of the sandstone's porosity value.

While X is used to represent the random variable, and x is the value of the random variable, we use x for both most of the time.

Sample Statistics

Only a finite population and discrete random variables are available in practice. The statistics calculated from these samples are called sample statistics.

Suppose we have samples from I = 1 to N, where N is the total number of samples. The sample mean is calculated as

$$\bar{x} = \frac{1}{N} \sum_{i=1}^{N} x_i. \tag{1.3.6}$$

The sample variance is given as

$$S^2 = \frac{1}{N-1}\sum_{i=1}^{N}(x_i - \overline{x})^2. \tag{1.3.7}$$

Eq. (1.3.7) uses $N - 1$ to make the estimate unbiased. If N is large, this correction is unnecessary. Besides, how large N is sufficiently large is undetermined since we do not know the population's size. Likewise, the normal PDF is widely used in many fields because of its mathematical simplicity and insufficient datasets to verify or disapprove the distribution. The following central limit theorem further supports this approach.

The Central Limit Theorem (CLT)

As a statistical premise, CLT states that when independent random variables are lumped together after being properly normalized, their distribution tends toward a normal distribution even if the original variables are not normally distributed. The central limit theorem has several variants. In its typical form, the random variables must be identically distributed. In variants, the convergence of the mean to the normal distribution also occurs for non-identical distributions or for non-independent observations. This theorem is a key in probability theory because it implies that probabilistic and statistical methods that work for normal distributions can apply to many problems involving other types of distributions.

1.4 Stochastic Process or Random Field

Instead of considering a variable, we often simultaneously consider many variables in time or space, leading to the adoption of the stochastic process or field concept. Formally, a stochastic process is a collection of an infinite number of random variables in space or time. Examples are the spatial distribution of porosity or hydraulic conductivity in a geologic formation or the spatial and temporal distribution of a lake's chemical concentration.

Consider a temporally varying concentration, $C(x,t)$, at a point, x, in space at the time t_1. If the concentration is not measured, and we guess it, we inevitably consider the concentration a random variable characterized by a probability distribution (PDF). If we guess the C at the time t_2, we again consider it a random variable at t_2. Guessing the concentration at that location for a period is treating the concentration as a stochastic process over time.

As articulated in the flipping coin experiment, we can conceive any known or deterministic event as a random event. Accordingly, a recorded concentration history at a location that has occurred (or deterministic in conventional thinking) in

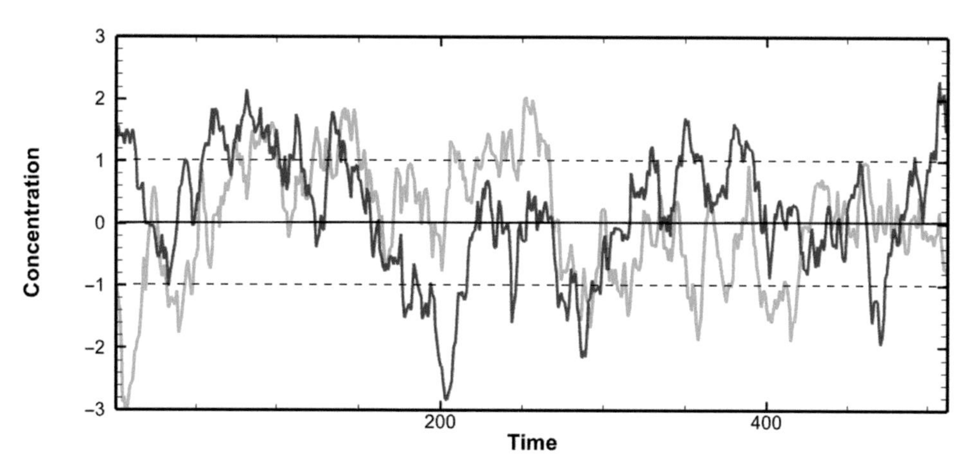

Figure 1.4 Illustration of two possible realizations of a concentration-time stochastic process, assuming jointly normal distribution with mean zero and variance one. The solid black line is the mean, and the two dashed lines are standard deviations.

time and space is a stochastic process or stochastic field. While this concept may be difficult to accept based on the conventional sense, it should become apparent as we apply it to many situations discussed in the book. The fact is that the stochastic process concept is abstract and exists in our imaginary space.

An observed record of concentration history at a location can be considered a stochastic process, consisting of infinite possible concentration-time series over the same period in our imaginary domain. Each possible concentration-time series is a **realization** (Fig. 1.4). The **ensemble** is the collection of all possible realizations (analogous to the population for a single random variable). The observed concentration-time series is merely one realization of the ensemble.

Determining the possibility (likelihood) of the occurrence of this observed concentration series demands the concept of joint probability density function (or a joint probability distribution, JPD). This JPD is different from the PDF of a random variable since the JPD considers the simultaneous occurrence of some concentration values at different times.

To explain the meaning of the JPD, we consider two random variables (e.g., porosity n and hydraulic conductivity, K) and assume that the logarithm of each of them ($\log K$ or $\log n$) has a normal distribution. Fig. 1.5 shows their JPD $f_{X,Y}(x, y)$. Here X and x denote the variable $\log K$, and its value, respectively. The Y is $\log n$, and y is its value, respectively. The JPD determines the probability of the simultaneous occurrence of porosity and hydraulic conductivity over a range of values. This JPD is different from the probability distribution of each variable individually. The individual probability distribution of X or Y is the marginal probability distribution of the JPD. The marginal probability distribution of X or Y can be determined from their joint probability distribution:

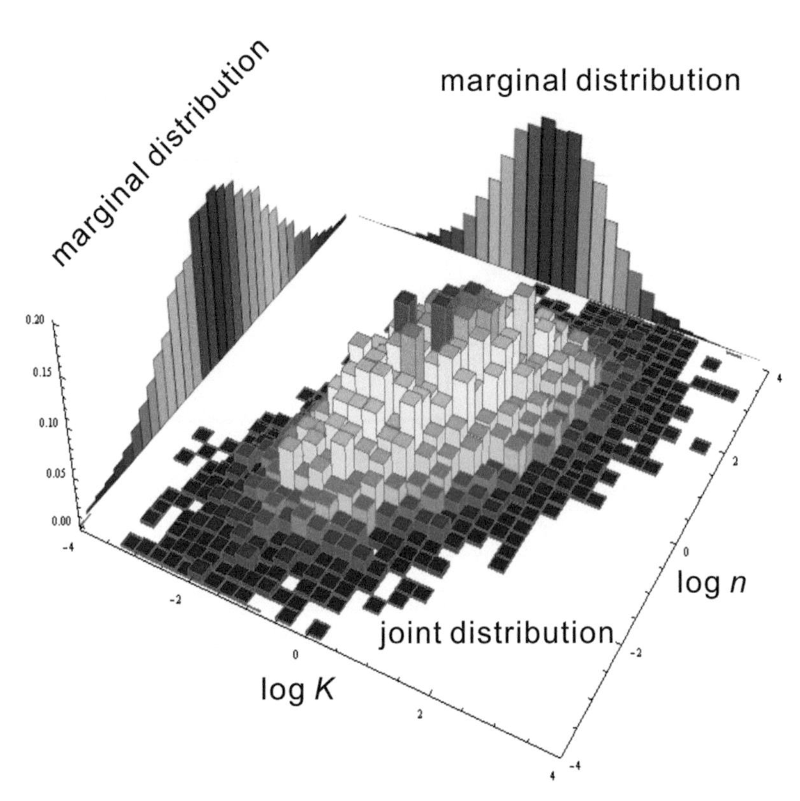

Figure 1.5 A joint probability density function of log K and log n. Modified from stackoverflow.com

$$f_X(x) = \int_{-\infty}^{\infty} f_{X,Y}(x, y)dy, \quad f_Y(y) = \int_{-\infty}^{\infty} f_{X,Y}(x, y)dx \tag{1.4.1}$$

The marginal probability is an orthogonal projection of the JPD to the X or Y-axis.

1.4.1 Joint Probability Density Function

A joint probability density function (JPDF) for the stochastic concentration process, $C_1, C_2, C_3 \ldots$, thus is represented as $f_{C_1C_2C_3} \cdots (C_1, C_2, C_3 \ldots)$, which satisfies the following properties:

(1) $$f_{C_1C_2C_3} \cdots (C_1, C_2, C_3 \ldots) \geq 0 \text{ for all } C_1, C_2, C_3 \ldots \tag{1.4.2}$$

This property states that the JPD must be greater than zero for all concentrations at any time.

(2)
$$\int_{-\infty}^{\infty}\int_{-\infty}^{\infty}\int_{-\infty}^{\infty}\cdots\int_{-\infty}^{\infty} f(C_1,C_2,C_3\ldots)dC_1dC_2dC_3\ldots = 1 \quad (1.4.3)$$

The sum of the JPD for all possible concentrations must equal 1.

(3)
$$P(A<C<B,\, D<C_2<E,\, F<C_3<G\ldots) = \int_{A}^{B}\int_{D}^{E}\int_{F}^{G}\cdots\int_{\ldots}^{\ldots} f(C_1,C_2,C_3\ldots)dC_1dC_2dC_3\ldots \quad (1.4.4)$$

The probability of occurrence of the concentration of certain intervals is the sum of all JPD over the given intervals.

1.4.2 Ensemble Statistics

Ensemble statistics assume that the ensemble (i.e., all possible realizations) is known. If the JPD of the process is Gaussian, the first and second moments are sufficient to characterize the stochastic process. Otherwise, this is the best we can do.

Mean (1st moment) is defined as

$$E[C] = \mu = \int_{-\infty}^{\infty}\int_{-\infty}^{\infty}\int_{-\infty}^{\infty}\cdots\int_{-\infty}^{\infty} C_1,C_2,C_3\ldots f(C_1,C_2,C_3\ldots)dC_1dC_2dC_3\ldots \quad (1.4.5)$$

The expectation value, E, is carried out over the entire ensemble. The mean, $E[C] = \mu$, thus is the average (the most likely) concentration value over the ensemble.

Covariance or Covariance function (2nd moment) is defined as

$$R(\xi) = E[(C(t)-\mu)(C(t+\xi)-\mu)] = \int_{-\infty}^{\infty}\int_{-\infty}^{\infty}\int_{-\infty}^{\infty}\cdots\int_{-\infty}^{\infty}(C_1-\mu),(C_2-\mu),(C_3-\mu)\ldots f(C_1,C_2,C_3\ldots)dC_1dC_2dC_3\ldots \quad (1.4.6)$$

In Eq. (1.4.6), ξ is the separation time between two concentrations (random variables) at two different times. As a result, the covariance measures the temporal relationship between the concentrations at a given time and other times, characterizing the joint probability distribution of a stochastic process. If the separation time is zero, the covariance collapses to the variance of the process. That is,

$$\sigma^2 = R(0) = E[(C(t) - \mu)(C(t) - \mu)] \tag{1.4.7}$$

The variance is always positive, and covariance is symmetric and could have negative values at some separation time, for instance, a periodic time series (see Fig. 1.10).

Autocorrelation or Autocorrelation function is the covariance normalized by the variance:

$$\rho(\xi) = \frac{R(\xi)}{\sigma^2} \tag{1.4.8}$$

It has the following properties:

1. The autocorrelation function is always symmetric.
2. $-1 < \rho(\xi) < 1$ and it is always a real number.
3. $\rho(0) = 1$.

The autocorrelation function is a statistical measure of the similarity of the concentration-time series offset by a given separation time. It often is used to detect the periodicity of the time series.

1.4.3 Stationary and Nonstationary Processes

A stationary process is a stochastic process where the JPDs or statistical properties do not vary with temporal or spatial locations. This concept is analogous to the spatial homogeneity of geologic media's hydraulic properties or the spatial representative elementary volume concept (spatial REV, Yeh et al., 2015, or Section 1.5). A second-order stationary process requires only the first and second moments of the JPD to be stationary. For a Gaussian JPDF, a second-order stationary process implies stationary. If a process does not meet this requirement, it is a nonstationary process. For instance, the time series, which has a large-scale trend (e.g., seasonal or annual trends), is nonstationary. In reality, natural processes are nonstationary since they have multi-scale variability. Nonetheless, they can always be treated as stationary since the large-scale trend can always be conceptualized as a stochastic process (an abstract concept).

Suppose one treats a nonstationary process as a stationary process. One implicitly includes the large-scale trend as the stochastic process. As such, the variance of this "stationary" process is large since it includes the variability of the large-scale trend. If the large-scale trend is known and characterized, it can be removed as a deterministic feature. The residual variable (after removing the trend)

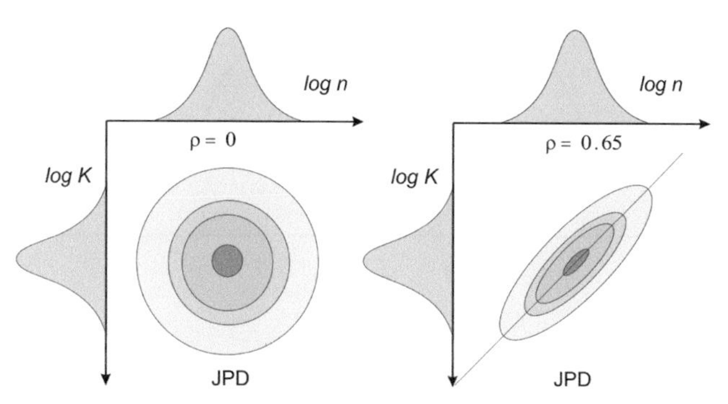

Figure 1.6 Effects of correlation on bivariate (log K and log n) JPD. ρ is the correlation between the two random variables.

can then be stationary if the residual statistics are invariant. In this case, the variance of the residual should be smaller than the original.

The above stochastic analysis assumes that the JPD is jointly Gaussian, like all statistical analyses. Therefore, a jointly Gaussian stochastic process is fully characterized if the mean and the autocovariance function are specified. An illustration of JPD of a Gaussian stochastic process is formidable. Nevertheless, we show in Fig. 1.6 a bivariate JPD with different correlations (covariances). As shown in Fig. 1.6, if two random variables are uncorrelated, their JPD is concentric bell-shaped. An increase in the correlation between the two variables squeezes the JPD's symmetric bell shape to an elliptic one. This correlation effect is further explained in Fig. 1.9.

1.4.4 Sample Statistics

As already discussed, the stochastic concept and theories are built upon the ensemble. In reality, we, however, can only observe one realization. Owing to this fact, we must adopt the ergodicity assumption as we apply stochastic theories to real-world problems. The **ergodicity** states that the statistics from the time series we observed represent the ensemble statistics as long as the series is sufficiently long. Of course, how long is sufficiently long is a question. "Up to one's judgment" is the answer since we never know the ensemble, and no one can discredit your assessment. In other words, ergodicity is our working hypothesis – the best we can do.

Furthermore, our observed time series always has a finite length, and the samples and records are not continuous but discrete. We, therefore, developed methods to calculate the sample statistics as follows:

Consider the case in which we take N concentration measurements, $C_i, i = 1 \ldots N$, at a regular time interval (Δt) at a lake's location. The mean of the concentration-time series is

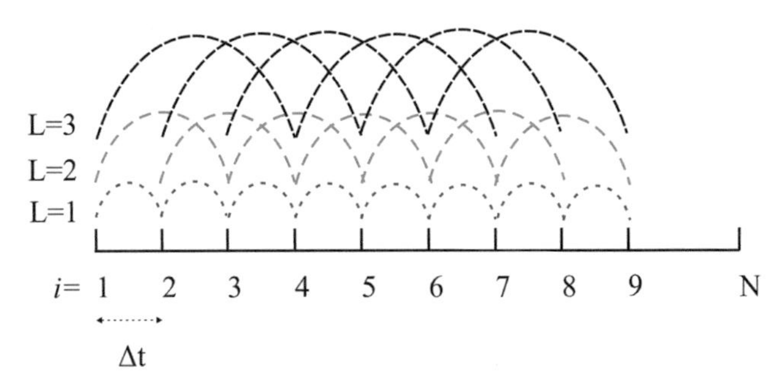

Figure 1.7 An illustration of the autocovariance algorithm (Eq. 1.4.10).

$$\overline{C} = \frac{1}{N}\sum_{i=1}^{N} C(t_i) \tag{1.4.9}$$

While this formula is identical to the sample mean of a random variable (Eq. 1.3.6), this concentration-time series is a stochastic process. Also, Eq. (1.4.9) is a temporal averaging, not ensemble averaging, implicitly invoking the ergodicity assumption.

To calculate the temporal relationship between the concentration value at t_i and at other times, we use the sample covariance function:

$$Cov(\xi) = Cov(t_i, t_i + \xi) = \frac{1}{N-L}\sum_{i=1}^{N-L}\left[C(t_i) - \overline{C}\right]\left[C(t_i + \xi) - \overline{C}\right] \tag{1.4.10}$$

In Eq. (1.4.10), $\xi = L\Delta t$ (separation time); Δt is the sampling time interval; L is the number of intervals, ranging from 1 to $N - 1$. Fig. 1.7 illustrates the operation of this equation. When $L = 1$, Eq. (1.4.10) calculates the sum of the product of the $N - 1$ pairs of concentration perturbations (concentration minus its mean) separated by one Δt, and the resultant is then divided by $N - 1$. As $L = 2$, it repeats the calculation for $N - 2$ pairs separated by twos. This procedure is repeated until $L = N - 1$, or only one pair of concentrations is left.

As $L = 0$, the covariance is the sum of the product of $C(t_i)$ and itself at all measurement times and divided by N. It becomes the variance, denoted as S^2.

$$S^2 = Cov(0) = Cov(t_i, t_i,) = \frac{1}{N}\sum_{i=1}^{N}\left[C(t_i) - \overline{C}\right]\left[C(t_i) - \overline{C}\right]. \tag{1.4.11}$$

Eq. (1.4.10), the sample covariance function, examines the relationship between pairs of concentration data separated at different numbers of Δt s by increasing the L value. Notice that increasing L decreases the number of pairs for the product operation, reflected by $N - L$'s value. For example, when $L = N - 1$, the formula has only one pair of concentrations (i.e., $c(t_1)$ and $c(t_N)$ for calculating the covariance. As L becomes large, the number of pairs for evaluating the

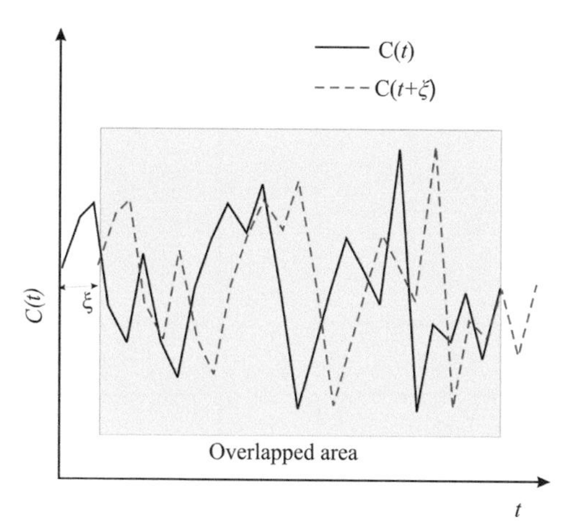

Figure 1.8 An illustration of the physical meaning of the autocovariance function.

covariance becomes small. The estimated covariance is thus not representative the time series's actual covariance and is often discarded.

Notice that some prefer to use $N - L - 1$ instead of $N - L$ for the denominator in Eq. (1.4.10) for unbiased estimates. However, statistics are merely a bulk description of a process, and such a theoretically rigorous treatment may not be necessary for a practical purpose.

Graphical interpretation of Eq. (1.4.10) is given in Fig. 1.8, where the solid black line denotes the observed time series of the concentration, and the red dashed line represents the observed series after being shifted by a separation time ξ. Eq. (1.4.10) sums up the products of the pairs of C values of the solid black and dashed red lines at every t inside the overlapped yellow area. It then divides the sum by the total number of pairs within the area to obtain the covariance at a given separation time interval (ξ). Subsequently, the two series are shifted by another ξ, and the above procedure is repeated till the overlapping area is exhausted.

Eq. (1.4.12) defines the sample autocorrelation function

$$\rho(\xi) = \rho(t_i, t_i + \xi) = \frac{1}{S^2(N - L)} \sum_{i=1}^{N-L} \left[C(t_i) - \overline{C}\right] \left[C(t_i + \xi) - \overline{C}\right] \tag{1.4.12}$$

which is simply the covariance (Eq. 1.4.10) normalized by the variance (Eq. 1.4.1). After the normalization, the maximum autocorrelation is 1, and the range of the correlation is bounded by -1 and $+1$. Eq. (1.4.12) can be implemented in Microsoft Excel (see homework 1.2).

Scatter plots in Fig. 1.9 explain the physical meaning of the autocorrelation function. Instead of summing the products of different pairs of time-series data

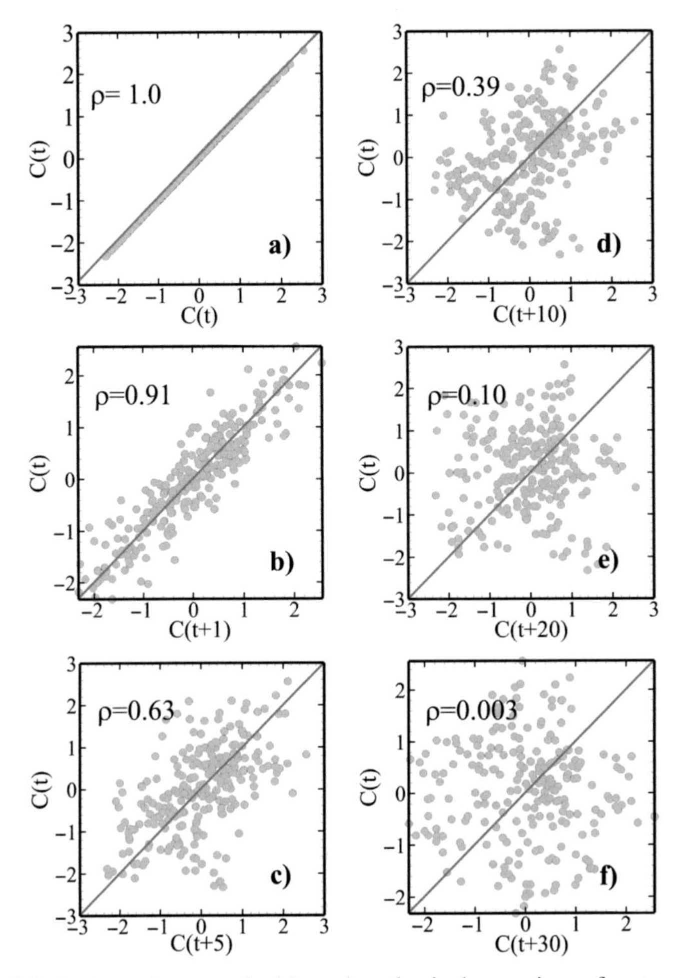

Figure 1.9 Scatter plots to elucidate the physical meaning of autocovariance function calculation. They show how the similarity between the original concentration time-series and those shifted with different time intervals.

using Eq. (1.4.12), we plot each pair of data series in the overlapped area with different shifts (separation times $\xi = 0$, 1, 5, 10, 20, and 30) on a X–Y plot and then carry out a correlation analysis (Eq. 1.4.12), as illustrated in Fig. 1.9a, b, c, d, e, and f, respectively. Each plot shows the correlation value ρ between the time series and that after a shift. When the shift is zero, the data pairs from the solid black and dashed red lines in Fig. 1.8 overlap on the 45-degree (1:1) red line, and the correlation value is one, indicating that these data pairs are perfectly correlated (identical). As the shift becomes large, the data pairs start scattering around the 1:1

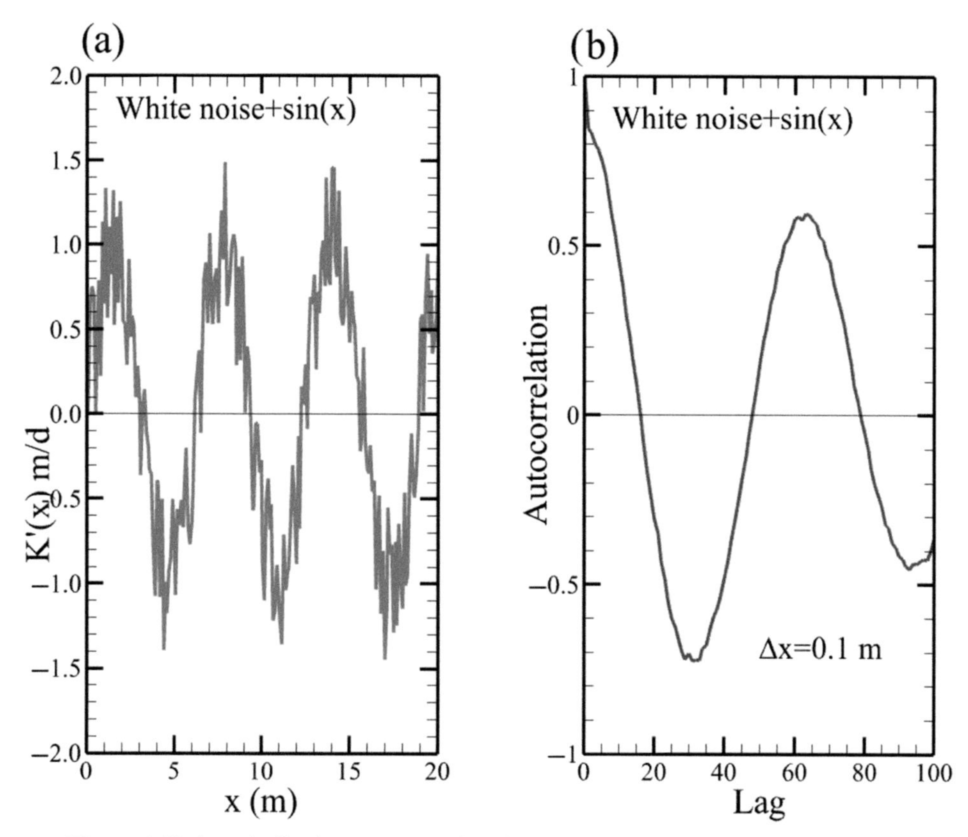

Figure 1.10 A periodic sine concentration time series corrupted with white noise is shown in (a). The time series autocorrelation function (b) indicates the periodicity of the sine function and noise. The sudden drop of the autocorrelation value at the origin reflects that the noise is non-periodic.

line, and the correlation value drops below 1, indicating that the shifted dashed red line is dissimilar to the solid black line in Fig. 1.8. As the shift becomes large, the dissmilarity becomes significant, and the correlation drops further.

These scatter plots manifest that an autocorrelation analysis compares the time series at different separation times for their similarity based on the linear statistical correlation analysis. Accordingly, autocorrelation analysis is a valuable tool for finding the time the recurrence of similar events in a time series (e.g., diurnal, seasonal, or annual variation in precipitation, temperature, or flood events of certain magnitudes).

Figure 1.10a shows a sinusoidal time series infested with white noises (i.e., uncorrelated random noise at a scale smaller than the sampling time interval). This series is nonstationary or has multi-scale variability. Its autocorrelation

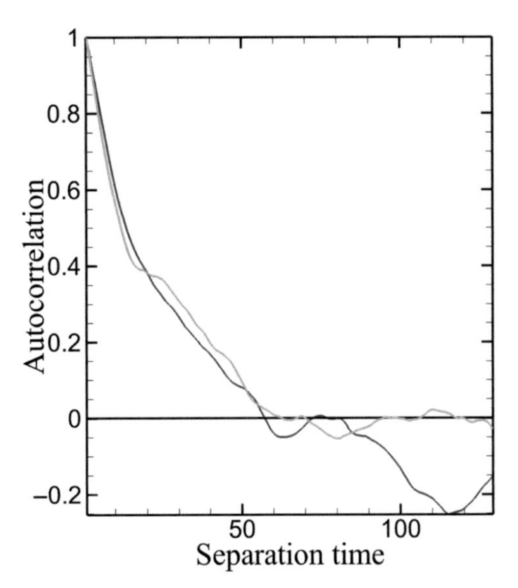

Figure 1.11 Estimated autocorrelation functions for the two realizations in Fig. 1.4. (file: molecular diffusion/console1/two realizations auto.lpk).

function is plotted in Fig. 1.10b. The autocorrelation function reveals that the peaks and troughs of the time series reoccur at a period of 60 and the effect of non-periodic white noise at the origin (also see stochastic representation in Chapter 9).

Likewise, Fig. 1.11 displays the sample autocorrelation functions of the two realizations of stochastic time series in Fig. 1.4, which does not have a noticeable large-scale periodic component as in Fig. 1.10a. First, the two realizations' autocorrelation functions decay as the separation time increases, but they are not identical. The difference manifests the non-ergodic issue of a single realization of a stochastic process of a short record. Second, the autocorrelations drop from 1 to 0 as the separation time increases beyond the separation time around 50, indicating that the pairs of concentration data, separated by the separation time greater than 50, are unrelated (or statistical independent). Statistical independence means that one value's occurrence does not affect the other's probability.

Further, the autocorrelation at the separation time greater than 50 becomes negative. A negative correlation value means that the time series are correlated negatively at this separation time – the series are similar but behave oppositely. Notice that the autocorrelations fluctuate at large separation times due to the small number of pairs available for the analysis. Thus, the autocorrelation values at these separation times are deemed unrepresentative.

This section introduces the basic concepts of the stochastic process, which will be utilized throughout this book to explain observations and theories of flow and solute transport in the environment. While the time series are used to convey the stochastic concept, the concept holds for spatial series, where the variable varies in space (to be discussed in Chapter 9).

In summary, a stochastic process comprises many possible realizations in our imaginary domain with a given joint probability density function. The distribution is generally considered a joint normal or Gaussian distribution (although not limited to). For this reason, a stochastic process can be described by its mean and covariance function. If the process's distribution is not Gaussian, the mean and covariance description is merely approximation – the best we can do and a working hypothesis for our scientific analysis.

1.4.5 Homework

The homework's exercises below intend to demonstrate the power of widely available Microsoft Excel for scientific and engineering applications and enhance understanding of stochastic processes and autocorrelation functions.

(1.1) Use Microsoft Excel to create 80 cells (time steps). Assign a constant temperature value 1 for the first 10 cells and a value of -1 to the next ten cells. Repeat this pattern until the end of the 80 cells (Fig. 1.12). Use Eq. (1.4.12). This equation can be implemented in Excel using the following formula.

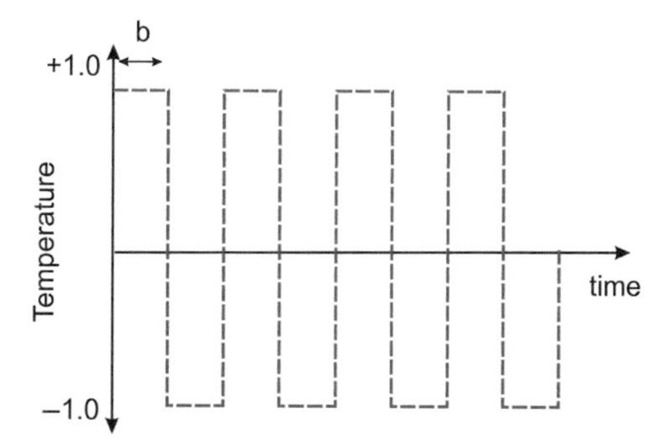

Figure 1.12 The graph shows the temperature variation as a function of time.

=SUMPRODUCT(OFFSET(data,0,0,n-lag,1)
-AVERAGE(data),OFFSET(data,lag,0,n-lag,1)-AVERAGE(data))/DEVSQ
(data). (A1)

Data in the above formula is the entire data set, n is the total number of data points (80), and lag is the L in the equation. The lag ranges from 0 to 20. Calculate the autocorrelation function of this series, following the instruction below:

- First, create a column for the data. Select "formulas" in the Excel manual, then select "Name Management" and define "data" as the name of the corresponding cell locations of all data in the column.
- Next, create a new column with name "lag" and assign 0, 1, through 20 to this new column. Use the name management to define the lag with the cell location of the first lag.
- Create another new column with the name "ACF". In the first cell below the name ACF, enter the formula in Eq. (A1).
- Drag this cell down to reach lag = 20, creating the autocorrelation values for ACF column.
- Plot the temperature data set vs time and the autocorrelation function vs. lag.

Determine the lag where the correlation value is close to zero. Is this lag close to the time step where the C value is constant? What does this result imply? Why does the correlation value drop at large time lags? (Hint: finite length)

(1.2) Generate autocorrelated random time series using Excel.

This homework exercise shows a simple Excel approach to generate an autocorrelated random time series, derive the sample autocorrelation function (ACF), and fit an exponential ACF to the sample ACF for estimating the correlation scale (lambda). Follow the steps below.

(1) Use "RAND()" function in Excel to generate a column of 500 random numbers (ranging from 0 to 1) vs row numbers in column A. Select the 500 random numbers and press CTRL C, then Shift F10, and then V to freeze the random number just generated. The series of the random number is a realization of an uncorrelated random field – Column A.
(2) Plot this realization of the random field and its histogram. Determine the mean and variance.
(3) Calculate and plot the autocorrelation function (ACF) vs. lag of Column A using the procedure in Problem 1. The ACF should drop from 1 rapidly after the first lag, indicating an uncorrelated random field.

(4) In the first cell of another column (e.g., Column B, cell B1), generate a random number using = RAND(). Then, set the next cell as "= B1") and drag this cell to the following eight cells. This step creates 10 cells with the random number, the same as the first one. Select the generated ten cells with the same random number and drag it down to yield 500 numbers, partitioned into 50 segments. Each segment has a different random value. Freeze this column, and this column of data will be Column B.

(5) Create another column (Column C), being the sum of Columns A and B, yielding a realization of an autocorrelated random field. Afterward, carry out the autocorrelation analysis of this realization, using the procedure in Problem 1.1 to obtain a sample ACF. Determine mean and variance and show the histogram. Plot the random fields of Columns A and C, and the sample ACF vs. lag of Columns C. Compare the ACFs of the two to see the differences between Columns A and C. Verify that this random field Column C correlates over more lags than in Column A.

Fitting an Exponential ACF Model to the Estimated ACF

(6) Set a cell as "lambda" (correlation scale) and assign a numerical value to the next cell. Then define lambda (correlation scale) with the cell using the name manager as in problem 1.

(7) Create a new column with a length of the lag (1–20). Enter "= EXP(-lag/ lambda)" at the first cell and drag it along the column to lag = 20, yielding the predicted theoretical ACF at each lag based on the exponential function with the given lambda value.

(8) Set up another column with the name "Error^2", which is the square of the difference between the sample ACF from the random field with the predicted theoretical ACF. Below the end of this column, create a cell, "sum error^2", as the sum of all cells of the Error^2 column.

A nonlinear regression solve must be used to find the optimal lambda value that best fits the exponential ACF to the sample ACF. We need to use Solver Add-in in Excel. This nonlinear regression tool is extremely helpful for the analyses and homework in the book.

Install Solver Add-In (If it is not installed).

Press "File" tab and then select "option".

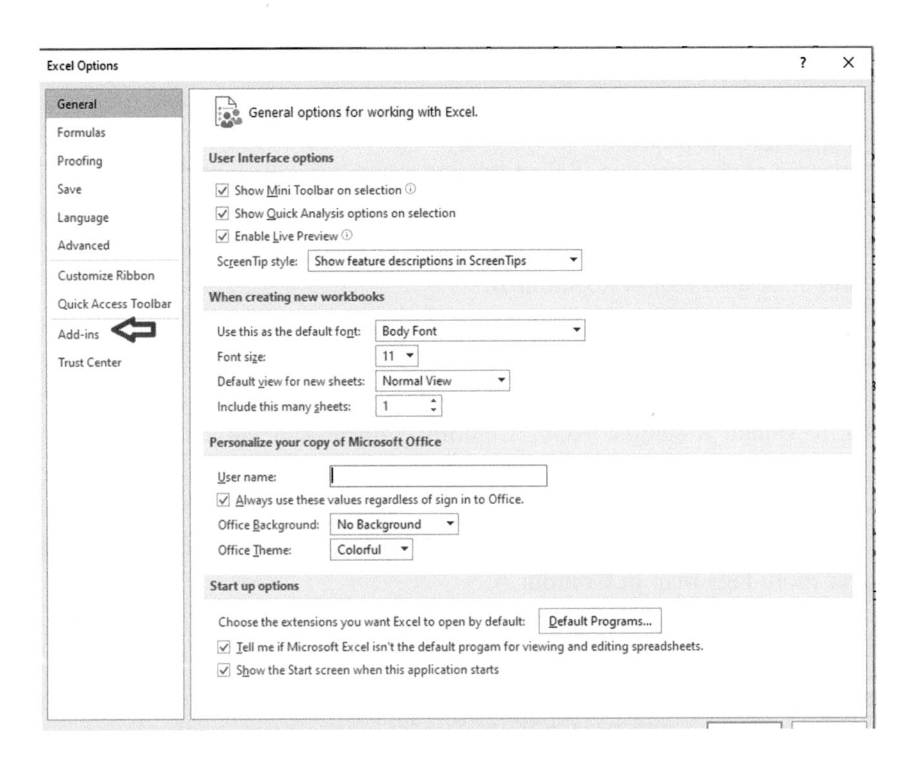
Excel Options
General
Formulas
Proofing
Save
Language
Advanced
Customize Ribbon
Quick Access Toolbar
Add-ins
Trust Center
General options for working with Excel.
User Interface options
Show Mini Toolbar on selection
Show Quick Analysis options on selection
Enable Live Preview
ScreenTip style: Show feature descriptions in ScreenTips
When creating new workbooks
Use this as the default font: Body Font
Font size: 11
Default view for new sheets: Normal View
Include this many sheets: 1
Personalize your copy of Microsoft Office
User name:
Always use these values regardless of sign in to Office.
Office Background: No Background
Office Theme: Colorful
Start up options
Choose the extensions you want Excel to open by default: Default Programs...
Tell me if Microsoft Excel isn't the default progam for viewing and editing spreadsheets.
Show the Start screen when this application starts

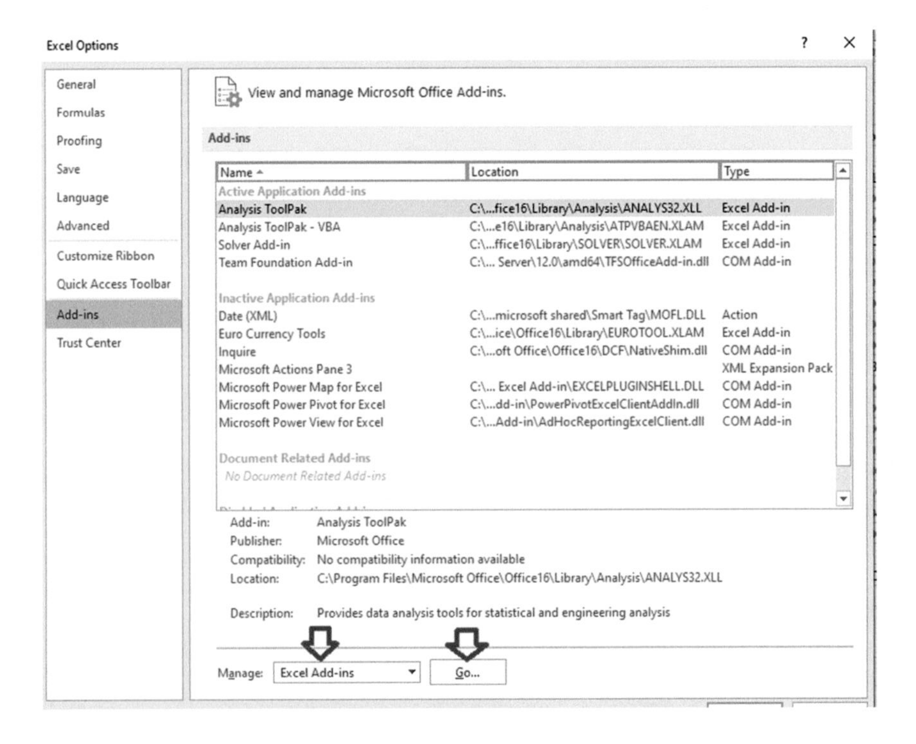
Excel Options
General
Formulas
Proofing
Save
Language
Advanced
Customize Ribbon
Quick Access Toolbar
Add-ins
Trust Center
View and manage Microsoft Office Add-ins.
Add-ins
Name
Location
Type
Active Application Add-ins
Analysis ToolPak C:\...fice16\Library\Analysis\ANALYS32.XLL Excel Add-in
Analysis ToolPak - VBA C:\...e16\Library\Analysis\ATPVBAEN.XLAM Excel Add-in
Solver Add-in C:\...ffice16\Library\SOLVER\SOLVER.XLAM Excel Add-in
Team Foundation Add-in C:\... Server\12.0\amd64\TFSOfficeAdd-in.dll COM Add-in
Inactive Application Add-ins
Date (XML) C:\...microsoft shared\Smart Tag\MOFL.DLL Action
Euro Currency Tools C:\...ice\Office16\Library\EUROTOOL.XLAM Excel Add-in
Inquire C:\...oft Office\Office16\DCF\NativeShim.dll COM Add-in
Microsoft Actions Pane 3 XML Expansion Pack
Microsoft Power Map for Excel C:\... Excel Add-in\EXCELPLUGINSHELL.DLL COM Add-in
Microsoft Power Pivot for Excel C:\...dd-in\PowerPivotExcelClientAddIn.dll COM Add-in
Microsoft Power View for Excel C:\...Add-in\AdHocReportingExcelClient.dll COM Add-in
Document Related Add-ins
No Document Related Add-ins
Add-in: Analysis ToolPak
Publisher: Microsoft Office
Compatibility: No compatibility information available
Location: C:\Program Files\Microsoft Office\Office16\Library\Analysis\ANALYS32.XLL
Description: Provides data analysis tools for statistical and engineering analysis
Manage: Excel Add-ins
Go...

Press go. Then select solver add-in, and then press OK.

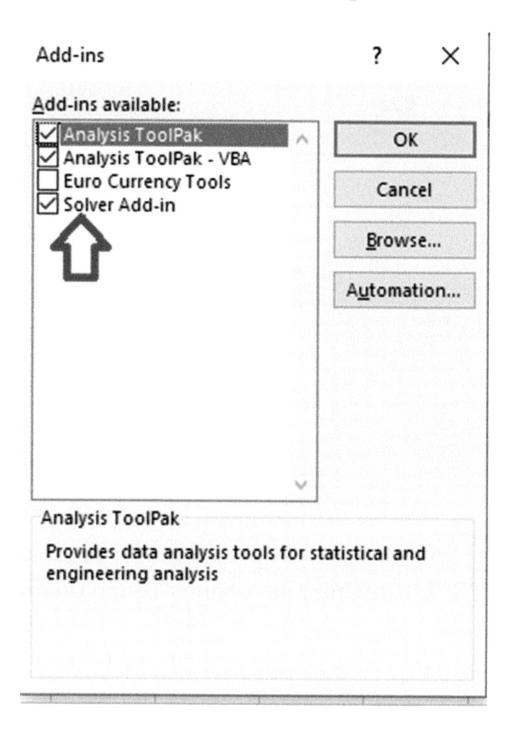

Back to excel sheet and select data and solver should appear.

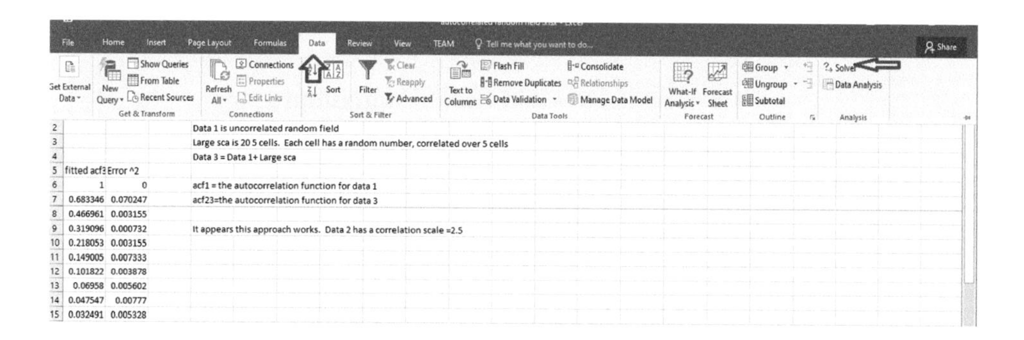

Click Solver. (1) define the objective, which is the cell numbere of the sum of error^2. (2) Then select Min (minimizing the sum of error^2). (3) Enter lambda (or the cell location of the paramter to be adjusted). (4) Select GRG Nonlinear solving method.

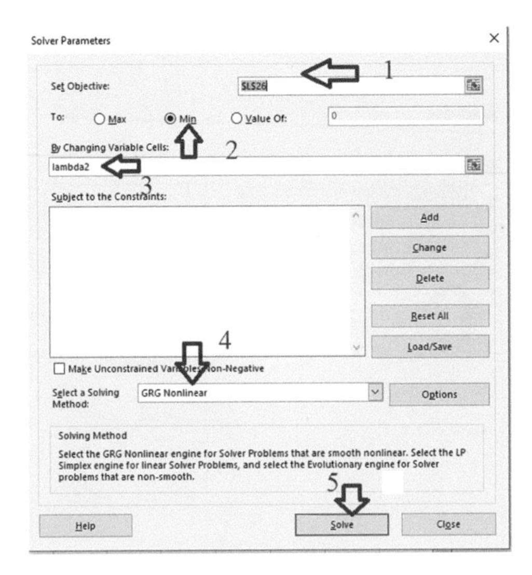

(5) Press solve. Excel will automatically change the lambda value to derive the minimal sum of error^2. Meanwhile, the column of the predicted theoretical ACF will be updated to be the best-fit ACF.

Plot the sample and the best-fit theoretical ACF. Discuss the relationship between the correlation scale (the lambda value) and the number of cells with the constant value. Is it close to 1/2 of the number of cells?

(1.3) Explain why the autocorrelation is one when the L is zero, but the covariance is the highest and decreases as the L becomes large, as the correlation becomes small?

1.5 Stochastic Fluid-Scale Continuum

After the basic concepts of stochastic processes, we are ready to illustrate their importance in the classical "deterministic" fluid properties in the fluid-scale-continuum.

1.5.1 Fluid Continuum and Control Volume

Like any scientific study of natural phenomena, fluid mechanics makes underlying assumptions about the materials under investigation. For example, the fluid properties in fluid mechanics refer to the properties of a point in the fluid. However, suppose we randomly pick a point (an infinitesimally small volume) in a fluid. Due to the fluid matter's discrete nature, that point might be within an atomic particle or in the space between atomic particles. Therefore, the fluid "properties" associated with a point in fluid mechanics would depend upon the measurement point's location. This point concept creates a discontinuous distribution of properties in space for the substance, prohibiting calculus and differential equations built on smoothness and continuity. To avoid this problem, scientists and engineers employ the ***continuum assumption***. In other words, they use the "volume-average" concept to idealize the fluid macroscopically continuous throughout its entirety; the molecules are being "smeared" or "averaged" to eliminate spaces between atomic particles and molecules. Similarly, the volume-averaged velocity in fluid mechanics conceals continuous collisions between molecules and omits molecular-scale velocities in fluid mechanics.

Indoctrinated by the continuum assumption, the fluid's physical or chemical attribute at a point in fluid mechanics' mathematics is the property in a small volume of the actual space. For example, a solute concentration is the mass of solute molecules (M) per volume of the solution (V):

$$C = \frac{M}{V} \tag{1.5.1}$$

This volume must be large enough to contain many molecules and is called a **control volume** (CV). This CV is our "**sample volume**" or "**sample scale**" from which we measure or define the fluid's attributes (such as force, temperature, pressure, energy, velocity, and other variables of interest in fluid mechanics or properties). Such a volume-average attribute ignores the distribution of molecules. For instance, a concentration informs us of the total solute mass, not the solute's spatial distribution in the volume.

The control volume could also be a volume over which we conceptualize the fluid's states, conduct mass or energy balance calculations, or formulate a law (e.g., Darcy's or Fick's law). Therefore, it can be our **model volume** or **model scale**.

1.5.2 Spatial Average, Ensemble Average, and Ergodicity

Suppose we define a volume-averaged fluid property in a volume of 1 m^3 fluid, using a small CV (0.01 m^3) at various fluid locations, and find the property is translationally (or spatially) invariant. This CV is then the property's **representative elementary volume (REV)**. The word **representative** implies that the fluid properties observed in this volume (0.01 m^3) are identical to those observed in other fluid locations (1 m^3). On the other hand, the word "**elementary**" means that the volume (0.01 m^3) is smaller than the entire fluid body and is an element of the entire fluid (1 m^3).

If this REV exists, the fluid is considered **homogeneous** regarding the property defined over this CV. Otherwise, the fluid is **heterogeneous**. Notice that the property defined over CVs smaller than the REV may still be heterogeneous – varying in space, implying homogeneity or heterogeneity definition is scale-dependent.

The REV concept is analogous to the size of the moving averaging window in signal processing. Suppose the window size (smaller than the entire record of signals) encompasses enough spatially varying signals to capture the representative characteristics of the entire signal record. This size of the window is a REV. Because this classical REV concept rests upon spatial invariance, it is most appropriately called the **spatial REV,** which is different from the **ensemble REV** to be explained next.

The following concentration examples explain the differences in **spatial average** and **ensemble average** concepts. Suppose we define the concentration at every point in a solution of $1 m^3$ by overlapping a small CV of $0.01\ m^3$ to obtain a continuous concentration distribution $C(\mathbf{x})$. The $\mathbf{x}$ **is** a location vector, x or (x, y) or (x, y, z) for one, two, or three-dimensional space.

If $C(\mathbf{x})$ is invariant in $\mathbf{x}$, the solution is well mixed at this CV, and this CV of $0.01\ m^3$ is the concentration spatial REV. Otherwise, the solution is poorly mixed, and this CV is not a spatial REV. Nonetheless, albeit $C(\mathbf{x})$ varies in $\mathbf{x}$, one may treat the solution of $1 m^3$ well mixed by defining the concentration as the total mass normalized by the total volume ($1\ m^3$), ignoring the spatially varying concentration defined at the CV of $0.01\ m^3$. This well-mixed proposition arises from our lack of interest in the detailed solute concentration spatial distribution within this volume ($1\ m^3$). This argument leads to the following discussion of CV and REV concepts for the concentration of a solute.

Consider a discrete solute molecule distribution everywhere in a one-dimensional solution domain, as illustrated in Fig. 1.13a. Suppose we adopt a

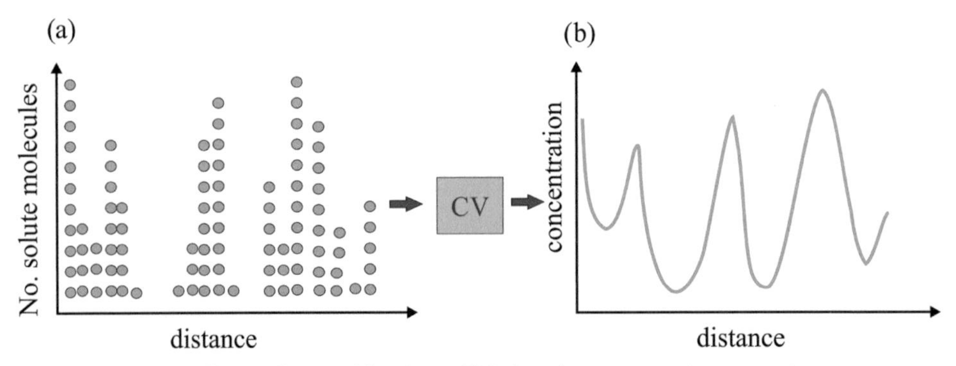

Figure 1.13 Effects of smoothing by a CV. A volume-averaging procedure over a CV converts a discrete and erratic solute spatial distribution into a continuous concentration distribution. The blue circles denote molecules.

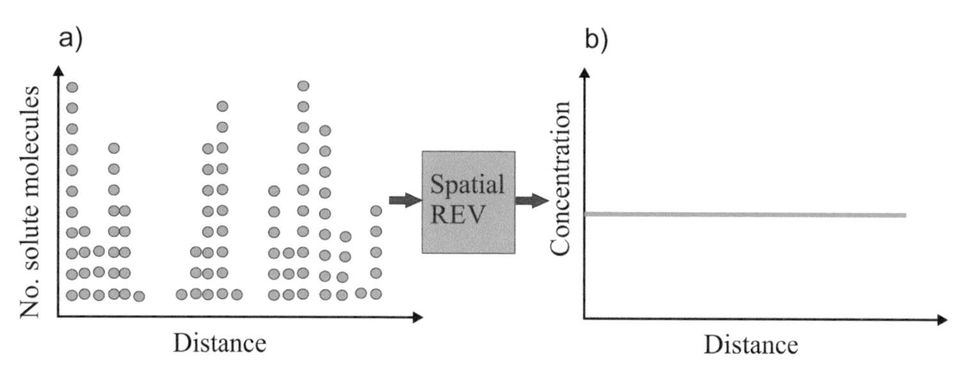

Figure 1.14 If a CV is sufficiently large and its average concentration is translationally constant, the CV is called a spatial REV and the solute is considered well mixed.

small CV to define the solute concentration and continuously overlap the CV at every point along this line, yielding a continuously but spatially varying concentration distribution, i.e., a poorly mixed solution (Fig. 1.13b). This distribution is characterized by a spatial mean and a spatial variance, representing the most likely concentration and the concentration's variability above the mean. The mean and variance calculated from the observed concentrations over the domain are **spatial statistics**.

By enlarging the CV's size yet smaller than the entire domain, we may obtain a constant concentration everywhere (equivalent to the spatial mean). These CVs are **spatial REVs**, as illustrated in Fig. 1.14b, and the solute defined over these sizes' CV is considered well-mixed.

There are situations where the concentration field, $C(\mathbf{x})$, remains varying in space even if the size of the CV is increased (e.g., Figs. 1.15 a and b): the $C(\mathbf{x})$ always exhibits a spatial trend (a large-scale variation). Despite this trend, one can employ the entire domain as a fixed CV to determine a volume-averaged concentration over the entire domain and treat the solution well-mixed (Fig. 1.16). This CV is fixed in space and is not spatially translational, as are those used to define $C(\mathbf{x})$. Such a well-mixed solution proposition based on the entire domain implies a spatial REV since the volume-averaged concentration over this fixed large CV is the mean concentration of the entire domain, satisfying the representativeness requirement of a REV. However, such a large CV is not the solution's elementary volume; therefore, it is not a spatial REV. The ensemble concept discussed next resolves this paradox.

The ensemble concept envisages that the observed spatially varying concentration field with a trend or not in the domain is merely a possible spatial distribution (realization) of many possible ones (ensemble), characterized by a JPD. In other words, an infinite number of CVs of the size as large as the entire domain exists,

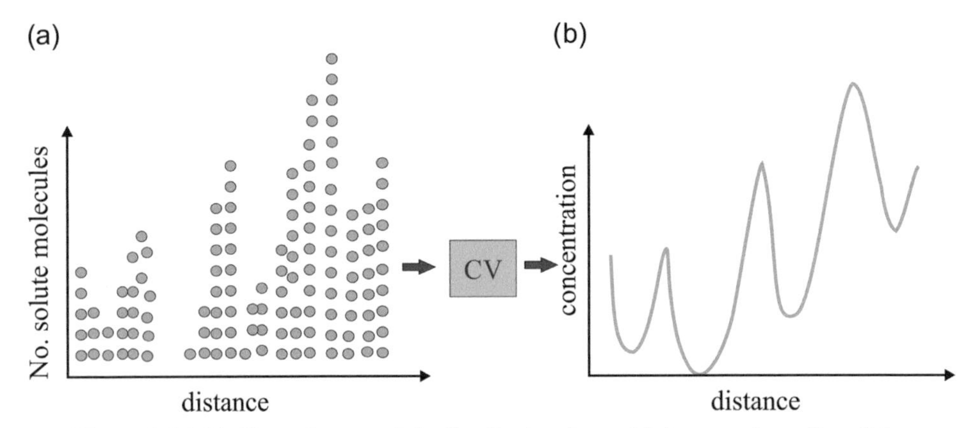

Figure 1.15 (a) The solute particle distribution has a higher number of particles toward the right. (b) The blue line represents the spatial distribution of concentration (% of solute particles over a small CV). The distribution is a continuously varying concentration field with an increasing trend toward the right. The solution is poorly mixed.

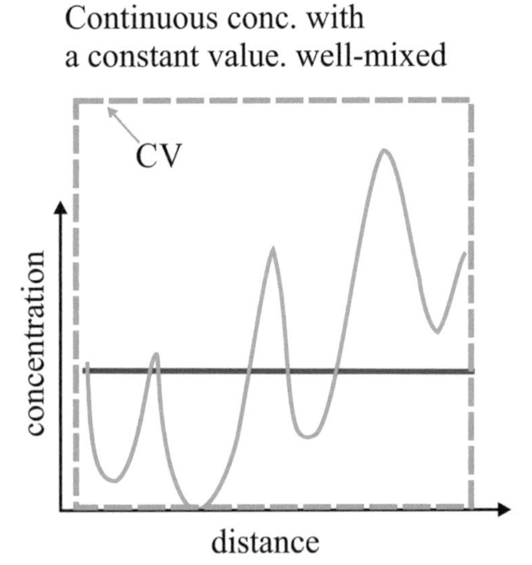

Figure 1.16 The blue line represents the concentration along a line, which exhibits variation and an increasing trend to the right. The red line is the volume-average concentration over a CV, covering the entire solution.

with its own spatially varying concentration field but the same spatial mean and variance, describing the concentration field's spatial variability in each CV (see Fig. 1.17). Such mean and variance are the **ensemble mean** and **variance**.

This concentration of the CV fixed in space as large as the entire domain thus possesses the same mean and variance as other realizations and is a realization of the

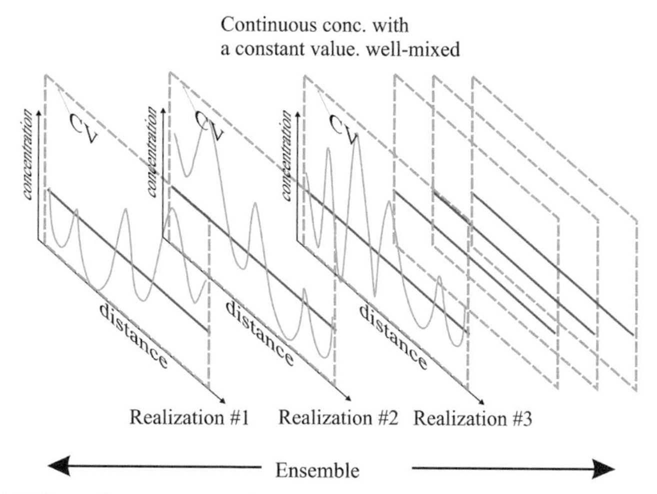

Figure 1.17 The volume-averaged constant concentration of a solution field can be viewed as the mean concentration of many possible solute concentration distributions over CVs of the same size in the ensemble space. In order to restrict the size of the ensemble, we constrain it by requiring that all the concentration realizations have the same spatial mean and variance.

ensemble. It satisfies the representativeness and elementary volume requirements of a REV. As a result, this large CV fixed in space qualifies as an **ensemble REV** (in our imaginary space), albeit it is not the **spatial REV**. Following this argument, any size CV is an ensemble REV at a point in space. That is to say, a volume-averaged concentration at a point is an ensemble-averaged concentration at that point. Ultimately, the ensemble REV is a more general definition of REV than the classic definition since it includes the spatial REV.

Once the ensemble REV is accepted, acquiring the JPD of the ensemble becomes a question since all the realizations in the ensemble are unknown. The "**Ergodicity**" assumption address this question, which states that the spatial statistics from our observations (one realization) are equivalent to the statistics of the ensemble. In plain English, we see one, and we see them all. As is, it may be counterintuitive because the ensemble is unknown. Nevertheless, we accept it as a working hypothesis – the best we can do.

At this point, readers may wonder about the need for these statistical concepts at the beginning of this book. This need arises because multi-scale heterogeneity exists in nature, they are unknown, and all our theories, laws, and observations inherently embrace the ensemble-average concept. For instance, a measured property may be deterministic, but it is an ensemble-averaged quantity – ignoring details within the volume. As we investigate the details within the volume, we find that the details are inherently stochastic (unknown). This stochastic nature of the property or process has often been ignored because it is beyond our observation scale and interest. As we extrapolate these theories and laws to field-scale problems much larger than our observation scales, these stochastic concepts become relevant to comprehending the stochastic nature of the classical theories for solute transport.

1.5.3 Flux Averages

Besides the volume and ensemble averages, flux averages are used to investigate solute transports in the environment, as discussed below (Fig. 1.18).

The classic solute transport analysis (e.g., Kreft and Zuber, 1978; Parker and Van Genuchten, 1984) has recognized two types of averaged concentrations.

1. Volume-Averaged (or resident) Concentration: the mass of a solute per unit volume

$$C = \frac{M}{V} = \left[\frac{M}{L^3}\right] \tag{1.5.2}$$

A typical example of volume-averaged concentration is illustrated in Fig. 1.17. If three point sampling boreholes (left-hand side of the figure) tap into dolomite, gravel, and sand layers of an aquifer, concentrations of some chemical at the three depths are sampled (namely, C_1, C_2, and C_3). The volume-averaged concentration is

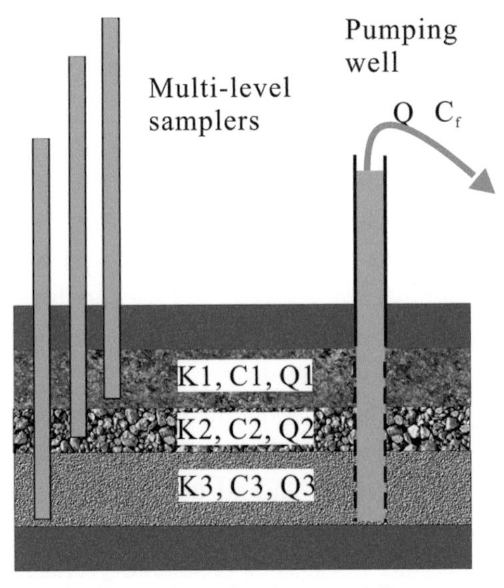

Figure 1.18 A schematic illustration of volume and flux averaged concentrations.

$$\overline{C} = \frac{C_1 + C_2 + C_3}{3}. \tag{1.5.3}$$

Such a volume-averaged approach is likely inappropriate in many applications. For example, as illustrated in the figure's right-hand side, a pumping well is fully screened over the entire aquifer. The concentration sampled from the discharge from the well is average. However, this average weighs heavily on the concentration from the gravel layer, which has the highest discharge to the total well discharge Q. As a result, a flux-averaged concentration is most appropriate to represent the sample concentration.

2. Flux-averaged (flux) concentration $\overline{C_f}$:

$$\overline{C_f} = \frac{Q_1 C_1 + Q_2 C_2 + Q_2 C_2}{Q} \quad \text{or } \overline{C_f} = \frac{\int Cq\,dA}{\int q\,dA} \tag{1.5.4}$$

Q_1, Q_2, and Q_3 are discharges to the well from layers 1, 2, and 3. Likewise, the concentrations in layers 1, 2, and 3 are denoted by C_1 C_2, and C_3. The right-hand of Eq. (1.5.4) is a continuous form of the flux-averaged concentration. In the equation, C and q are the concentration and specific discharge at any point in the aquifer. A denotes the total surface area of the screening interval.

Kreft and Zuber (1978) and Parker and Van Genuchten (1984) pointed out that the resident or flux concentration could lead to different analytical solutions for the advection–diffusion equation. Nevertheless, the importance of the differences would depend on our scales of observation, interest, and model. Therefore, we will adopt the residence concentration throughout the rest of this book.

Units of Concentration: The commonly used unit for concentration is mass per volume (mg/liter). However, it is convenient to express the solute concentration in mass per mass unit as described below for many situations where fluid density varies in time and space (e.g., compressible fluids).

$$C = \frac{\text{Mass of } \alpha/\text{Vol}}{\text{Mass of solution}/\text{Vol}} = \frac{\rho_\alpha}{\rho} = \frac{\text{Mass of } \alpha}{\text{Mass of solution}}$$
$$= \text{ppm(parts per million)} = 1 \times 10^{-6}$$

For water at room temperature, 1 liter (l) of water is about 10^6 mg. Therefore $1\,\text{mg}/l \approx 1\,\text{ppm}$ (part per million). This allows the conversion of volume-averaged concentration to the mass per mass basis concentration definition.

1.6 Summary

- A random variable has a probability distribution characterized by a mean and variance.
- A stochastic process is a collection of many random variables. It has a joint probability distribution, characterized by a mean and autocovariance function, which quantifies the statistical relationship between the random variables.
- If the property in a CV at every point is known, spatial or volume average and autocovariance can be defined. Otherwise, they are strictly ensemble statistics. The spatial and ensemble statistics are equivalent if ergodicity is met or the number of samples is sufficiently large.
- All properties in the fluid continuum are ensemble means over a given CV.
- The ensemble REV concept is more general than the classical spatial REV since it applies to spatially variable properties or processes with a trend or not.

2

Well-Mixed Models for Surface Water Quality Analysis

2.1 Introduction

After introducing the volume-average and stochastic concepts in Chapter 1, widely used, parsimonious, practical well-mixed models are presented in this chapter for lakes water quality analyses. In addition, the volume-average, ensemble average, and stochastic concepts implicitly rooted in these models are explained and emphasized.

2.2 Well-Mixed Systems

A well-mixed (or lumped) model is the fundamental building block of a mathematical model for any natural system. It aims to balance an entire system's mass or energy as a function of time, ignoring the states or properties' spatial variation in the system. Specifically, the model treats the states or properties of the system as well-mixed.

When applying a well-mixed model to a non-well-mixed system, we invoke the ensemble REV and derive the system's ensemble-averaged behaviors since the spatial behavior is likely unknown. We reiterate that if the spatial distribution of the states or properties is known, the word "volume average" is appropriate. Otherwise, the ensemble average is preeminent. For example, applications of the model to study the temporal variation of water quality of bays, estuaries, or lakes inadvertently assume a completely mixed system. Large-scale velocity variations in the lake by wind and tide or small-scale velocity variations (such as diffusion or other processes) have been the justification for this well-mixed assumption, but it is likely deceiving. Despite this faulty assumption, like our presumption of well-mixed chemical concentration from a bottle of the water sample, this model, at best, yields the ensemble-averaged water quality of these systems. It is an averaged water quality over many possible spatially varying water quality distributions. In other words, the model adopts the ensemble REV concept (Chapter 1). As such, the well-mixed model is an **ensemble mean model**.

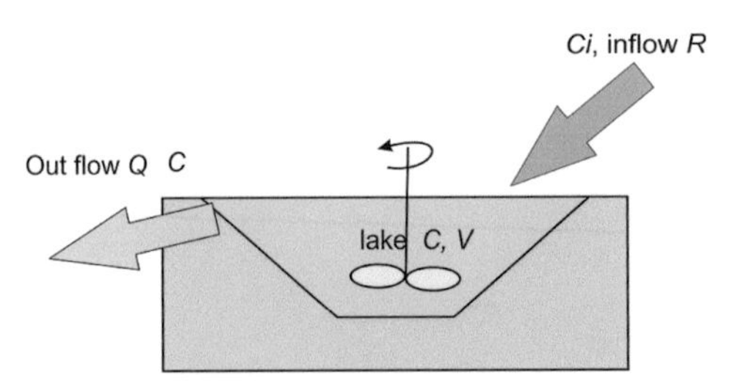

Figure 2.1 A schematic illustration of a well-mixed reservoir.

2.3 Mathematical Model for Water Quality of Surface Water Reservoir

As illustrated in Fig. 2.1, consider a lake where a river flows in with a flow rate of R with a chemical concentration C_i. The lake's initial concentration instantaneously mixes with that of the inflow and reaches a new concentration $C(t)$. Meanwhile, the lake discharges Q amount of water with $C(t)$ to an outflow stream. In the figure, the variables are

V = total volume of water in the lake [L^3]
Q = outflow rate from the lake [L^3/T]
R = inflow rate to the lake [L^3/T]
C_i = inflow concentration [M/M_w] (mass of solute/mass of water, ppm)
C = concentration of the outflow from lake [-] (ppm)
ρ = density of water [M_w/L^3].

The figure's propeller represents the lake's mixing mechanisms (e.g., diffusion, disturbance of winds, currents, turbulences, volume or ensemble averaging).

We next translate the above conceptual model to a mathematical model to quantify the temporal evolution of the lake water and chemical concentration. Recognizing that water carries chemicals, we must consider the water balance for the lake first.

The water balance in the lake states that the change of water mass equals the mass inflow minus the mass outflow of the water:

$$\frac{d\rho V(t)}{dt} = \rho[R - Q]. \tag{2.3.1}$$

If the density of water, ρ, is time-invariant, Eq. (2.3.1) becomes

$$\frac{dV}{dt} = R - Q. \tag{2.3.2}$$

Next, the mass balance for the chemical in the lake is:

$$\frac{d\rho CV}{dt} = \rho C_i R - \rho CQ + \rho SV \tag{2.3.3}$$

where $\rho = [M_w/V_w]$, $C = [M/M_w]$ and the last term in Eq. (2.3.3) represents a chemical sink (e.g., chemical degradation, precipitation, or other reactions) or source (production or growth). Assuming that ρ is constant in Eq. (2.3.3), and recognizing C and V could be a function of time, we have

$$V\frac{dC}{dt} + C\frac{dV}{dt} = C_iR - CQ + SV. \tag{2.3.4}$$

The first term on the left-hand side of Eq. (2.3.4) denotes the change in the solute mass in a given volume of water due to a change in concentration. The second term describes the change in the solute mass by changing water volume with a given concentration. Since the change in water volume, *dV/dt,* is the water balance, substituting Eq. (2.3.2) into Eq. (2.3.4) yields

$$V\frac{dC}{dt} + C(R - Q) = C_iR - CQ + SV \tag{2.3.5}$$

or

$$V\frac{dC}{dt} + CR = C_iR + SV \tag{2.3.6}$$

Notice that V is outside the time derivative, and the outflow rate, Q, vanishes in Eq. (2.3.6) after the inclusion of the water balance equation. In other words, Eq. (2.3.6) has implicitly considered the water balance in the lake. Further, the well-mixed assumption: the outflow concentration is the same as the reservoir's concentration contributes to the vanish of the product of C and Q. For these reasons, Eq. (2.3.6) is valid for either a steady-state or transient flow. Specifically, the reservoir water volume, V, water inflow, R, in Eq. (2.3.6) could vary with time.

Next, consider the sink or source (chemical reaction) term in Eq. (2.3.6). First, we assume that the sink term in Eq. (2.3.6) follows a linear isotherm reaction model for the sake of simplicity.

$$S = -kC \tag{2.3.7}$$

in which the negative sign denotes the sink (losing mass). Eq. (2.3.7) states that the chemical loss amount is linearly proportional to its concentration in the lake water. The parameter k is a chemical reaction rate (constant), [1/T].

2.3.1 Applications of the Well-Mixed Model

The following three examples illustrate applications of the well-mixed model developed above to different environmental problems.

2.3.1.1 The Half-Life of Radioactive Chemicals

Consider a radioactive chemical in a closed chamber (i.e., $Q = R = 0$) of a volume V where the chemical decays with time. The chemical mass balance equation, Eq. (2.3.6), becomes

$$V\frac{dC}{dt} = SV \qquad \frac{dC}{dt} = -kC \tag{2.3.8}$$

The analytical solution to Eq. (2.3.8) is

$$C(t) = C_o e^{-kt} \tag{2.3.9}$$

where C_o = initial concentration. Taking the logarithm of the solution yields

$$ln\,\frac{C(t)}{C_o} = -kt \tag{2.3.10}$$

If we set $C(t_{1/2})/C_o = 1/2$, then we have

$$t_{1/2} = \frac{ln\,2}{k} \tag{2.3.11}$$

Eq. (2.3.11) defines the half-life of the chemical: the time required for the chemical to decay to half its original concentration.

2.3.1.2 Point Dilution Method for Determining Groundwater Velocity

The well-mixed model is also the basis of the popular point dilution method for estimating groundwater velocity (Drost et al., 1968). The method involves injecting a certain mass of a tracer into a well and monitoring the change in tracer concentration over time. It assumes an instantaneous mixing of the tracer with the original water in the well (Fig. 2.2) and the ambient clean groundwater entering the well, leading to the decay of the tracer concentration in the well. Based on the rate of decay of the tracer concentration, the rate of groundwater entering the well can be determined.

To formulate a mathematical model for this method, consider a screened section of a well as the CV (Fig. 2.2) and assume that there is no storage effect in the well, i.e., $Q = R$, inflow equals outflow: flow is steady.

Suppose that the tracer's concentration in the ambient groundwater is zero, $Ci = 0$, and the tracer is well mixed in the well at $t = 0$. That is,

$$C(t) = C_o \quad \text{at} \quad t = 0 \tag{2.3.12}$$

Based on Eq. (2.3.5) and the initial condition above, the following equation describes the temporal change of the concentration in the well:

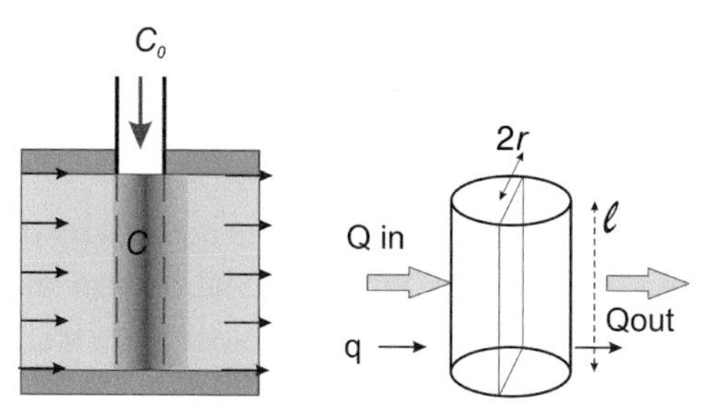

Figure 2.2 A schematic illustration of a point dilution method for determining groundwater velocity.

$$V\frac{dC}{dt} + CR = C_iR + SV \tag{2.3.13}$$

Suppose that the tracer is non-reactive ($S = 0$). Eq. (2.3.13) becomes

$$V\frac{dC}{dt} = -CR. \tag{2.3.14}$$

Knowing that $V = \pi r^2 l$ (r is the radius of the well and l is the length of the screen (Fig. 2.2), assuming that q (specific discharge of groundwater) is uniform and that the cross-sectional area, A, approximates the surface area where groundwater enters the well, we can calculate the total inflow or outflow using

$$R = q \cdot A = q \cdot 2rl. \tag{2.3.15}$$

Thus, an ordinary differential equation for the average tracer's concentration in the well is

$$\frac{dC}{dt} = -\frac{q \cdot 2rl}{\pi r^2 l}C = -\frac{2q}{\pi r}C \tag{2.3.16}$$

Let $C = C_o$ at $t = 0$, we solve Eq. (2.3.9) for C and we have

$$C(t) = C_o\, exp\left(-\frac{2qt}{\pi r}\right) \quad or \quad ln\left(\frac{C}{C_o}\right) = -\frac{2qt}{\pi r}. \tag{2.3.17}$$

This equation inspires us to plot the natural log of the measured *C(t)* normalized by the well-mixed concentration at $t = 0$ as a function of time. This plot should form a straight line, and the slope of this line is $2q/\pi r$, or the specific discharge of the groundwater at this location is

$$q = -\frac{\pi r}{2t}\; ln\left(\frac{C}{C_o}\right). \tag{2.3.18}$$

Afterward, the linear average velocity, u^*, (magnitude) of the groundwater can then be obtained by

$$u^* = \frac{q}{n\alpha}, \tag{2.3.19}$$

where α is a correction factor to account for factors ignored by our assumptions, and n is the porous medium's porosity. A similar equation can also be formulated for injecting a heat pulse to determine groundwater velocity in a well (see Section 2.3.1.3).

Question: The velocity determined from the point dilution method is just the magnitude. How would the direction of the groundwater flow be determined?

2.3.1.3 Power Plant Cooling Pond Design

A cooling pond is an artificial water body primarily for cooling heated water and storing and supplying cooling water to a power plant or industrial facilities (e.g., a petroleum refinery, pulp, paper mill, chemical plant, and steel mill or smelter.) As an alternative to discharging heated water to a nearby river or coastal bay, causing thermal pollution of the receiving waters, cooling ponds are used where the climate is cool and sufficient land is available. Cooling ponds are also sometimes used with air conditioning systems in large buildings.

The pond receives thermal energy in the water from the plant's condensers during energy production. The thermal energy is then dissipated mainly through evaporation and convection in the cooling pond. Once the water has cooled in the pond, it is reused by the plant.

If the cooling pond's temperature or thermal energy is well-mixed, an energy balance model is readily applicable to the cooling pond's design. Before formulating the energy balance for the cooling pond, we define some terms. First, kilocalorie (kcal) is the energy required to raise the temperature of 1 kg of water by 1 °C. British thermal unit (BTU) is the energy required to raise the temperature of 1 lb of water by 1 °F (1 kcal = 3.968 BTU) Thermal energy (E) of the water in the pond is defined as

$$\text{Thermal energy}\,(E) = \int_V \rho\hat{c}T dV = \frac{[M]}{[L^3]}\cdot\frac{\text{cal}}{\text{gm }^\circ\text{C}}\,^\circ\text{C}\,[L^3], \tag{2.3.20}$$

where ρ is the density of water, and $\hat{c}$ is the specific heat of water, which is the energy required to raise the temperature of a unit mass of material by one degree.

$$\hat{c} = \frac{1\,\text{cal}}{\text{gm}^\circ\text{C}} = \frac{1\,\text{BTU}}{\text{lb}^\circ\text{F}} \tag{2.3.21}$$

The conceptual model for the energy balance equation is illustrated in Fig. 2.3, in which V is the pond's volume, A is its area, and T is its water temperature. Q is the

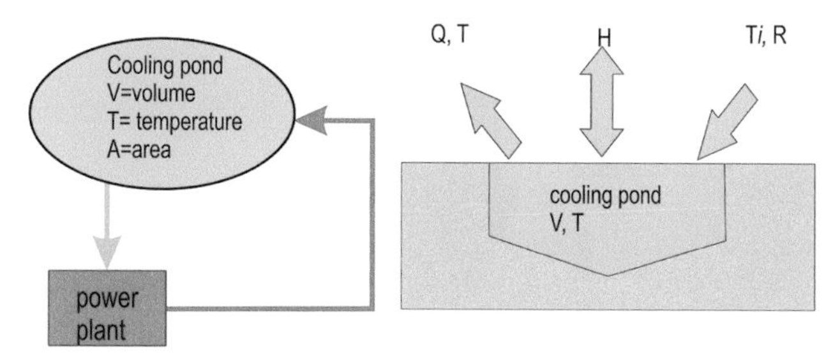

Figure 2.3 Illustrations of a well-mixed model for a power plant cooling pond.

water flow rate from the pond to the power plant, R is the discharge of the heated water from the plant, and T_i is its temperature. H represents the heat exchange between the pond and the atmosphere.

Using the definitions and the conceptual model, we formulate an energy balance equation for the cooling pond:

$$\frac{d}{dt}(\rho \hat{c} T V) = \underbrace{\rho \hat{c} T_i R}_{\substack{\text{Energy}\\ \text{inflow}\\ \text{rate}}} \underbrace{- \rho Q T \hat{c}}_{\substack{\text{Energy}\\ \text{outflow}\\ \text{rate}}} \underbrace{+ H}_{\substack{\text{Net heat}\\ \text{exchange}\\ \text{rate}}} \tag{2.3.22}$$

Assuming inflow to and outflow from the cooling pond are equal (i.e., steady flow) and that the density of the water remains constant within the temperature range considered, we will then let $R = Q$, and V and $\rho =$ constant. The energy balance equation for the cooling pond becomes

$$\frac{dT}{dt} = \frac{R}{V}(T_i - T) + \frac{H}{\rho \hat{c} V} \tag{2.3.23}$$

Here, no seepage into or from groundwater below is assumed, heat exchange occurs only near the pond's surface, and the heat conduction with the surrounding soils is negligible.

Next, we define the net heat transfer term (the second term on the right-hand side of Eq. (2.3.23). To do so, we first define $\Phi_n = H/A$, the rate of net heat-transfer/ unit area, which can be evaluated using

$$\Phi_n = \frac{H}{A} = \Phi_r - (\Phi_{br} + \Phi_e + \Phi_c) \tag{2.3.24}$$

where $\Phi_r =$ net incident radiation (long and short waves)/area, $\Phi_e =$ surface heat flux by evaporation (latent heat)/area, $\Phi_c =$ surface heat flux by conduction (sensible heat)/area, and $\Phi_{br} =$ the surface back radiation/area.

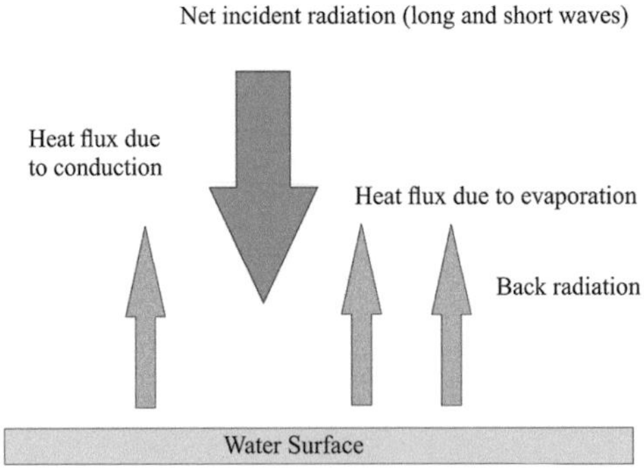

Net incident radiation (long and short waves)

Heat flux due to conduction

Heat flux due to evaporation

Back radiation

Water Surface

Figure 2.4 An illustration of heat fluxes on the water surface.

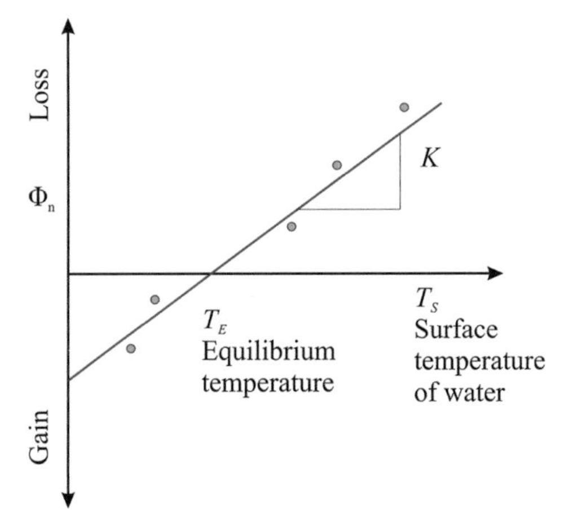

Figure 2.5 A simplified conceptual model of the relationship between heat transfer and temperature of a cooling pond's surface water temperature. Green dots are data points.

Formulas for each term are available but complicated (see Ryan et al., 1974, or Thomann, p. 602). These terms are a function of wind speed, water surface temperature, air temperature, and relative humidity (Fig. 2.4). For simplicity, we will use a simple empirical relationship to determine the heat transfer. That is, we assume that there exists a linear relationship between the net heat exchange and the surface temperature of the water in the cooling pond (Fig. 2.5):

$$\Phi_n = -K(T_S - T_E). \tag{2.3.25}$$

In Eq. (2.3.25), T_s is the water temperature at the surface, and T_E is the equilibrium temperature, the temperature at which net heat exchange between the atmosphere and the water is zero. Its value varies with locality, solar radiation, wind speed, and meteorological conditions, and K is a heat transfer coefficient with the unit $BTU/(ft^2 day^\circ F)$ or $cal/m^2 sec^\circ C$.

If $T_S = T$ (i.e., a well-mixed system), Eq. (2.3.23) becomes

$$\frac{dT}{dt} = \frac{R}{V}(T_i - T) - \frac{AK(T - T_E)}{\rho \hat{c} V} \tag{2.3.26}$$

If we define the water retention rate constant as $k_r = R/V$, [1/T] and the thermal rate constant as $k_T = AK/\rho \hat{c} V$, [1/T]. The energy balance equation takes a new form:

$$\frac{dT}{dt} = k_r(T_i - T) - k_T(T - T_E) \tag{2.3.27}$$

Suppose we consider a steady-state energy system:

$$dT/dt = 0 \quad t \to \infty; \quad \text{at} \quad \text{which} \quad T = T_\infty.$$

Assuming that T_i is constant, we have

$$k_r(T_i - T_\infty) = k_T(T_\infty - T_E) \tag{2.3.28}$$

$$k_r[(T_i - T_E) - (T_\infty - T_E)] = k_T(T_\infty - T_E) \tag{2.3.29}$$

$$\frac{T_\infty - T_E}{T_i - T_E} = \frac{k_r}{k_r + k_T} = \frac{1}{1 + r} \tag{2.3.30}$$

where $r = k_T/k_r$, which is the characteristics of a cooling pond. Generally, the parameter, k_r, can be determined if R and V are known. The parameter, k_T, which involves K, and T_E may be obtained by calibrating Eq. (2.3.27) to the site's available temperature data.

Notice that a similar energy balance approach can also be applied to geothermal reservoirs, hot springs, and other related problems such as determining groundwater velocity (Section 2.3.1.2) using a heat pulse.

The above examples illustrate the applications of the well-mixed model to practical problems. We emphasize that the variables in the application equations are ensemble-averaged quantities: the average of all possible permutations of spatial point variables in the system. If the system's size is sufficiently large to contain all possible spatial permutations of the system variables, the ergodicity condition is satisfied. The ensemble one then is equal to the point variables' spatial average. That is, the output of the well-mixed model will be close to the spatially averaged quantity over the system.

2.4 Analytical Solutions

In the case of the well-mixed model for water quality in lakes, in which the chemical follows a first-order decay reaction, the governing equation is

$$\frac{dC}{dt} + C\left(\frac{R}{V}\right) = C_i\left(\frac{R}{V}\right) + S \quad \frac{dC}{dt} + k_r C = k_r C_i - \underbrace{kC}_{\substack{\text{1st Order} \\ \text{Decay}}} \tag{2.4.1}$$

or

$$\frac{dC}{dt} + (k_r + k)C = k_r C_i. \tag{2.4.2}$$

Again, the water retention rate constant is $k_r = R/V$, [1/T]. Its reciprocal is the **water retention time** (t_r), the time required to fill the reservoir with an inflow rate of *R*. This well-mixed model is a first-order differential equation, which generally has a closed-form analytical solution. We present solutions to some problems below.

2.4.1 Steady-State Solution

Suppose the temporal evolution of the concentration in a lake reaches a steady state:

$$\frac{dC}{dt} = 0 \rightarrow C_i \text{ constant} \quad \text{or} \quad t \rightarrow \infty$$

Eq. (2.4.2) becomes $(k_r + k)\,C = k_r\,C_i$, which has a solution

$$C(\infty) = \left(\frac{k_r}{k_r + k}\right) C_i \tag{2.4.3}$$

If the lake has the characteristic *V/R* = 10 days (retention time) = $1/k_r$. $k_r = 0.1\,day^{-1}$ and $k = 0$, (i.e., conservative chemical). Accordingly, we have

$$C = \frac{0.1}{0 + 0.1} C_i = 1\,C_i \tag{2.4.4}$$

Eq. (2.4.4) shows that the reservoir's chemical concentration equals the inflow concentration when $t \rightarrow \infty$ (i.e., at the equilibrium stage). The conservative chemical fully replaces the original water in the lake. On the other hand, if the chemical is reactive, undergoing the first-order decay with a rate constant $k = 1\,\text{day}^{-1}$, the concentration at the extensive time is

$$C(\infty) = \frac{k_r}{k + k_r}\,C_i = \frac{0.1\,C_i}{1 + 0.1} = 0.09\;C_i$$

That is, the chemical concentration decays substantially in this lake due to chemical reactions.

2.4.2 Transient Solutions

The input type and initial condition dictate the system's response in the transient situation (before the system reaches equilibrium). We examine impulse, step, periodic, and arbitrary inputs in the following sections.

2.4.2.1 Impulse Input

Consider that a mass of contaminant, M, is suddenly discharged into a lake at $t = 0$, where the lake is initially free of the contaminant.This input can be expressed as

$$\rho R C_i = M\delta(t) \qquad C_i = \frac{M\delta(t)}{\rho R} \tag{2.4.5}$$

where $\delta(t)$ is the Dirac delta function, [1/T]. Mathematically, the impulse function can be expressed as

$$\int_{-0}^{+0} \delta(t)\,dt = 1 \quad \text{and} \quad \delta(t) = 0 \quad t \neq 0 \tag{2.4.6}$$

The upper and lower bounds of the integral reflect that the function exists only close to the origin. In Eq. (2.4.5), the dimensions of the concentration, discharge rate, and density are

$$C_i = \frac{[M]\left[\frac{1}{T}\right]}{\left[\frac{M}{L^3}\right]\left[\frac{L^3}{T}\right]} = \left[\frac{M}{M}\right], \quad R = \left[\frac{L^3}{T}\right], \quad \text{and} \quad \rho = \left[\frac{M}{L^3}\right]$$

Then the mass balance equation (Eq. 2.4.2) becomes:

$$\frac{dC}{dt} + k_e C = \frac{R}{V} C_i = \frac{\cancel{R}}{V} \cdot \frac{M\delta(t)}{\rho \cancel{R}} = \frac{M\delta(t)}{V\rho} \tag{2.4.7}$$

If we let

$$k_e = \left(\frac{R}{V} + k\right) = (k_r + k) \tag{2.4.8}$$

we have

$$\frac{dC}{dt} + k_e C = \frac{M\delta(t)}{V\rho} \tag{2.4.9}$$

The solution to Eq. (2.4.9) is

$$C(t) = C_0 e^{-k_e t} \quad \text{where} \quad C_o = \frac{M}{\rho V} \tag{2.4.10}$$

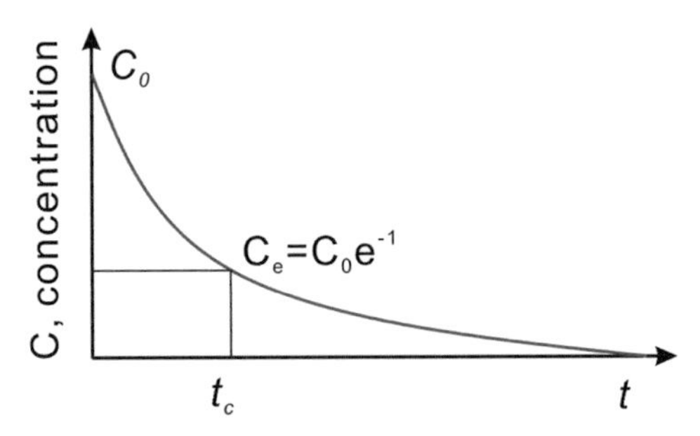

Figure 2.6 An illustration of the output concentration due to an impulse input and the chemical response time.

If $M/(\rho V) = 1$, the input is the unit impulse input. At time $t = 1/k_e$, the concentration at this time becomes

$$C_e = C_0\, exp\left[-k_e\left(\frac{1}{k_e}\right)\right] = C_0 e^{-1} \tag{2.4.11}$$

Accordingly, the **solute (or chemical) response time** ($t_e = 1/k_e$) of this reservoir is defined as when the solute concentration in the reservoir reaches C_e level. When this definition applies to non-reactive chemicals (i.e., $k = 0$), $k_e = k_r$, the solute response time is the same as the **water retention time.** At this water retention time, the mixing process in the reservoir has diluted the input solute concentration to e^{-1} level. If $k>0$, the solute loses its mass due to its decay, and the solute response time is reduced, reflecting the effects of both mixing and decay processes' effects.

To elucidate the distinction between water retention and chemical response time (Fig. 2.6), we consider the situation where blue water flows continuously into and discharges from a tank until a steady-state flow is reached. A slug of yellow water is then suddenly injected into the tank. We consider this experiment under two possible cases:

Case 1 is illustrated in Fig. 2.7, where there is no mixing between the blue and yellow water. The continuous inflow of blue water pushes the slug of yellow water through the tank as a piston-type displacement with sharp fronts (advection of the yellow slug). Thus, the time to move the yellow slug entirely from the reservoir is identical to the time required to fill the tank with any water. This time is the **water retention time** (or the travel time of the yellow slug of water through the tank), which is finite:

$$t_r = \frac{1}{k_r} = \frac{V}{R}$$

Case 2 (Fig. 2.8) is that the yellow water slug is instantaneously well mixed or diluted with the blue water. Consequently, the water's color in the tank becomes

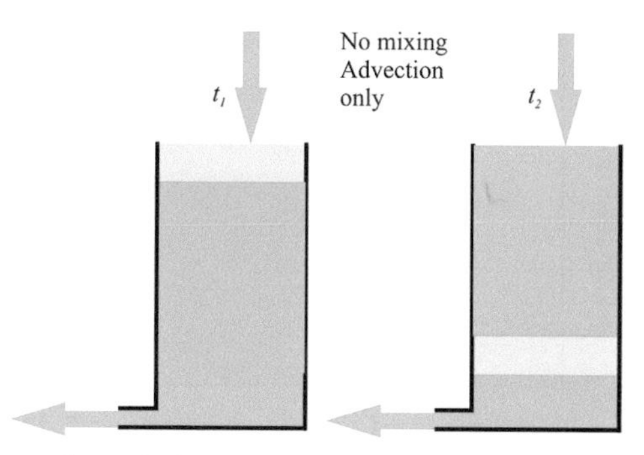

Figure 2.7 An illustration of piston displacement without mixing in Case 1.

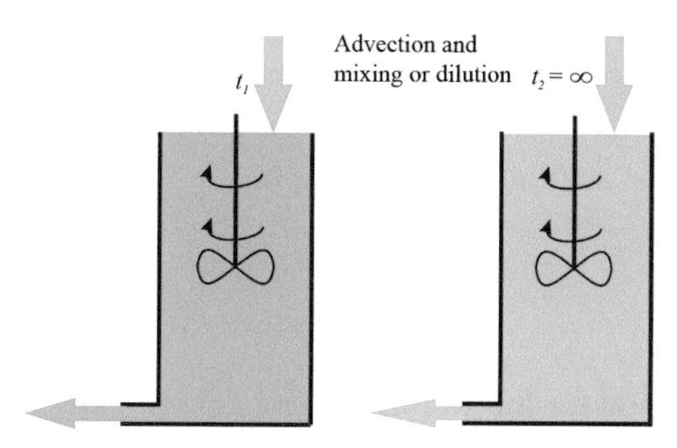

Figure 2.8 Illustration of effects of advection and mixing or dilution.

greenish at first and eventually blue. Moreover, the time to completely replace the yellow slug (i.e., the water retention time) is infinity in contrast to Case 1. For this reason, we use the solute or chemical response time: the time the yellow concentration $C(t)$ in the tank reaches a e^{-1} level of the input concentration C_0. Thus, this chemical response time is finite, equal to the retention time in Case 1. That is to say, the chemical response time can be used to estimate the retention time if the chemical is conservative.

Notice that Case 1 involves only advection (no mixing), and Case 2 considers mixing or dilution and advection (i.e., continuous inflow to and outflow from water from the tank). Behaviors in these two cases are analogous to that in a CV of the advection–diffusion/dispersion equation (ADE) in the later chapter of this book. However, the size of the CV in well-mixed models is much larger than in ADE. So is the scale of our interest.

Thinking Points 1. How would one estimate the retention time and the decay rate constant using tracers? Can one estimate the size of the reservoir? If the reservoir is not well-mixed, how would one estimate them?

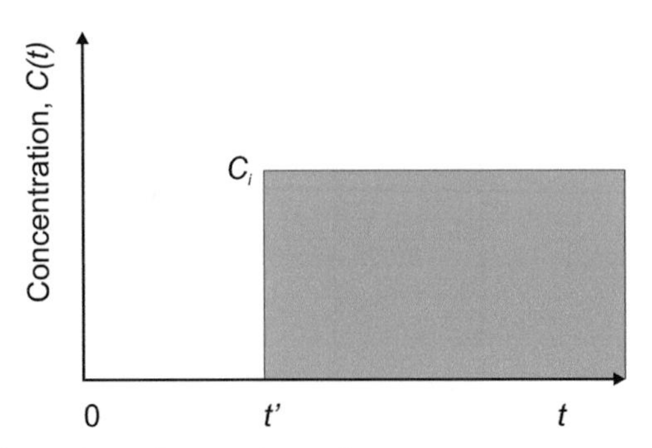

Figure 2.9 An illustration of a step input form.

2.4.2.2 Step Input

This input form (Fig. 2.9) represents a sudden discharge of a constant and continuous concentration C_i from a source to the reservoir.

Mathematically, the step input is defined as

$$C = 0 \text{ at } t < t'$$

$$C = C_i \text{ at } t \geq t'$$

The governing equation for the well-mixed model with this input is

$$\frac{dC}{dt} + (k_r + k)C = k_r C_i \tag{2.4.12}$$

where C_i = constant. Let $k_r + k = k_e$, we have

$$\frac{dC}{dt} + k_e C = k_r C_i. \tag{2.4.13}$$

To solve this equation for the step input, we recall

$$y' + f(x)y = r(x) \neq 0 \tag{2.4.14}$$

and the formula for the solution

$$y(x) = e^{-h}\left[\int e^h r\, dx + A\right] \tag{2.4.15}$$

where $h = \int f(x)\, dx$

$$f(x) \Rightarrow k_e,\ h = \int k_e\, dt = k_e t,\ r(x) \Rightarrow k_r C_i. \tag{2.4.16}$$

Then, the general solution becomes

$$C(t) = e^{-k_e t}\left[\int e^{k_e t} k_r C_i\, dt + A\right] = e^{-k_e t}\left[\frac{k_r C_i}{k_e} \cdot e^{k_e t} + A\right]. \tag{2.4.17}$$

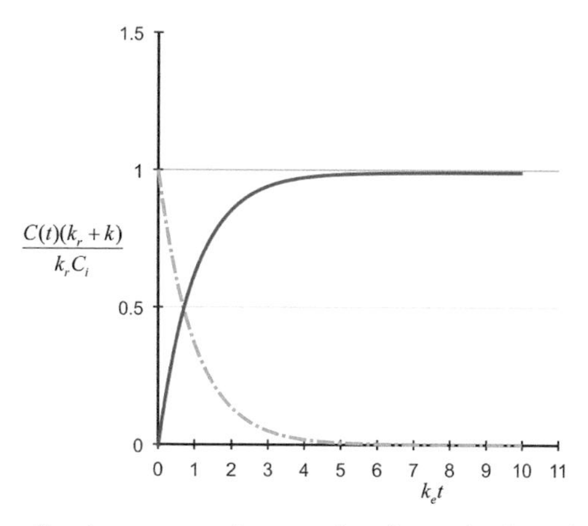

Figure 2.10 Normalized concentration vs. the dimensionless time. The red line denotes $C(t)$, and the blue line is the exponential decay term in Eq. (2.4.20).

To derive the unknown constant A, we apply an initial condition, $t < t', C_o = 0$

$$0 = \frac{k_r C_i}{k_e} + A e^{-k_e t} \tag{2.4.18}$$

the constant is

$$A = -\frac{k_r C_i}{k_e}. \tag{2.4.19}$$

Substituting the constant to the general solution, we have the final particular solution,

$$C(t) = \frac{k_r}{k_e} C_i \left(1 - e^{-k_e t}\right). \tag{2.4.20}$$

The solution graph is given as the red line in Fig. 2.10, where the vertical axis is the normalized concentration. The solution Eq. (2.4.20) consists of two parts. The first part $(k_r/k_e)C_i$ reflects the effect of the chemical reaction. If the chemical is non-reactive ($k = 0$, $k_e = k_r$), the concentration from the first part equals the input concentration (the black line at 1). The second term $C_i\left(e^{-k_e t}\right)$ accounts for mixing with the initial concentration in the lake (effect of dilution), as indicated by the blue dashed line in the figure.

Effects of the Initial Condition Now consider if at t $= 0$, $C = C_0 \neq 0$.

$$C_o = \frac{k_r}{k_e} C_i + A \tag{2.4.21}$$

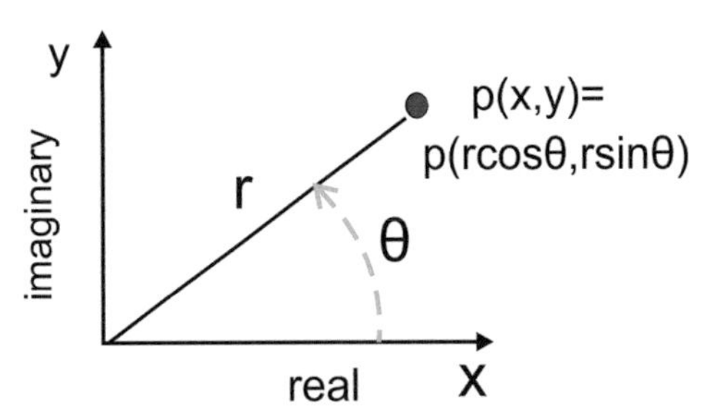

Figure 2.11 A plot of a complex variable in a complex plane.

The constant associated with the initial condition is

$$A = C_o - \frac{k_r}{k_e} C_i. \tag{2.4.22}$$

The particular solution takes the form

$$C(t) = \frac{C_i k_r}{k_r + k}\left(1 - e^{-k_e t}\right) + C_o e^{-k_e t}. \tag{2.4.23}$$

The first term in Eq. (2.2.23) represents the inflow concentration's mixing and chemical reaction effects. The second term denotes the change in initial concentration due to mixing and chemical reaction.

2.4.2.3 Harmonic Input

Harmonic or periodic input represents processes subject to diurnal (e.g., DO in water due to photosynthesis), monthly, or seasonal variation. We shall review some fundamentals of the complex variable before we can derive the solution for the harmonic input.

A complex variable can be expressed as the sum of a real part and an imaginary part:

$$z = \underbrace{x}_{\text{real}} + \underbrace{iy}_{\text{imaginary parts}} \tag{2.4.24}$$

This complex variable z can be plotted on a complex plane (Fig. 2.11).

$$\begin{gathered} x = r\cos\theta, \quad y = r\sin\theta \\ \therefore z = r(\cos\theta + i\sin\theta) \end{gathered} \tag{2.4.25}$$

Recall Euler's formula:

$$z = re^{i\theta} \quad \therefore \quad e^{i\theta} = \cos\theta + i\sin\theta \tag{2.4.26}$$

$$\theta = \tan^{-1}\left(\frac{y}{x}\right) \tag{2.4.27}$$

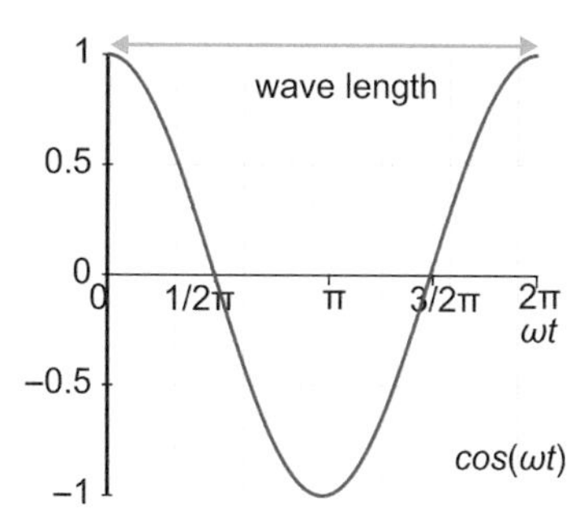

Figure 2.12 A plot of $\cos \omega t$ as a function of ωt.

Suppose the harmonic input concentration to a reservoir is given as

$$C_i(t) = a \cos \omega t \tag{2.4.28}$$

where a is the amplitude [M/L^3], ω is the angular frequency referring to the angular displacement per unit time (e.g., in rotation) or the rate of change of the phase of a sinusoidal waveform (e.g., in oscillations and waves), or as the rate of change of the argument of the sine function. Note that one revolution is equal to 2π radians:

$$\omega = \frac{2\pi}{T} = 2\pi f \quad [\text{radians}/\text{T}]$$

where T is the period [T], and f is the ordinary frequency [1/T], the number of occurrences of a repeating event per unit of time. The period is the duration of one cycle in a repeating event, so the period is the reciprocal frequency. Fig. 2.12 shows the behavior of a cosine function.

Now, we consider that the initial concentration of the chemical in a reservoir is zero: $C\ (0) = 0$. The chemical mass balance equation takes the following form:

$$\frac{dC}{dt} + k_e C = k_r C_i(t) = k_r a \cos(\omega t) \tag{2.4.29}$$

Using Euler's formula for $\cos(\omega t)$, we can rewrite Eq. (2.4.29) as

$$\frac{dC}{dt} + k_e C = k_r C_i(t) == k_r a e^{i\omega t}.$$

To derive the solution, we assume that the solution takes this form:

$$C(t) = \phi e^{i\omega t} \tag{2.4.30}$$

in which ϕ is the unknown variable to be determined. If the assumed solution form is correct, it must satisfy Eq. (2.4.29). Specifically, after we substitute Eq. (2.4.30) into Eq. (2.4.29), we should have

$$\phi i\omega e^{i\omega t} + k_e \phi e^{i\omega t} = k_r a e^{i\omega t}$$

$$i\omega\phi + k_e\phi = k_r a$$

$$\phi = \frac{a k_r}{i\omega + k_e}. \tag{2.4.31}$$

Then, substituting (2.4.31) into (2.4.30) to yield

$$C(t) = \frac{a k_r}{i\omega + k_e} e^{i\omega t} \tag{2.4.32}$$

Now, use the complex variable $z = k_e + i\omega = re^{i\theta}$ where $r = \left(k_e^2 + \omega^2\right)^{½}$ and $\theta = \tan^{-1}(\omega/k_e)$. Finally, multiply the denominator and numerator of Eq. (2.4.32) with the conjugate of the complex variable $z^* = k_e - i\omega = re^{-i\theta}$. Thus, we have the final solution, which takes the form:

$$C(t) = \frac{a k_r}{\left(k_e^2 + \omega^2\right)^{½}} e^{i(\omega t - \theta)} = \frac{a k_r}{\left(k_e^2 + \omega^2\right)^{½}} \cos(\omega t - \theta) + i \sin(\omega t - \theta) \tag{2.4.33}$$

The imaginary part in Eq. (2.4.33) represents the transient evolution of the sinusoidal input starting from the initial condition. Specifically, once the cosine input enters the reservoir, the reservoir responds slowly to reach a steady form of cosine wave oscillation. At this time, this imaginary term becomes zero. Ignoring the transient portion, we can express the steady oscillating concentration as

$$C(t) = aA_m \cos(\omega t - \theta) \tag{2.4.34}$$

where $A_m = k_r/\left(k_e^2 + \omega^2\right)^{½}$ is the attenuation effect on the amplitude (amplitude effect) and

$$\theta = \tan^{-1}(\omega/k_e) \tag{2.4.35}$$

is the delay on phase (phase lag) and is bounded $0 \leq \theta \leq \pi/2$.

The above analysis shows that attenuation and phase effects depend on: $k_{r=}$retention characteristics of the system, k, rate constant of the chemical, and ω, the input frequency.

Noticeably, the output lags behind the input signal by a phase lag, and the output frequency remains the same as the input frequency, as shown in Fig. 2.13. The effects of attenuation as a function of frequency are in Fig. 2.14, while the phase lag is displayed in Fig. 2.15.

Plots in Figs. 2.14 and 2.15 demonstrate that the well-mixed model significantly attenuates and delays high-frequency (large ω values) inputs (such as noise or short-term temporal variations). On the other hand, it only slightly modifies the

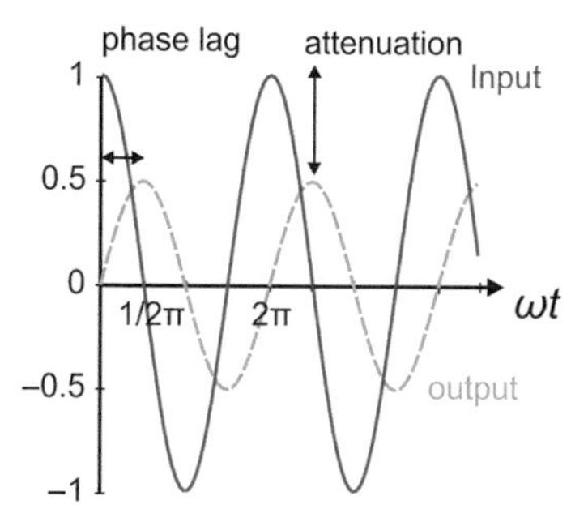

Figure 2.13 An illustration of sinusoidal input and output and attenuation and phase lag.

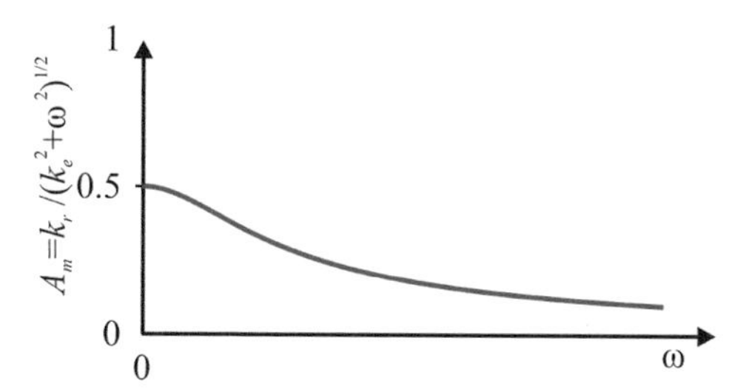

Figure 2.14 A graph of the attenuation as a function of frequency.

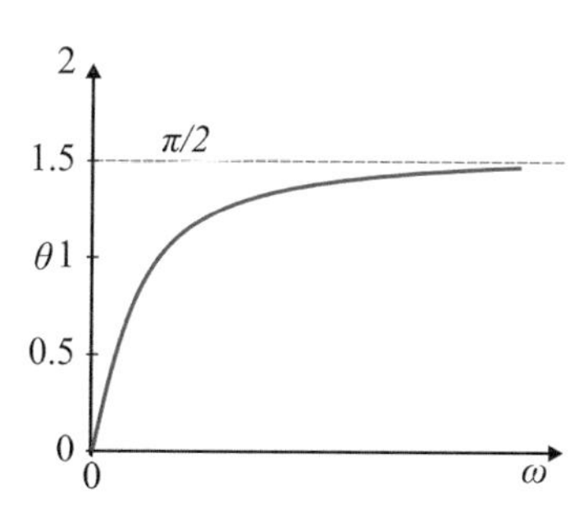

Figure 2.15 A plot of the phase lag as a function of ω.

characteristics of low-frequency inputs (such as long-term temporal variations). In addition, the analysis shows that the output retains the frequency of any input. Therefore, it should be possible to identify inputs (e.g., precipitation, industrial input, and municipal input) from the output record if the input's frequency characteristics are known.

Questions: 1. Plot the attenuation and delay as a function of the chemical response time $(1/k_e)$, which composes water retention and chemical reaction times. Speculate its effects on the attenuation and phase lag. How could the attenuation and phase lag be used to determine these characteristics?

2. In real-world scenarios, the inputs and outputs are likely corrupted by many different signals. How would one determine the attenuation and delay (hint: cross-correlation analysis)?

Remarks: The most common misuse of the well-mixed model is the failure to recognize the model's well-mixed assumption (i.e., ensemble-average concept). Specifically, the model results have often been compared with one or two spatial point observations in a poorly mixed reservoir. Likewise, a single spatial point observation is often used to estimate a well-mixed model's parameter value. To be consistent with the theory of the well-mixed model, one should take as many point samples at different locations as possible and then compare the output from the model against the observed values to check the bias of the model output. Similarly, estimating the parameters of a well-mixed model should calibrate the model against many spatially observed data.

2.4.2.4 Arbitrary Input

A simple mathematical function in the previous sections rarely describes natural inputs to a lake or reservoir. Specifically, the C_i term on the right-hand side of Eq. (2.2.36)

$$\frac{dC}{dt} + k_e C = k_r C_i(t) \tag{2.4.36}$$

has no exact math function (e.g., Fig. 2.16). In this case, one can solve the equation using a numerical or a semi-analytical approach.

Numerical Solution (Finite Difference) The following forward finite difference approximates the time derivative (see Chapter 5):

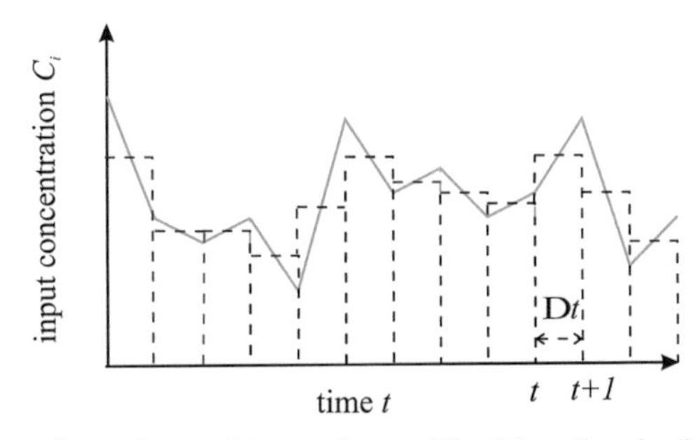

Figure 2.16 Illustration of an arbitrary input. The blue line is the arbitrary input, and the black dashed line is the finite difference temporal discretization.

$$\frac{dC}{dt} \approx \frac{\Delta C}{\Delta t} = \frac{C^{t+1} - C^t}{\Delta t} \tag{2.4.37}$$

where superscript denotes the time step: $t + 1$ is the next step and t is the current step. The term $k_e C$ in Eq. (2.4.36) is approximated by $k_e\left(\left(C^{t+1} + C^t\right)/2\right)$. Therefore, Eq. (2.4.36) becomes:

$$\frac{C^{t+1} - C^t}{\Delta t} + k_e\left(\frac{C^{t+1} + C^t}{2}\right) = k_r C_i^t \tag{2.4.38}$$

where Δt = the time interval between $t + 1$ and t, C^{t+1} denotes the concentration at time $t = t + \Delta t$ and is the unknown to be solved, and C^t = concentration at current time step t, is known. C_i^t can be obtained by discretizing the input concentration record. Subsequently, the finite difference solution for C^{t+1} is

$$C^{t+1} = \left[C^t\left(1 - \frac{k_e \Delta t}{2}\right) + k_r \Delta t\, C_i^t\right] \Big/ \left(1 + \frac{\Delta t k_e}{2}\right) \tag{2.4.39}$$

This numerical solution requires the time step size, Δt, to be small to obtain an accurate solution.

Semi-Analytical Approach This approach discretizes the input record into many input time intervals and then takes advantage of the solution for step inputs for each time interval. We then treat the input of each time interval as a step input. According to the solution in Section 2.4.2.2, the concentration at the time over this time interval is

$$C(t) = \frac{k_r}{k_e} C_i\left(1 - e^{-k_e t}\right) + C_o e^{-k_e t} \tag{2.4.40}$$

where C_o is the initial concentration. C_i is the inflow concentration at the beginning of the time interval, $C(t)$ is the reservoir or outflow concentration at time t. Afterward, we use the calculated concentration at the end of this interval as the initial concentration for the next time interval. Finally, we repeat the above procedures for the rest of the time series.

To illustrate this approach, we use the example shown in Fig. 2.17, where the input record is discretized into two intervals (0–10 yrs) and (10–15 yrs). Then, the concentration from 0 to 10 yrs is calculated using the concentration at year 0 as the initial concentration and use the input concentration $C_i(0)$:

$$C(t) = \frac{k_r}{k_e} C_i(0)\left(1 - e^{-k_e t}\right) + C_o e^{-k_e t}. \tag{2.4.41}$$

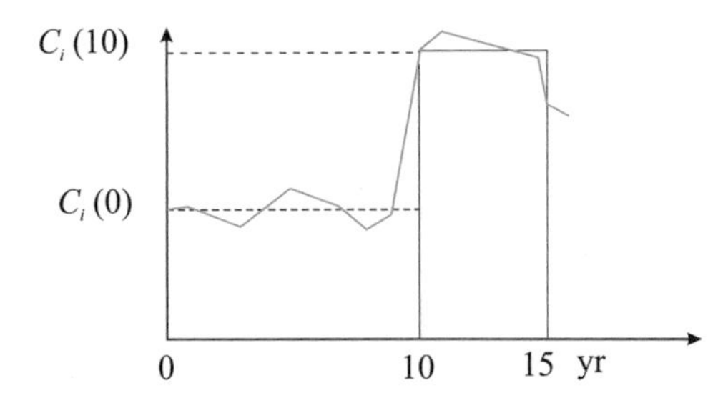

Figure 2.17 A illustration of the semi-analytical solution approach. The blue line is the recorded input concentration series.

At the end of the 10th year, the outflow concentration from the reservoir is:

$$C_{10} = \frac{k_r}{k_e} C_i(0)\left(1 - e^{-k_e(10)}\right) + C_o e^{-k_e(10)}. \tag{2.4.42}$$

Then, use this concentration C_{10} as the initial concentration and $C_i(10)$ as input for the second time interval (5 years from year 10 to year 15):

$$C_{15} = \frac{k_r}{k_e} [C_i(10)]\left(1 - e^{-k_e(5)}\right) + C_{10} e^{-k_e(5)}. \tag{2.4.43}$$

Then repeat this procedure for the next time interval till the end of the time series.

2.5 An Application to Water Quality Analysis

O'Connor and Mueller (1970) used connected well-mixed reservoirs to investigate chloride concentration evolution in the Great Lakes (Lake Superior, Michigan, Huron, Lake St. Clair, Erie, and Ontario). They assumed each lake is a well-mixed reservoir, and the five lakes are mutually connected, as shown in Fig. 2.18. A 1-year simulation time interval was used in the model to justify the well-mixed assumption since the 1 year's time scale was large enough to meet the well-mixed assumption. This assumption is unnecessary if the ensemble REV concept is recognized.

The inflow to each reservoir consisted of runoff and precipitation on the lake, while its outflow included evaporation from the lake and discharge to other lakes and oceans. The chloride sources include (1) its natural background concentrations (initial concentrations), (2) municipal inputs (i.e., chlorides introduced by human and domestic wastes, and various commercial usages), as well as industrial discharges and road salt (de-icing salt). Finally, unaccounted sources were considered as other sources, contributing to the misfit of the prediction to the observed chloride concentrations.

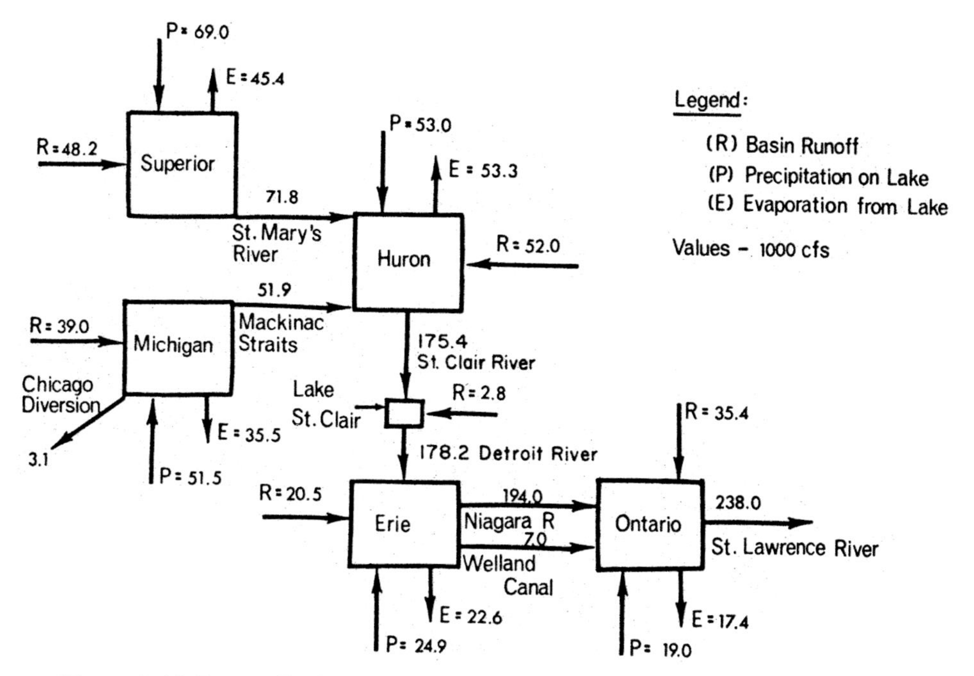

Figure 2.18 The well-mixed conceptual model for the hydrology of the Great Lakes. (O'Conner and Mueller, 1970).

According to Fig. 2.19, the agreement between the calculated and observed values is reasonably good despite incomplete data from various sources. As illustrated in Fig. 2.20, the well-mixed models also serve as a tool for quantifying the magnitude of various sources of potential pollutants to evaluate their relative importance. This information could be used to study the effects of imposing controls on waste load inputs.

2.6 Homework

1. Fig. 2.21 shows a result of some laboratory experiments on the flushing of a small lake. The model was an elliptical pond 6 × 10′ in plan (2–4" deep) which was initially filled with a fluorescent dye solution of concentration C_0. At time $t = 0$, the flow of clean water through the pond was started. The dye is not degradable.

Compare this experimental result with the corresponding solution for a completely mixed steady flow system. Comment on the differences. The nonlinear regression tool in excel can be applied here.

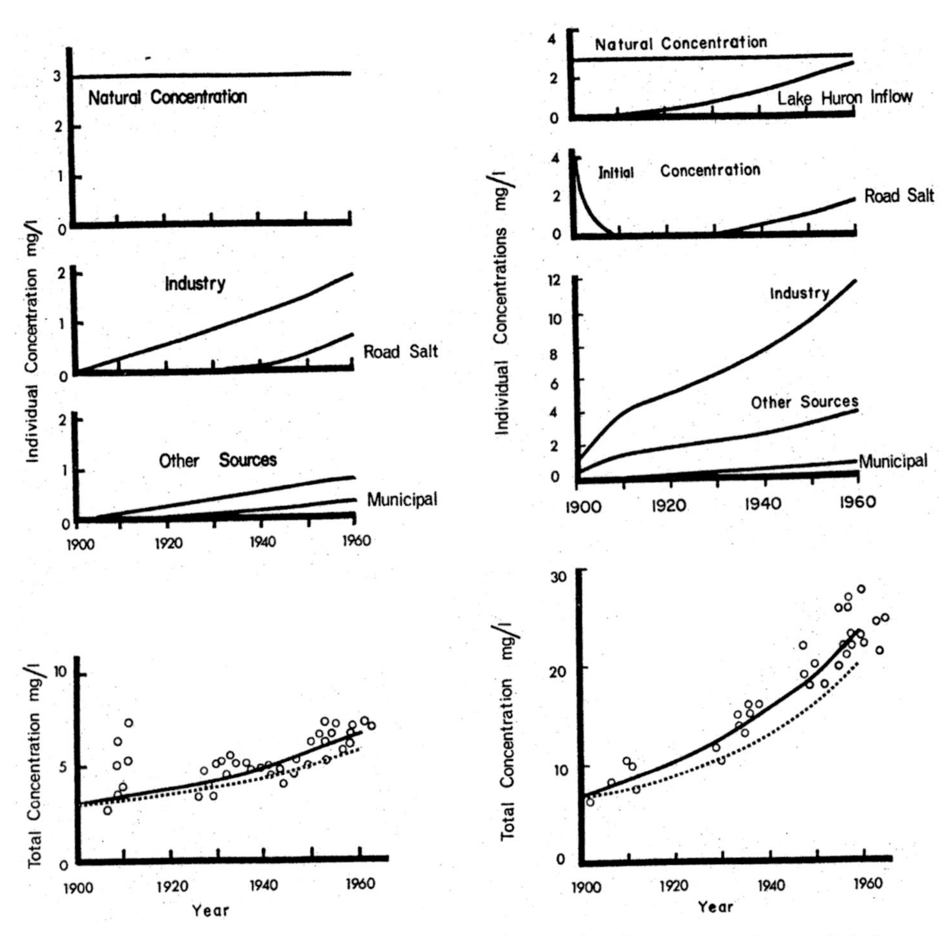

Figure 2.19 Sources (top six plots) and calculated concentrations of Lake Michigan (left bottom) and Erie (right bottom) from 1900 to 1960. (The circles are the observed concentration, the dashed lines are the calculated concentration without considering other sources, and the solid lines are the calculated concentration with other sources.) (O'Conner and Mueller, 1970).

2. The graph given in Fig. 2.22 shows the observed behavior of a model cooling pond (40 $\times$ 22 feet, approximate depth of 0.5 feet). Note that the temperature notation differs from that used in class. T_i refers to the plant intake (outflow from the pons) and T_o the outflow from the plant. The time is normalized by the retention time t_r based on $Q = 5$ gpm throughout the length of the set.

During the experiment, the inflow temperature to the pond was nearly constant. The only change in conditions is the change in the flow through the pond. Using the data after steady temperature conditions have developed, estimate the parameter $r = k_t/k_r$ and the results to predict the transient response assuming a

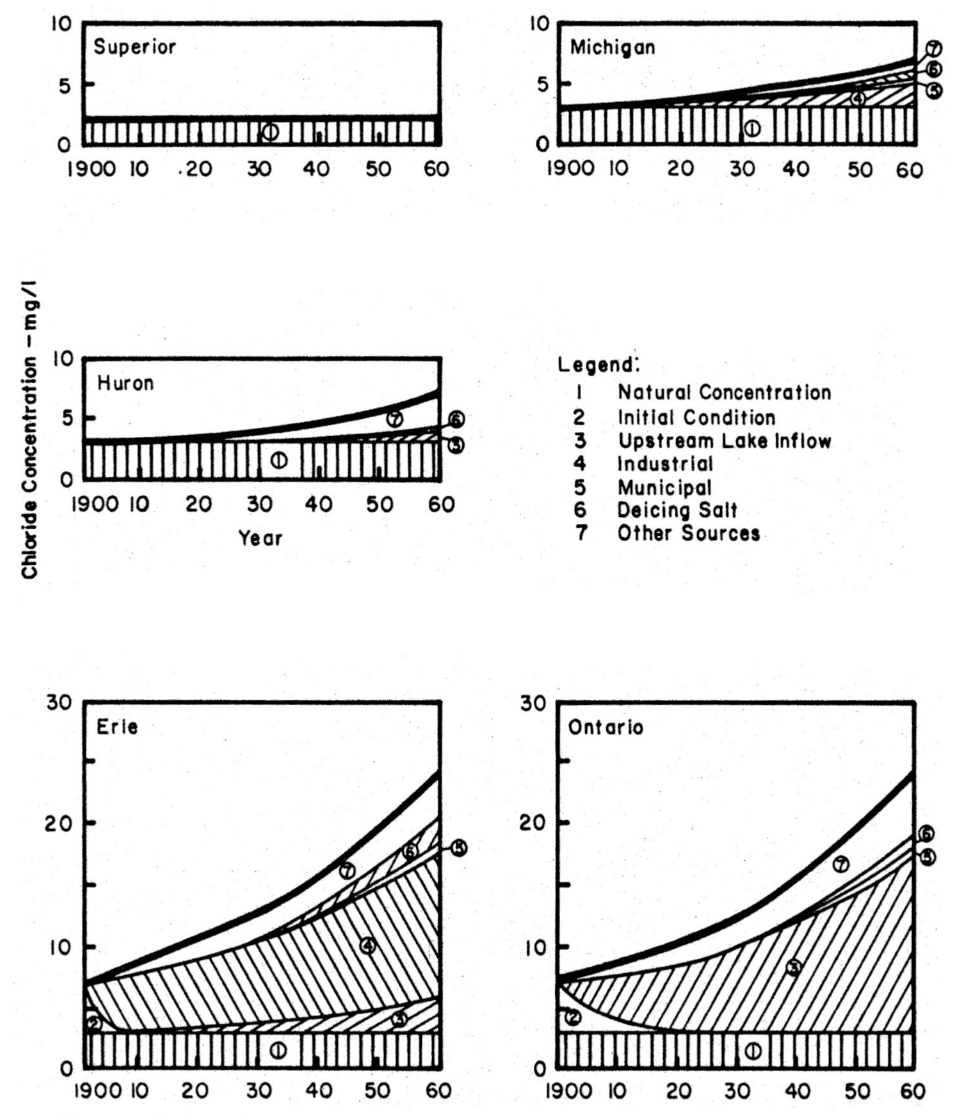

Figure 2.20 Contributions of different chloride sources on the concentration in the five lakes. (O'Conner and Mueller, 1970).

completely mixed system. Compare your results with the experiments and discuss any differences. The nonlinear regression tool in excel can be applied her.

3. Develop the equations describing water quality in a series and parallel two-component mixed reservoir systems, as shown in Fig. 2.23. A contaminant of concentration C_0 and degradation rate constant k is introduced with a steady flow Q into each system. Develop the solution for the outflow concentration C if

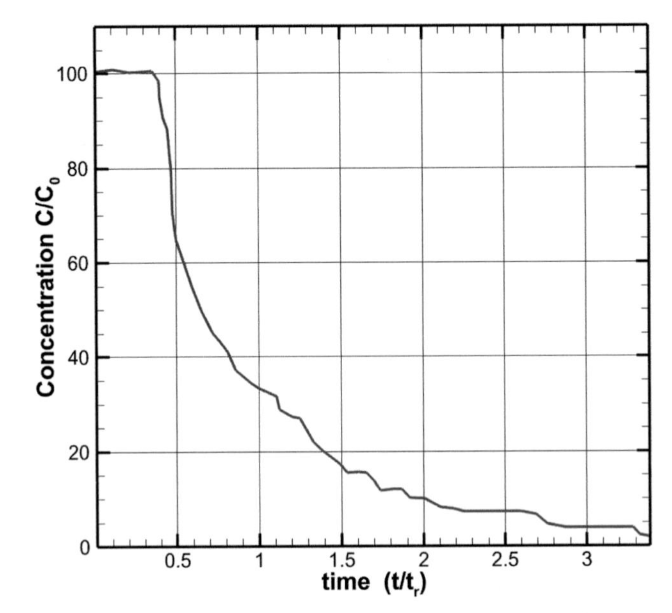

Figure 2.21 Typical outflow hydrograph, C/C_0 versus t/t_r, t_r is the retention time. (Lomax and Osborn, 1970).

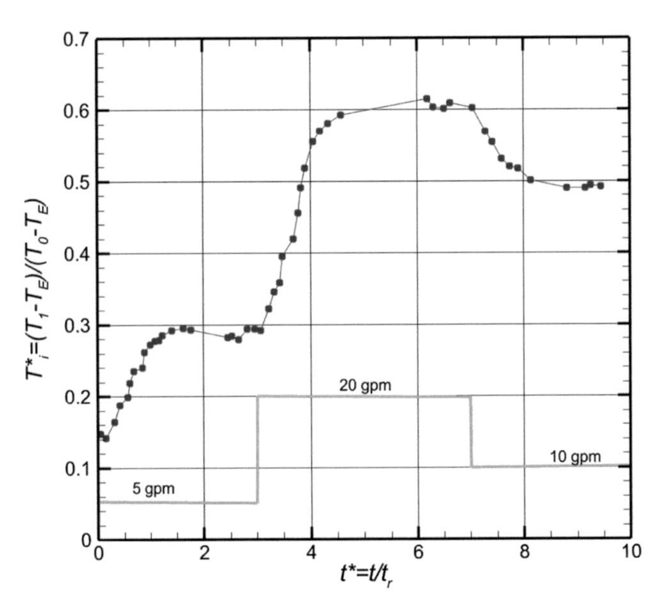

Figure 2.22 Response of Pond to Loading Changes (Laboratory) , t_r is the retention time. (Ryan and Herleman,1974).

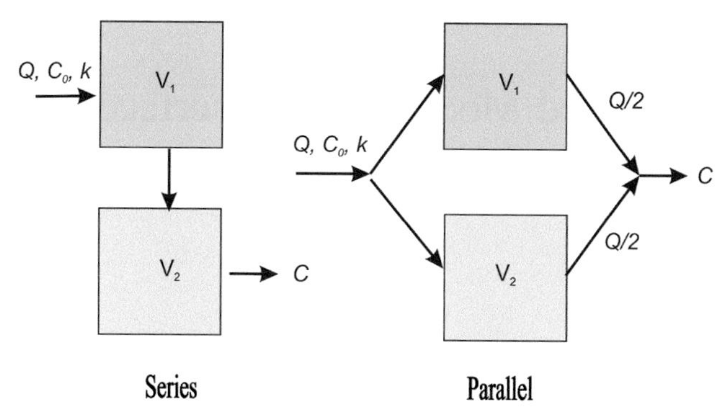

Figure 2.23 An illustration for problem 3.

the initial concentration within the systems is zero. Compare the ultimate steady-state and the transient responses for the two configurations given $k = 1$ day^{-1}, $Q/V_1 = 1$ day^{-1}, $Q/V_2 = 2$ day^{-1}. Suggest some examples of natural flow systems to which these configurations would apply.

3

Well-Mixed Models for Subsurface Water Quality Analysis

3.1 Introduction

This chapter mainly follows the work by Gelhar and Wilson (1974). It first develops the water balance equation and its solutions for various inputs to study temporal water fluctuations in the groundwater system. Application of the model to a real-world situation to estimate the model's parameters and recharge comes next. This chapter then presents the well-mixed solute model for groundwater reservoirs and derives analytical solutions for various input forms. It explains the hydraulic response, water retention, and chemical response time, pertinent to understanding energy propagation, solute advection, and mixing concepts in this book's later chapters. Subsequently, managing the highway de-icing salt application demonstrates the well-mixed model's practical utility.

The formulation of the well-mixed model with reactive chemical reactions is presented next, including chemical decay, first-order equilibrium, and nonequilibrium reactions. Their effects on the output from the well-mixed reservoirs are discussed.

The last part of the chapter introduces the Monte Carlo simulation, sensitivity analysis, and first-order analysis based on the well-mixed model to address uncertainty in the outputs due to unknown parameters and inputs. These approaches are subsequently applied to the study of the role of groundwater reservoirs on buffering acid rains in a lake.

3.2 Water Balance Equation

Consider a cross-sectional view of a stream-water table aquifer system, as shown in Fig. 3.1a. The lower boundary and right-hand side of the aquifer are impermeable. The aquifer connects to a stream on the left-hand side. This figure assumes that the same cross-section view extends infinitely into and out of the figure. The associated

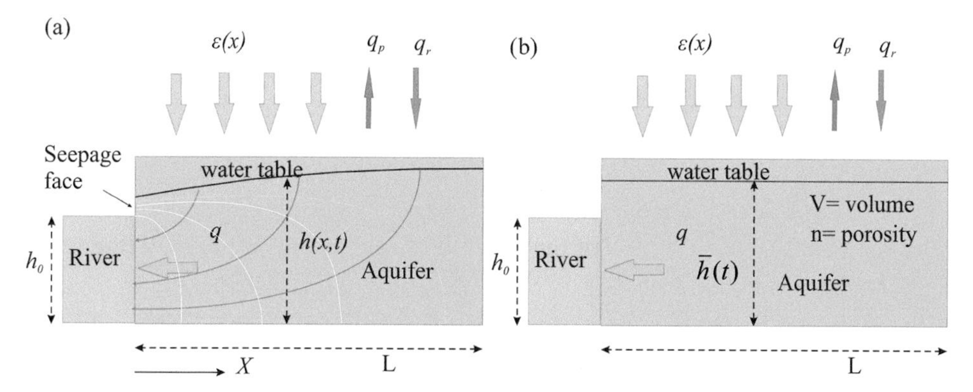

Figure 3.1 A schematic illustration of (a) a stream-water table aquifer system. The white lines are the equal potential lines, and the blue lines are streamlines. (b) The well-mixed conceptual model.

well-mixed model is displayed in Fig. 3.1b, which replaces spatial varying heads and discharges by their averaged values without considering the effects of boundaries. These averages again invoke the ensemble-average concept (Chapter 1).

First, we consider the water balance for the aquifer mathematically as

$$\frac{dV}{dt} = R - Q, \tag{3.2.1}$$

where V, R, and Q are volume [L^3], inflow, and outflow rates [L^3/T], stating that the rate of the change in the aquifer's water volume equals the inflow rate minus the outflow rate.

Assuming that the aquifer extends uniformly in the direction perpendicular to the cross-section in Fig. 3.1b (i.e., y-axis), a well-mixed flow model mathematically expressed in terms of per unit area (on an x–y horizontal plane) is:

$$n\frac{d\bar{h}}{dt} = (\bar{\varepsilon} + q_r) - \left(q + q_p\right), \tag{3.2.2}$$

where n is the porosity (or specific yield), a dimensionless quantity or [—], $\bar{h}$ is the average thickness of the saturated thickness [L] along the x-direction, i.e., the saturated thickness per unit area of the entire aquifer.

$$\bar{h}(t) = \int_0^L h(x, t)dx/L \tag{3.2.3}$$

Here L is the total length of the aquifer in the x-direction and

$$\bar{\varepsilon}(x, t) = \int_o^L \varepsilon(x, t)dx/L \tag{3.2.4}$$

which represents the average recharge rate/area, [L/T]. q_r is the artificial recharge rate/area, [L/T], q_p is the pumping rate/area, [L/T], and q denotes the natural outflow rate from the aquifer/area, [L/T]. The area means the horizontal (x–y) plane of the aquifer.

Next, we assume that the inflow and outflow between the aquifer and steam obey a linear reservoir relationship: they are linearly proportional to the difference between the average thickness of the saturated zone and the river stage (the water level in the stream),

$$q = a(\overline{h} - h_o) \tag{3.2.5}$$

The proportionality a is an outflow constant, [1/T], which is related to aquifer properties (Eq. (3.2.13)) and h_0 is the water stage in the stream, which could be a function of time.

While dropping the overhead bar for all the variables for convenience, we should remember that they are of volume-average quantities. If we consider that artificial recharge and groundwater withdrawal are zero: $q_r = q_p = 0$, the water balance equation (3.1.2) for the aquifer becomes

$$n\frac{dh}{dt} + a(h - h_o) = \varepsilon \tag{3.2.6}$$

3.2.1 Verification of the Linear Reservoir Assumption

We now explore the physical meaning of the outflow constant, a, in the linear reservoir model. Adopting the Dupuit assumption (Fig. 3.2), a one-dimensional steady flow for the stream-water table aquifer system can be written as

$$\frac{d}{dx}\left(Kh(x)\frac{dh(x)}{dx}\right) = -\varepsilon(x) \tag{3.2.7}$$

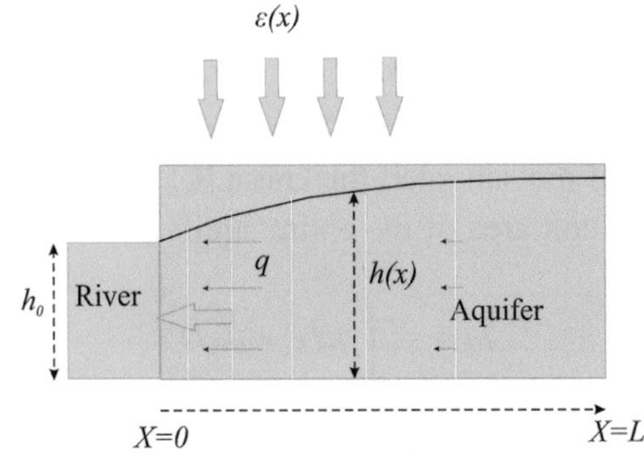

Figure 3.2 A cross-sectional view of a Dupuit aquifer, where flow is assumed to be horizontal and the seepage face does not exist. The white lines are equal potential lines.

with boundary conditions: $x = 0$, $h = h_o$ and $x = L$, $Kh(L)\frac{dh(L)}{dx} = 0$. The solution can be obtained by integrating Eq. (3.2.7) over x and is

$$h(x)^2 - h_0{}^2 = \varepsilon x(2L - x)/K \tag{3.2.8}$$

and can be expressed as

$$(h - h_o)(h + h_o) = \varepsilon x(2L - x)/K \tag{3.2.9}$$

If $(h - h_o) \ll h_o$, we may say $h \approx h_o$ we have an approximate solution:

$$h(x) - h_o = \frac{\varepsilon x(2L - x)}{2T} \tag{3.2.10}$$

where $T = Kh_o$. Subsequently, the average thickness of the saturated zone is obtained by

$$\overline{h} = \frac{1}{L}\int_0^L \left(\frac{\varepsilon x(2L - x)}{2T} + h_o\right) dx = \frac{\varepsilon L^2}{3T} + h_o \tag{3.2.11}$$

Recall that the flow is steady: $\varepsilon = q$ and the linear reservoir assumption used in the well-mixed model takes the following form,

$$\varepsilon = q = a(h - h_o) \tag{3.2.12}$$

In comparison with Eq. (3.2.11), the constant for the linear reservoir model is

$$a = \frac{3T}{L^2} \quad \left[\frac{1}{T}\right] \tag{3.2.13}$$

where T is the transmissivity [L^2/T]. The above derivations show that the linear reservoir model is consistent with the flow process described by the governing equation for unconfined groundwater based on the Dupuid assumption.

3.2.2 Analytical Solutions

This section derives analytical solutions to Eq. (3.2.6) for impulse, step, and sinusoidal inputs.

3.2.2.1 Impulse Input

Suppose a sudden recharge is injected into the aquifer:

$$\varepsilon = \varepsilon_0 \delta(t) \tag{3.2.14}$$

where ε_0 denotes the depth of recharge [L], and $\delta(t)$ is a Dirac delta function ($\delta(t) = 1$, when $t = 0$, and zero otherwise) [1/T]. The solution associated with this input function is

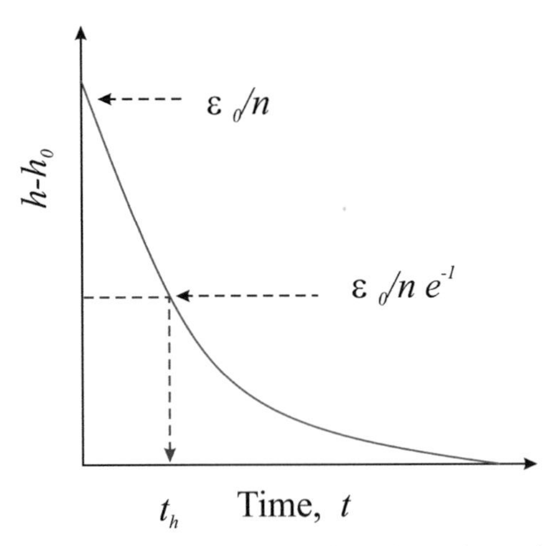

Figure 3.3 Response of the aquifer under an impulse recharge input.

$$h - h_o = \frac{\varepsilon_0}{n} \exp\left(-at/n\right) = \frac{\varepsilon_0}{n} \exp\left(-t/t_h\right) \tag{3.2.15}$$

where $t_h = n/a$, the **hydraulic response time** [T] of the aquifer. It represents the time required to drain the aquifer to a $(\varepsilon_0/n)\exp(-1)$ level for a given impulse input of ε_0. It is a characteristic of the aquifer, depending on the values of n and a. h_o is the initial head at $t = 0$. The behavior of the solution is shown in Fig. 3.3.

3.2.2.2 Step Input

Suppose a constant recharge rate is injected into the aquifer at a time equal and greater than t_1. This injection can be expressed mathematically as:

$$\varepsilon(t) = \begin{cases} 0 & t < t_1 \\ \varepsilon_o & t \geq t_1 \end{cases} \tag{3.2.16}$$

The solution for this input is

$$h(t) - h_o = \frac{\varepsilon_o t_h}{n}\left[1 - e^{-(t-t_1)/t_h}\right] \qquad t > t_1. \tag{3.2.17}$$

The behavior of the solution is plotted in Fig. 3.4.

3.2.2.3 Sinusoidal Input

Consider a groundwater recharge rate that has a sinusoidal fluctuation, described by

$$\varepsilon(t) \quad = \quad \varepsilon_o(1 + \alpha \sin \omega t) \tag{3.2.18}$$

where ε_0 is the base recharge, [L], α is the amplitude of the sine wave, [L], and ω is its frequency, [1/T]. The corresponding solution is

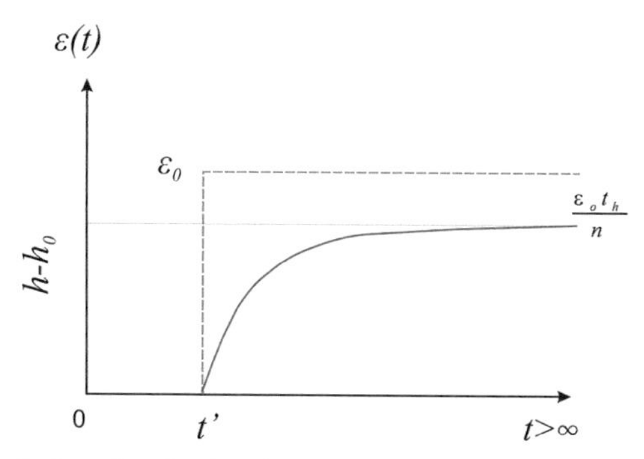

Figure 3.4 The solution for the head vs. time (red line) due to a step input (blue line). The steady-state solution is $\varepsilon_o t_h / n$.

$$h(t) - h_o = \frac{\varepsilon_o t_h}{n} \left[1 + \frac{\alpha}{\sqrt{1 + (t_h \omega)^2}} \right] \sin(\omega t - \phi) \tag{3.2.19}$$

where ϕ is the phase shift and $\tan\phi = t_h\omega$. The aquifer's response to the sinusoidal input is similar to the surface water reservoir in Chapter 2. The output reflects the phase delay and attenuation of the amplitude. The high-frequency input is generally filtered and low frequency preserved.

3.3 Application to the Recharge Estimation

Updegraf (1977) applied a well-mixed model to groundwater recharge estimation for the Mesilla Valley, NM. The water from the Rio Grande was diverted for irrigating farms in the valley. The irrigation water becomes the groundwater and then returns to farm drainage ditches. The data available for this application include monthly water levels in 39± wells over several years, monthly total drainage flow (acre-feet/month), and monthly irrigation diversion (acre-feet/month) in the valley.

The well-mixed model for the water balance for the study was

$$S\frac{dh}{dt} + a(h - h_0) = r_{ij} + u_i. \tag{3.3.1}$$

In the equation, h is the water level averaged over a month and all the wells, [L]. S is the averaged storage coefficient [ ], a is the linear reservoir constant, [1/T], and h_0 is the linear outflow drainage datum. The first term on the right side of the equation, r_{ij}, is the temporally varying irrigation recharge from river diversion averaged over a month (i and j are the year and month index, respectively). u_i is the leakage (constant during a given year, i) to deeper aquifers or boundaries. The objective is to estimate the model parameters, S, a, h_0, r_{ij}, and u_i.

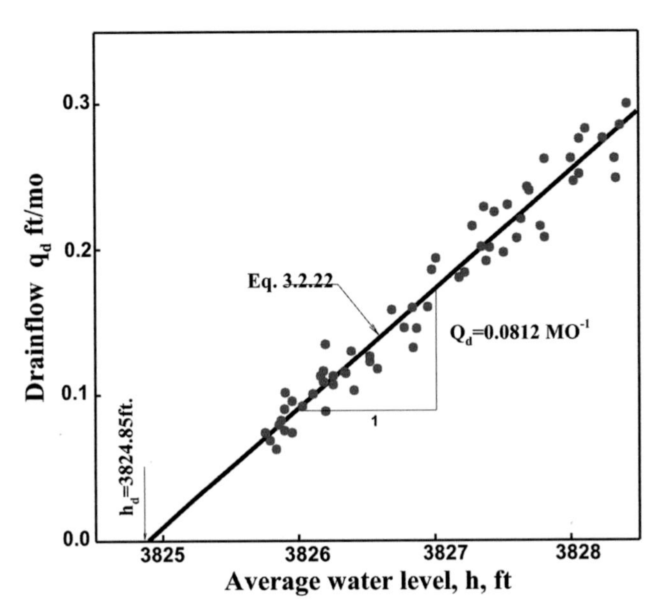

Figure 3.5 Regression analysis of drain flow and average water level data in Mesilla Valley, New Mexico. Updegraf (1977).

3.3.1 Parameter Estimation Procedure

The estimation procedure takes the following steps:

(1) Apply linear regression to estimate the parameters, a, and h_o, using drain flow and water level data and the linear reservoir model, $q = a(h - h_0)$, by minimizing

$$\sum_{ij}\left(\hat{q}_{ij} - q_{ij}\right)^2 = \text{minimal} \tag{3.3.2}$$

where $\hat{q}_{ij}$ is the observed drain flow averaged over the basin and q is the simulated, using the well-mixed model (Fig. 3.5), where i and j are the year and month index, respectively.

(2) Find S and u from the well hydrograph's recession limbs when the recharge is zero ($r = 0$). To accomplish this step, we first integrate Eq. (3.3.2) over the month when irrigation is not applied (Fig. 3.6).

$$\int_{j-1}^{j+1} S\frac{dh}{dt}dt + a\int_{j-1}^{j+1}(h - h_o)\,dt = \int_{j-1}^{j+1} u\,dt. \tag{3.3.3}$$

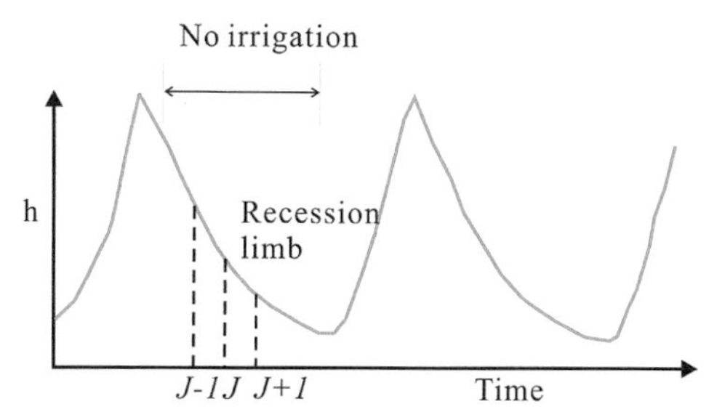

Figure 3.6 Averaged water level variation in the valley. J is the month index for the integration. Updegraf (1977).

Then apply a finite difference approach to approximate Eq. (3.3.3), and we have

$$S\left(\frac{h_{j+1}-h_{j+}}{2\Delta t}\right)2\Delta t + a\int_{j-1}^{j+1} h\,dt - a\int_{j-1}^{j+1} h_o\,dt = \int_{j-1}^{j+1} u\,dt \tag{3.3.4}$$

where $\Delta t = t_j - t_{j-1}$. After rearranging Eq. (3.3.4), we obtain

$$S\left(\frac{h_{j+1}-h_{j-1}}{2\Delta t}\right)2\Delta t + a\left(\frac{h_{j-1}+4h_j+h_{j+1}}{3} - h_o\right)2\Delta t = u_j 2\Delta t. \tag{3.3.5}$$

Rearranging Eq. (3.3.5), we have

$$S\left(h_{j+1}-h_{j-1}\right) + a\left[\frac{\left(h_{j-1}+4h_j+h_{j+1}\right)}{3} - h_o\right]2\Delta t = u_j\,2\Delta t. \tag{3.3.6}$$

Let

$$x_j = \left(h_{j+1}-h_{j-1}\right) \text{ and } y_j = a\left[\frac{\left(h_{j-1}+4h_j+h_{j+1}\right)}{3} - h_o\right]2\Delta t. \tag{3.3.7}$$

Eq. (3.3.6) is simplified to a linear equation

$$y = \alpha x + \beta_j \tag{3.3.8}$$

where $\alpha = -S$ and $\beta_j = 2u_j\Delta t$. A linear regression determines their values since x and y are known from data and previously estimated parameter a. This step yields $S = 0.21$, $a = 0.0812$/month ($T_h = 2.59$ month). The comparison of the observed and simulated average water level in the valley is displayed in Fig. 3.7.

(3) Find recharge r from the integrated equation with $r \neq 0$

$$S\left(h_{j+1}-h_{j-1}\right) + a\int_{j-1}^{j+1}(h-h_o)dt - 2u\Delta t = \frac{\Delta t}{3}\left(r_{j-1}+4r_j+r_{j+1}\right). \tag{3.3.9}$$

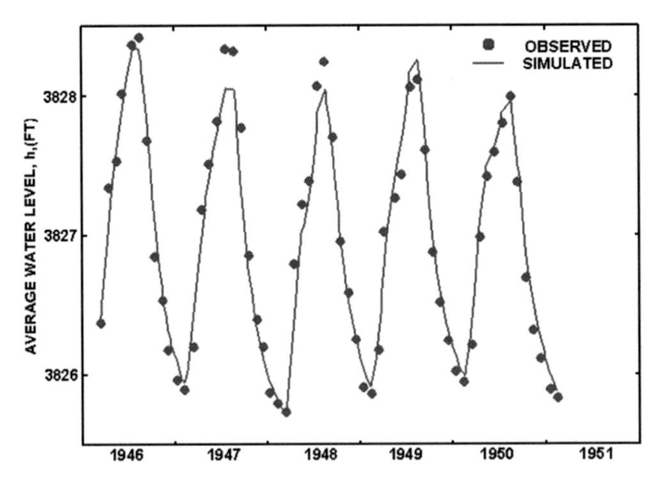

Figure 3.7 The observed and simulated (calibrated) water level in the valley. Updegraf (1977).

The left-hand side of Eq. (3.3.9) is known because of the previously estimated S, a, u, and h at different times. The right-hand side contains the unknowns, representing the recharge at different months. As a result, we have a system of linear equations for r_j at different months, which forms a tri-diagonal matrix.

$$\begin{bmatrix} 4\Delta t/3 & \Delta t/3 & & & & \\ \Delta t/3 & 4\Delta t/3 & \Delta t/3 & & & \\ & \Delta t/3 & 4\Delta t/3 & \Delta t/3 & & \\ & & \cdots & \cdots & \cdots & \\ & & & \cdots & \cdots & \cdots \\ & & & & \Delta t/3 & 4\Delta t/3 \end{bmatrix} \begin{Bmatrix} r_1 \\ r_2 \\ r_3 \\ r_4 \\ r_{..} \\ r_n \end{Bmatrix} = \begin{Bmatrix} S(h_3 - h_1) + a\int_1^3 (h_2 - h_o)dt - 2u\Delta t \\ S(h_4 - h_2) + a\int_2^4 (h_3 - h_o)dt - 2u\Delta t \\ \cdots \\ \cdots \\ \cdots \\ S(h_{n+1} - h_{n-1}) + a\int_{n-1}^{n+1} (h_n - h_o)dt - 2u\Delta t \end{Bmatrix} \tag{3.3.10}$$

The tri-diagonal matrix equations Eq. (3.3.10) can be solved to yield the recharge estimated (see Fig. 3.8).

The above application of the well-mixed model to the valley-wide groundwater balance demonstrates that the model can reproduce the observed long-term groundwater level. It also yields reasonable estimates of the recharge pattern, consistent with the historical records of river diversion. Specifically, the well-mixed model is a valuable yet straightforward first-cut analysis tool for estimating

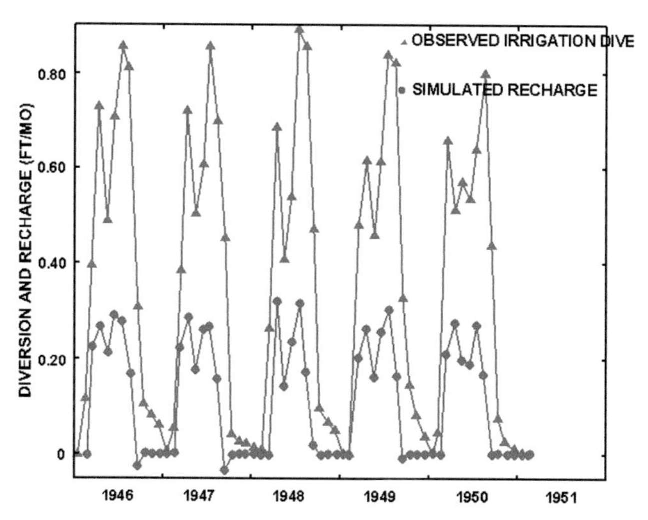

Figure 3.8 Observed irrigation diversion vs. the estimated groundwater recharge. Updegraf (1977).

groundwater recharge over the valley or a watershed, which can be a general knowledge of the recharge pattern for further calibration of a multidimensional distributed surface/groundwater model.

3.4 Water Quality Formulation

Next, we will consider the chemical mass balance for a stream-aquifer system. Suppose we focus on spatially averaged concentration:

$$\overline{C}(t) = \int_o^{h(x)} \int_o^L C(x, z, t)dx \; dz / \int_o^L h(x)\,dx, \tag{3.4.1}$$

a well-mixed model can be formulated as

$$\frac{d(nhC)}{dt} = \varepsilon C_\varepsilon + q_r C_r - q_p C - qC - nhS, \tag{3.4.2}$$

where S is the sink or source term per unit area of the aquifer [1/T], C, $C\varepsilon$, and Cr are the concentration the groundwater reservoir, natural rescharge, and artifical recharge, respectivel. See Fig. 3.9.

The last term in Eq. (3.4.2) is the sink or source term (i.e., chemical reaction term). After decomposition of the left-hand side term, Eq. (3.4.2) becomes

$$nh\frac{dC}{dt} + nC\frac{dh}{dt} = \varepsilon C_\varepsilon + q_r C_r - q_p C - qC - n\,hS. \tag{3.4.3}$$

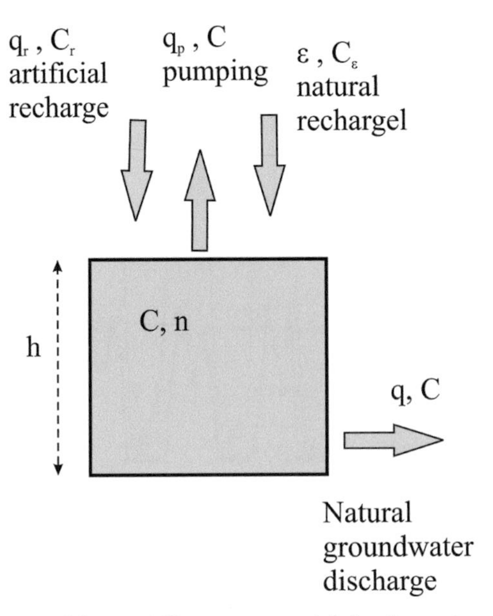

Figure 3.9 A conceptual lumped Parameter model for the water chemistry.

The first term on the left-hand side of Eq. (3.4.3) denotes the rate of change in solute mass due to the rate of change in solute concentration under a given groundwater volume (nh). The second term represents the change rate in solute mass due to the rate of change in the aquifer's saturated thickness, given a concentration C. We notice that the water balance equation dictates the rate of change in the saturated thickness for the aquifer system:

$$n\frac{dh}{dt} = \varepsilon + q_r - q_p - q. \tag{3.4.4}$$

Substituting Eq. (3.4.4) to replace ndh/dt (the second term on the left-hand side of Eq. (3.4.3)) leads to

$$nh\frac{dC}{dt} = \varepsilon C_\varepsilon + q_r C_r - \varepsilon C + q_r C - nhS. \tag{3.4.5}$$

Notice that the term $-q_p C - qC$ in Eq. (3.4.3) vanished from Eq. (3.4.5), and h could vary with time, although it is outside the time derivative. Specifically, Eq. (3.4.5) is valid for a steady or transient flow situation.

To consider chemical reaction, we assume that the sink or source term is a first-order decay process, $S = kC$. Eq. (3.4.5) becomes

$$\frac{dC}{dt} + \frac{(\varepsilon + q_r + knh)}{nh}C = \frac{\varepsilon C_\varepsilon + q_r C_r}{nh}. \tag{3.4.6}$$

Suppose we let

$$k_e = \frac{\varepsilon + q_r}{nh} + k = k_r + k \tag{3.4.7}$$

be the chemical response constant, consisting of the water retention constant $k_r = (\varepsilon + q_r)/nh$ and chemical decay constant k. Note that retention constant can change with time since h changes under transient flow conditions. Under steady-state flow, the reciprocal of the retention constant is the **retention time, t_r,** while the **chemical response time** of the aquifer system is defined as

$$t_c = k_e^{-1}. \tag{3.4.8}$$

This time is when the chemical concentration drops to e^{-1} level due to mixing and chemical reaction in the aquifer (Section 3.4.3).

Albeit Eq. (3.4.6) is valid for steady flow or transient flow situations, we first consider the steady flow condition, in which the saturated thickness of the aquifer is time-invariant. Then the water balance equation,

$$n\frac{dh}{dt} + a(h - h_o) = \varepsilon + q_r - q_p \tag{3.4.9}$$

becomes

$$a\,(h - h_o) + q_p = \varepsilon + q_r. \tag{3.4.10}$$

Eq. (3.4.10) states that outflow (natural discharge and groundwater withdrawal) equals the inflow (natural recharge and artificial recharge). From this equation, the steady thickness of the saturated zone can be determined, which could be used in the following chemical mass balance equation:

$$\frac{dC}{dt} + \frac{(\varepsilon + q_r + knh)}{nh} = \frac{\varepsilon C_\varepsilon + q_r C_r}{nh}. \tag{3.4.11}$$

For simplicity, let $q_r = 0$, and we have

$$\frac{dC}{dt} + \left(\frac{\varepsilon}{nh} + k\right)C = \frac{\varepsilon}{nh}C_\varepsilon. \tag{3.4.12}$$

Next, we derive analytical solutions for different scenarios using Eq. (3.4.12) in the following sections.

3.4.1 Analytical Solutions

The mathematical solution technique for this section is similar to those in Section 2.4.2 of Chapter 2. Please see that section for the details. For an impulse input of a concentration C_ε into an aquifer with an initial concentration, $C_o = 0$ and a steady-state flow, the solution is

$$C(t) = C_\varepsilon e^{-k_e t} = C_\varepsilon e^{-t/t_C} \tag{3.4.13}$$

where $t_C = k_e^{-1}$, the chemical response time of the aquifer.

For a step input, the solution assuming $C_o = 0$ is

$$C(t) = C_\varepsilon \frac{k_r}{k_e}\left(1 - e^{-t/t_C}\right). \tag{3.4.14}$$

For a linear increase in input concentration, $C_\varepsilon = \alpha t$, where α is a constant, the solution is given as

$$C(t) = \alpha\left[(t - t_C) + t_C e^{-t/t_C}\right]. \tag{3.4.15}$$

For large $t \gg t_c$, the solution becomes

$$C(t) = \alpha(t - t_C). \tag{3.4.16}$$

This equation shows that the output concentration lags behind the input concentration by the response time t_c.

3.4.2 Numerical Solutions

The chemical response under transient flow conditions can also be studied by numerically solving the following water and chemical mass balance equations.

$$n\frac{dh}{dt} + a(h - h_o) = \varepsilon + q_r - q_p \tag{3.4.17}$$

$$\frac{dC}{dt} + \frac{(\varepsilon + q_r + knh)}{nh} C = \frac{\varepsilon}{nh} C_\varepsilon + \frac{q_r}{nh} C_r. \tag{3.4.18}$$

A more general approach suitable for any arbitrary input form is a finite difference approach. If we use the forward finite difference approach to discretize the flow equation in time (Chapter 5), we then have

$$n\frac{(h_{i+1} - h_i)}{\Delta t} + a\left(\frac{h_{i+1} + h_i}{2} - h_o\right) = \varepsilon_i + q_{ri} - q_{pi} \tag{3.4.19}$$

where the subscript i is the time step index, and the solution is given by

$$\left(1 + \frac{1}{2}\frac{\Delta t}{t_h}\right) h_{i+1} = h_i - \frac{\Delta t}{2 t_h} h_i + \frac{\Delta t}{t_h} h_o + \Delta t \frac{\left(\varepsilon_i - q_{ri} - q_{pi}\right)}{n} \tag{3.4.20}$$

where $t_h = n/a$ is the hydraulic response time. Similarly, we can discretize the chemical mass balance equation to obtain

$$C_{i+1} - C_i = -\frac{\Delta t}{n}\left[(\varepsilon_i + q_i)\left(\frac{C_{i+1} + C_i}{2}\right) - (\varepsilon_i C_{\varepsilon i} + q_{ri} C_{ri})\right]\left(\frac{2}{h_{i+1} + h_i}\right). \tag{3.4.21}$$

Once the heads at different times are known, they can be used in Eq. (3.4.21) to calculate the concentration at various times. Because of the forward difference, small time steps are necessary to control the solutions' stability and accuracy.

3.4.3 Hydraulic Response, Water Retention, and Chemical Response Time

Definitions of these times are similar to those in the surface water reservoirs in Chapter 2. However, they differ because of the involvement of aquifer characteristics (i.e., porosity, n, and outflow constant, a). In the groundwater system, **hydraulic response time**, t_h, is when a sudden increase of $h(\Delta h = h - h_0 = \varepsilon_0/n)$ due to recharge ε_0 dissipates to $\Delta h e^{-1}$ level (Fig. 3.3). It is related to the dissipation of energy, and it depends on the aquifer characteristics, n and a.

The chemical response time of a groundwater reservoir is defined as the time when the chemical concentration C_ε of a slug input is diluted to $C_\varepsilon e^{-1}$ level due to mixing and chemical decay in a steady flow system. The chemical response time is the reciprocal of k_e,

$$t_C = \frac{1}{k_e} \tag{3.4.22}$$

where

$$k_e = \frac{\varepsilon}{nh} + k. \tag{3.4.23}$$

If the chemicals are conservative, $k = 0$, the chemical response time becomes

$$t_C = \frac{1}{k_r} = \frac{nh}{\varepsilon} = t_r \tag{3.4.24}$$

or the water retention time t_r, which physically represents the time a solute slug advects through an aquifer a saturated thickness of h without mixing or the time to fill the empty groundwater reservoir to a thickness of h. On the other hand, the chemical response time denotes the time when the input conservative solute concentration has reached its e^{-1} level due to mixing and dilution. While the physical meanings of these times are different, the result indicates that the retention time can be estimated via a tracer test in a steady flow system. See the explanation of Figs. 2.7 and 2.8 in Section 2.4.2.1).

Notice that the water retention time differs from the hydraulic response time ($t_h = n/a$, see Eq. 3.2.15). The former characterizes the groundwater reservoir's size and the advection (inflow magnitude). The latter quantifies the reservoir's ability to propagate or dissimilate an excitation (recharge or energy) to its e^{-1} level under a steady flow system.

Albeit both the chemical and hydraulic response times designate the time to dissipate the input to its e^{-1} level, the chemical response time refers to a solute and the hydraulic response time to energy. The hydraulic response time t_h is much shorter than the chemical response time because energy dissipation is much faster than the solute advection or mixing of solutes. These concepts are fundamental to the understanding of groundwater flow and solute transport.

3.4.4 Application to Highway De-Icing Salts

Below we present an application of the well-mixed model to investigate highway de-icing salts' effects on Eastern Massachusetts aquifers (Gelhar and Wilson, 1974). The study solved the water balance model in time first,

$$n\frac{dh}{dt} + a(h - h_o) = \varepsilon \tag{3.4.25}$$

then solve the chemical mass balance model,

$$nh\frac{dC}{dt} + \varepsilon C = \varepsilon C_\varepsilon. \tag{3.4.26}$$

The study assumed that $q_r = q_p = 0$, and $C_r = 0$ since groundwater withdrawal is approximately 2 inches/yr and recharge through septic tanks in the area is minimal. Solutions for the two equations were obtained by using finite difference analogs of Eqs. (3.4.25) and (3.4.26) with $\Delta t = 1$ month. Required input parameters, ε, C_ε, n, and a, are estimated as follows:

(1) Recharge rate, ε, (inches/month). Determining the monthly recharge rate assumes that the annual recharge $\bar{\varepsilon}$ is linearly proportional to the annual precipitation $\overline{P}$, which is $\bar{\varepsilon} = 0.35\overline{P}$. Then, the annual recharge is distributed over January through May, according to Fig. 3.10.
(2) Salt Input:

The city government's records indicated that 12 tons/lane mile/year road salt had been used, and the width of the highway is assumed to be 15 ft. This information led to an estimate of the amount of road salt used per square highway area.

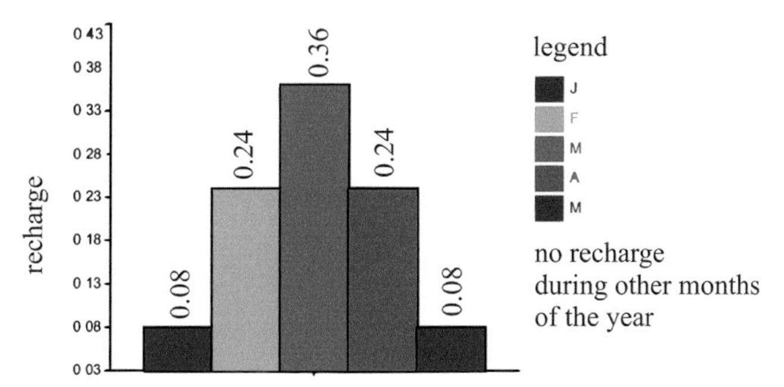

Figure 3.10 The distribution pattern for the annual recharge. (Gelhar and Wilson, 1974).

$$\frac{12 \times 2000\,\text{lb}}{15\,\text{ft} \times 5280\,\text{ft}} = \frac{0.3\,\text{lb Salt}}{ft^2\,\text{Highway}}.$$

The highway density is 5 % of the aquifer area. The road salt per aquifer area becomes

$$\frac{0.3\,\text{lb Salt}}{\text{ft}^2\ \text{Highway}} \times \frac{5\ (\text{area of highway})}{100\ (\text{area of Aquifer})} = 1.5 \times 10^{-2} \frac{\text{lb Salt}}{\text{ft}^2\ \text{Aquifer}}.$$

One could multiply this with the fraction of salt that reaches the water table.

If annual precipitation is

$$\overline{P} = 4\ \text{ft/yr} \quad \bar{\varepsilon} = 0.35 \times 4 = 1.4\ \text{ft/yr}$$

$$C_\varepsilon = \frac{\text{Mass of Salt}}{\text{Mass of Water}} = \frac{1.5 \times 10^{-2}\ \text{lb/ft}^2}{1.4f \times 62.4\text{lb/ft}^3} = 1.72 \times 10^{-4} = 172 \underbrace{\times 10^{-6}}_{\text{ppm}} = 172\ \text{ppm}$$

(3) Aquifer Parameters.

The study used $n = 0.25$, $h_o = 25$, 100, 400 ft, and $L = 500$, 5000, and 50000 ft and found the hydraulic response time ranging from 0.08 to 52 years and chemical response time from 6.25 to 100 years, indicating a slow change in water quality.

The above steps lead to the simulated and observed water at well 78, Willimington, Massachusetts, as illustrated in Fig. 3.11. Overall, the simulated water level matches well with the observed at the well satisfactory. Discrepancies are expected since the comparison was based on one well only, and the model yields the ensemble mean.

Likewise, the comparison between observed and simulated concentrations (Fig. 3.12) shows that the well-mixed model also predicts the averaged chloride concentration over the basin at different years.

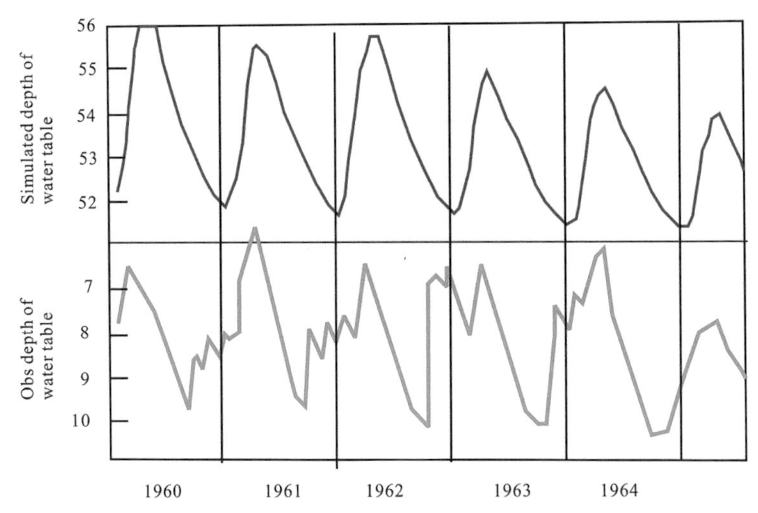

Figure 3.11 The simulated and observed water at well 78, Willimington, Massachusetts. (Gelhar and Wilson, 1974).

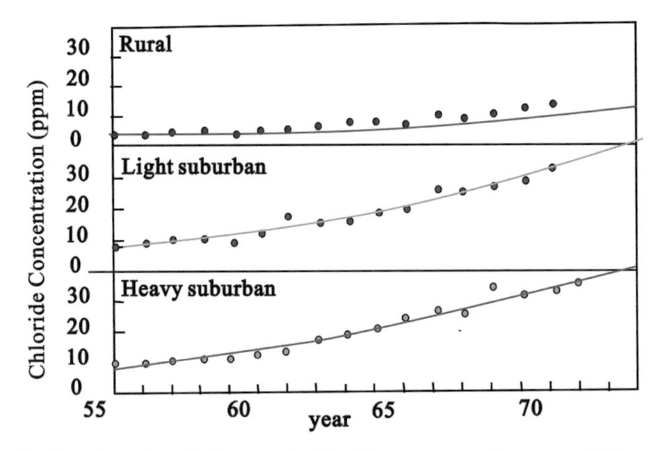

Figure 3.12 Simulated and observed chlorite concentrations in different areas in the basin. The solid lines represent the simulated, and circles are the spatially averaged observed data. (Gelhar and Wilson, 1974).

These results suggest that the model could be a viable and straightforward first-cut tool for decision-makers to investigate the effects of different applications of highway de-icing salts on the groundwater quality in the basin. As suggested, Fig. 3.13 plots the effects of different strategies for applying highway de-icing salts on the basin's groundwater quality.

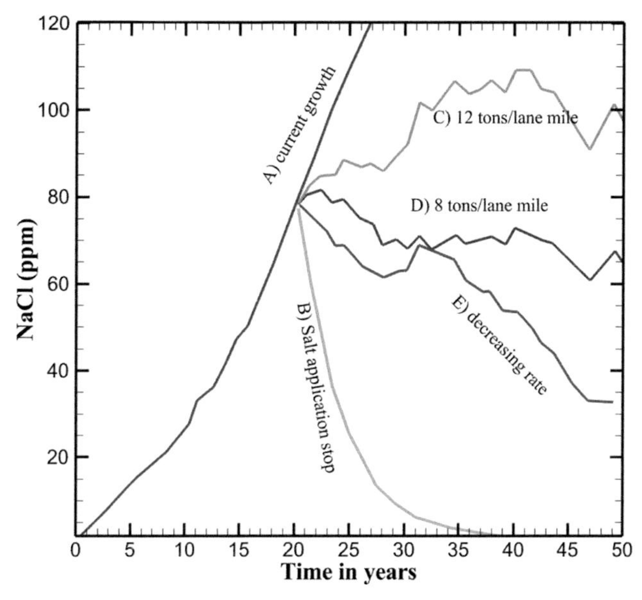

Figure 3.13 An illustration of the utility of the well-mixed model for management of highway de-icing salt application. (Gelhar and Wilson, 1974).

3.5 Well-Mixed Models with Reactive Chemicals

A chemical reaction is a process that leads to the chemical transformation of chemical substances to another. Chemical reactions encompass changes that only involve electrons' positions in the forming and breaking chemical bonds between atoms, with no change to the elements present. Without exploring the detailed mechanisms, kinetics describes the rate of change in a chemical concentration in the volume-averaged sense. Thus, reaction kinetics is an ensemble mean approach.

In the ensemble sense, the rate depends on various factors, such as:

Reactant concentrations usually make the reaction faster if raised through increased collisions per unit time. Some reactions, however, have rates that are independent of reactant concentrations.

Surface area available for contact between the reactants, particularly the solid surface area, plays a critical role. Larger surface areas lead to higher reaction rates.

Pressure–increasing the pressure decreases the volume between molecules and increases molecules' collision frequency.

Activation energy defines the energy required to make the reaction start and carry on spontaneously. Higher activation energy implies that the reactants need more energy to start than a lower activation energy reaction.

Temperature hastens reactions if raised since higher temperature increases the energy of the molecules, creating more collisions per unit time,

The presence or absence of **catalysts**, substances that change the pathway (the mechanism) of a reaction, increases the reaction's speed by lowering the activation energy needed to occur.

3.5.1 Reaction Kinetics Models

The study of reactive solute migration in groundwater focuses on the interaction between solutes in the groundwater and the geologic media. A conceptual model in Fig. 3.14 explains the interaction. The model views the reactive solute's migration in the aquifer as two well-mixed systems: one for the solute in the groundwater (the right in the figure) and the other for the aquifer matrix (the left in the figure). The former involves groundwater flow due to natural discharge and recharge, artificial recharge, and pumping. The latter is the solute staying on the aquifer matrix but undergoing chemical exchange with the migrating groundwater solute. For this reason, we say the solute in the groundwater is the water phase solute concentration, and the solute in the matrix is the solid phase solute concentration. The solute exchange between the water and solid phases (below the scale of the fluid-dynamic-scale velocity) is considered to be dictated by molecular diffusion (Chapter 4) and is called the mass transfer process.

The mass transfer process leads to changes in the solute concentration in the water and solid phases. As such, we must consider the chemical mass balances in each phase.

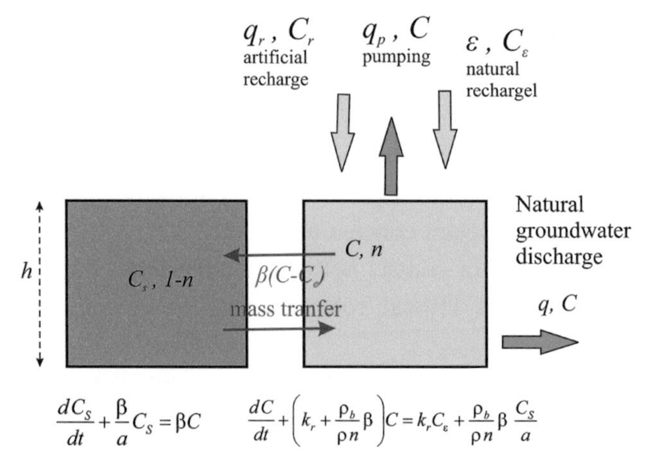

Figure 3.14 The conceptual model for interaction between solutes in the water and solid phases in an aquifer. The equations at the bottom are those to be formulated at the end of this section.

3.5.2 Chemical Mass Balance for Water Phase

Recall the chemical mass balance equation for the water phase in the previous section:

$$n\frac{dhC}{dt} = C_{\varepsilon}\varepsilon - Cq - S \quad \text{if} \quad q_p = q_r = 0 \tag{3.5.1}$$

and ρ = constant; S = sink or source term due to chemical reactions, representing ion exchange (adsorption and desorption) or mass transfer between the chemical in water and solid phases. It will be described by a kinetic model below.

Consider a unit volume of porous media in which the solid occupies a fraction (volume fraction of solid = $1 - n$, where n is the porosity.) The solid density is $\rho_S = M_S/V_S$, where M_S and V_S are the mass and volume of the solids, respectively. The bulk density is $\rho_b = M_S/V_T$, where M_S and V_T are the solid mass and total volume of the porous medium, consisting of solids and voids. The relationship between the bulk density and the solid density is

$$\rho_b = \frac{M_S/V_S}{V_T/V_S} = \frac{\rho_S}{1/(1-n)} = (1-n)\rho_S. \tag{3.5.2}$$

These definitions are to be used in the subsequent formation of the mass transfer process.

3.5.3 Chemical Mass Balance for Solid Phase

Let C_s be the concentration of a chemical X in the solid phase. Mass balance states that the mass change in the solid phase per V_T due to chemical reactions equals the net mass of α removed from or added to the fluid per V_T. For a unit volume of a porous medium, this change in the mass in the solid phase is

$$\frac{d\rho_S(1-n)\,C_S}{dt} = n\rho\gamma \tag{3.5.3}$$

where γ is the reaction rate constant, [1/T]. The equation can also be written in terms of the bulk density as

$$\frac{d\rho_b C_S}{dt} = n\rho\,\gamma \tag{3.5.4}$$

Dimension analysis of the equation yields

$$\left[\frac{M_S}{V_T}\right]\left[\frac{M_\alpha}{M_S}\right]\left[\frac{1}{T}\right] = \left[\frac{V_w}{V_T}\right]\left[\frac{M_w}{V_w}\right]\left[\frac{M_\alpha}{TM_w}\right]$$

Eq. (3.5.4) can be rewritten as

$$\frac{\rho_b}{n\rho}\frac{dC_S}{dt} = \gamma \tag{3.5.5}$$

The next step is to relate the solid phase concentration to the liquid phase's concentration. Two chemical reaction models (i.e., equilibrium and nonequilibrium models) are considered.

3.5.4 Equilibrium Model

This model assumes that the chemical reaction between the solid and liquid phases is "instantaneous." Specifically, the reaction time is shorter than the groundwater retention time ($t_r = nh/\varepsilon$), which is the time for groundwater traveling through the CV – the entire aquifer in the well-mixed model. Under this situation, the equilibrium model generally adopts the linear isotherm model:

$$C_S = aC \left[\frac{M_\alpha}{M_S}\right] = \left[\frac{M_w}{M_S}\right]\left[\frac{M_\alpha}{M_w}\right] \tag{3.5.6}$$

This linear model assumes that the chemical X concentration in the solid phase C_s is linearly proportional to the X in the fluid phase C. **Notice that the dimensionless partition coefficient a is not the same a as the linear reservoir model, Eq. (3.2.5).** It is the ratio of X concentration in the fluid phase to in the solid phase. This value is equal to or greater than zero. If it is zero, the solute is none-reactive to the chemical on the solid and otherwise. If it is nonzero, the rate of the change in both phases becomes

$$\frac{dC_S}{dt} = a\frac{dC}{dt}. \tag{3.5.7}$$

Eq. (3.5.7) indicates that the rate of change in solid-phase concentration is linearly proportional to that of the water-phase concentration. Therefore, the solid phase mass balance equation can be related to the concentration in the liquid phase:

$$\gamma = \frac{\rho_b}{n}\frac{a}{\rho}\frac{dC}{dt} = \frac{\rho_b}{n}K_d\frac{dC}{dt} \tag{3.5.8}$$

where K_d is the distribution coefficient, i.e.,

$$K_d = \frac{a}{\rho} = \left[\frac{M_w}{M_S}\right]\left[\frac{V_w}{M_w}\right] = \left[\frac{L^3}{M}\right] \tag{3.5.9}$$

Thus, the sink or source term in the mass balance equation, Eq. (3.5.1), for the chemical X is the product of the volume of water (nh) and the rate constant (γ): $S = nh\gamma$. Eq. (3.5.1) becomes

$$\frac{dC}{dt} + \frac{\varepsilon}{nh}C = \frac{\varepsilon}{nh}C_{\varepsilon} - \frac{\rho_b}{n}K_d\frac{dC}{dt} \tag{3.5.10}$$

since the flow is steady and h does not change with t. Rearrange the equation yields

$$\left(1 + \frac{\rho_b}{n}K_d\right)\frac{dC}{dt} + \frac{\varepsilon}{nh}C = \frac{\varepsilon}{nh}C_{\varepsilon}. \tag{3.5.11}$$

If we define the retardation factor as

$$R = 1 + \frac{\rho_b}{n}K_d \tag{3.5.12}$$

which is a dimensionless parameter, then we have

$$\frac{dC}{dt} + \frac{\varepsilon}{nhR}C = \frac{\varepsilon}{nhR}C_{\varepsilon}. \tag{3.5.13}$$

If we further define a chemical response constant,

$$k_{cr} = \frac{\varepsilon}{nhR}. \tag{3.5.14}$$

Its reciprocal is the chemical response time, t_{cr},

$$t_{cr} = \frac{nhR}{\varepsilon}. \tag{3.5.15}$$

This chemical response time is the water retention time multiplied by R. Recall that the water retention time is the travel time of a slug of water advecting through the aquifer without mixing. In the case of a well-mixed conservative solute model, it is the time when the concentration drops to e^{-1} level of the input concentration. As shown in Eq. (3.5.15), the effect of the equilibrium reaction is manifested by R. If $R = 1$, a conservative chemical, its chemical response time is the same as the water retention time. If $R > 1$, the chemical reaction slows down the solute advection velocity and increases the water retention time. A real-life analogy is that stopping at rest areas for resupply and relief during a trip (although the stopping time is short or "instantaneous") delays the trip while the number of travelers remains the same. More specifically, the reaction prolongs the time for the input concentration to drop to the e^{-1} level or the water retention time. Therefore, R is called the retardation factor.

For chemicals that can undergo both decay, adsorption, and desorption, the mass balance equation is

$$\frac{dC}{dt} + (k_{cr} + k)C = k_{cr}C_{\varepsilon} \tag{3.5.16}$$

where k is the decay constant. Notice that the chemical reaction between the solid and liquid phases increases the chemical response time, but the chemical decay decreases it (see Section 2.3.2.1). Notice that the decay reduces the solute mass, and the equilibrium reaction does not.

3.5.5 Non-Equilibrium Model

This model considers chemicals that do not react with solids "instantaneously," or the chemical equilibrium time is much longer than the groundwater retention time. In other words, the reaction has not reached equilibrium after the chemical exits the aquifer. As a result, the transient effect must be considered, and therefore, the mass balance for the solid phase, assuming the first-order reaction, is

$$\frac{dC_S}{dt} = \beta(C - C_e) \tag{3.5.17}$$

where β is the first-order reaction rate, [1/T], and C_e denotes the equilibrium concentration in the liquid phase. That is,

$$C_e = \frac{C_S}{a} \qquad or \qquad C_S = aC_e \tag{3.5.18}$$

Subsequently, the mass balance equation for the chemical in the liquid phase is

$$\frac{dC}{dt} + \frac{\varepsilon}{nh}C = \frac{\varepsilon}{nh}C_\varepsilon - \frac{\rho_b\beta}{n\rho}\left(C - \frac{C_S}{a}\right) \tag{3.5.19}$$

Eq. (3.5.19) contains solute concentrations in water and solid phases, which cannot be solved unless the water or solid phase's concentration is known. We need to couple the mass balance equations for the solid and liquid phases to determine the chemical concentration in either the solid or liquid phase. Specifically, we must solve the following equations simultaneously, considering the chemical interaction in both phases.

$$\begin{aligned} &\frac{dC}{dt} + \left(k_r + \frac{\rho_b}{\rho n}\beta\right)C = k_r C_\varepsilon + \frac{\rho_b}{\rho n}\beta\,\frac{C_S}{a} \\ &\frac{dC_S}{dt} + \frac{\beta}{a}C_S = \beta C \end{aligned} \tag{3.5.20}$$

Before solving the equations, we investigate the meaning of the nonequilibrium model (1st Order). Let's consider Eq. (3.4.17),

$$\frac{dC_S}{dt} = \beta(C - C_e) \tag{3.5.21}$$

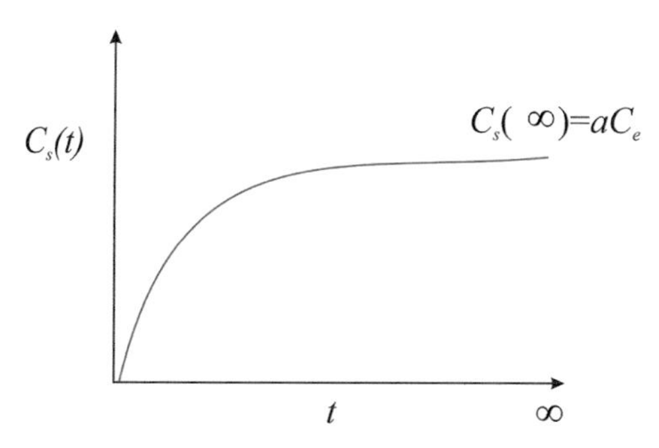

Figure 3.15 Evolution of the solid concentration as a function of time, based on a first-order nonequilibrium reaction model.

where $C_e = C_S/a$. Using this relationship, we rewrite the mass balance equation as

$$\frac{dC_S}{dt} + \frac{\beta}{a} C_S = \beta C. \tag{3.5.22}$$

The solution to this equation yields the concentration in the solid phase with a given fixed chemical concentration in the liquid phase is

$$C_S(t) = aC\left(1 - e^{-\frac{\beta}{a}t}\right) \tag{3.5.23}$$

if the initial solid concentration is zero, $C_S(0) = 0$. The solid phase concentration behavior is graphed in Fig. 3.15, showing that the model becomes an equilibrium model, $C_S(\infty) = aC$, when the time approaches infinity.

If we examine the problem differently, we observe that at a given $t = t^*$, the solid concentration is related to any liquid concentration, C, by

$$C_S(t^*) = a\left(1 - e^{-\frac{\beta}{a}t^*}\right)C \tag{3.5.24}$$

which can be rewritten as

$$C_S(t^*) = A(t^*)C. \tag{3.5.25}$$

The above equation suggests that the reaction constant, A, is a time-dependent variable. However, the reaction between chemicals in the solid and the liquid phases remains linear at any given time, as illustrated in Fig. 3.16, demonstrating the linear reaction at any time.

Questions: How to design laboratory batch experiments to determine if the solute of a chemical behaves as a first-order equilibrium model or a first-order nonequilibrium model?

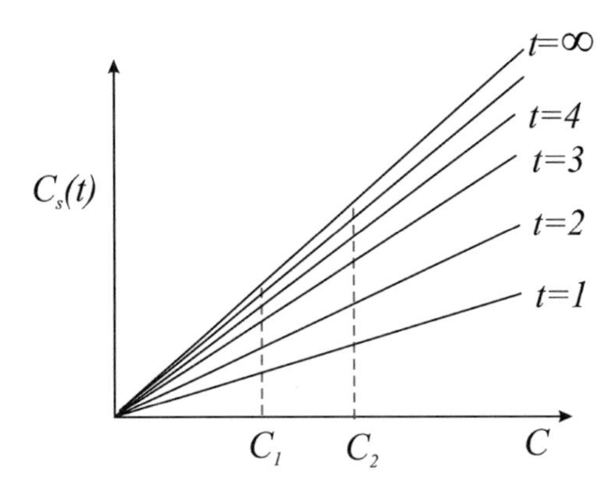

Figure 3.16 Nonequilibrium linear reaction model still retains the linear relationship between the chemical concentration in the solid phase $C_s(t)$ and different liquid chemical concentrations C at any given time.

3.6 Sensitivity and Uncertainty Analysis

A model's sensitivity is the amount of change in the model output due to a unit change of a model parameter. Accordingly, one derives the output of a model based on a set of model parameter values and then obtains another output using a new parameter set in which one of the parameters is perturbed by a unit. The difference between the outputs presents the model's sensitivity to a unit change in that parameter. This approach to sensitivity analysis is the most straightforward perturbation analysis.

On the other hand, uncertainty analysis quantifies the variation of the output of a model due to the parameter's unknown values. Specifically, the parameter's unknown values compel the parameter to be a random variable, characterized by a probability distribution with a given mean and variance (the normal PDF is optimal). Intuitively, using the statistics information, one creates many possible parameter values and, in turn, generates many model outputs. This procedure is called **Monte Carlo simulation**. Statistical analysis of all the outputs yields the mean and variance of the output, representing the most likely output and possible deviations from the most likely one.

Alternatively, a first-order analysis can be applied to the sensitivity and the uncertainty analysis. It takes advantage of the Taylor series in mathematics. The Taylor series expresses a nonlinear function as an infinite sum of terms expressed in terms of the function's derivatives at a single point. For example, a function is evaluated at a point $x = \bar{x} + x'$. Taylor Series states that $f(\bar{x} + x')$ is

$$f(\bar{x}+x') = f(\bar{x}) + \underbrace{x'\frac{df(\bar{x})}{dx}}_{\textit{first order}} + \underbrace{\frac{x'^2}{2}\frac{d^2f(\bar{x})}{dx^2} + \cdots}_{\text{higher order}} \tag{3.6.1}$$

For mathematical simplicity, a first-order approximation neglects the higher-order terms of the series but retains the zeroth and first-order terms such that

$$f(\bar{x}+x') \approx f(\bar{x}) + x'\,df(\bar{x})/dx \tag{3.6.2}$$

a first-order approximation of $f(\bar{x}+x')$. Precisely, the function value can be approximated by a linear extrapolation of $f(\bar{x})$ by adding the increment x' multiplied by the slope at the function evaluated at $\bar{x}$. If the function is linear, the approximation is exact, and Eq. (3.6.2) can be expressed as

$$\Delta f = f(\bar{x}+x') - f(\bar{x}) \approx x'\,df(\bar{x})/dx. \tag{3.6.3}$$

Specifically, the change in the function's value Δf due to a change in the parameter value x' is approximated as the product of x' and the function's first derivative evaluated at the original parameter value $\bar{x}$. This derivative is the function value change due to a unit change of the function parameter at the current parameter value (i.e., the function's first-order sensitivity).

Let us apply the first-order analysis to the uncertainty analysis of the well-mixed model. Suppose k is an input parameter to a well-mixed model and has some uncertainty due to measurement error or unknown value of the parameter. We like to address the effects of this uncertainty on the model's prediction. For this purpose, first assume that k is a random variable with a PDF, expressed as the mean plus the perturbation:

$$k = \bar{k} + k' \tag{3.6.4}$$

where $\bar{k} = E[k]$ and E denotes the expected value (Chapter 1). The perturbation k' is characterized by a PDF with a zero mean and a variance of σ_k^2. The output (e.g., outflow concentration from a well-mixed model, which varies in time) becomes

$$C(t) = f(t, k) = f\left(t, \bar{k} + k'\right) \tag{3.6.5}$$

where $f(t, k)$ represents the mathematical formulation of the well-mixed model. Since k is a random variable, the output $C(t)$ is a random variable, which can be decomposed into mean and perturbation parts.

$$C(t) = E[C(t)] + c'(t) = \overline{C}(t) + c'(t) \tag{3.6.6}$$

The term $c'(t)$ is a perturbation. Taking the expected value on both sides of the equation leads to the mean value of the output

$$E[C(t)] = E[E[C(t)]] + E[c'(t)] = E[C(t)] = \overline{C}(t). \tag{3.6.7}$$

Next, we approximate that $\overline{C}(t)$ equals the function (the well-mixed model) evaluated using the mean parameter value, $\bar{k}$:

$$\overline{C}(t) \approx f\left(t, \overline{k}\right). \tag{3.6.8}$$

This approximation is exact if f is a linear function. We can now use the first-order analysis to determine the uncertainty of output associated with the uncertain input.

$$C(t) = f\left(t, \overline{k} + k'\right) \approx f\left(t, \overline{k}\right) + \left.\frac{df}{dk}\right|_{\overline{k}} \left(k - \overline{k}\right) \tag{3.6.9}$$

or

$$C(t) = \overline{C}(t) + c'(t) \approx \overline{C}(t) + \left.\frac{df}{dk}\right|_{\overline{k}} k' \tag{3.6.10}$$

The vertical bar with subscript $\overline{k}$ indicates that the derivative of f is evaluated at the mean parameter value $\overline{k}$. It is called the sensitivity of the output to a unit change in the parameter value.

Subtracting $\overline{C}(t)$ from both sides of Eq. (3.6.10) leads to the relationship between output perturbation $c'(t)$ and input perturbation k'. That is,

$$c'(t) \approx \left.\frac{df}{dk}\right|_{\overline{k}} k'. \tag{3.6.11}$$

Then, the variance of the output is

$$E\left\{\left[c'(t)\right]^2\right\} = \left(\left.\left(\frac{df}{dk}\right)\right|_{\overline{k}}\right)^2 E\left[\left(k'\right)^2\right] \tag{3.6.12}$$

and can be simplified to

$$\sigma^2_{c'(t)} = \left(\left.\left(\frac{df}{dk}\right)\right|_{\overline{k}}\right)^2 \times \sigma^2_{\overline{k}} = J^2_{ck}\sigma^2_{k}. \tag{3.6.13}$$

The term J_{ck} is the sensitivity evaluated at $\overline{k}$. Based on this first-order analysis, first, find the derivative of the function (the sensitivity), evaluate it at the mean parameter value, and then use Eq. (3.6.13) to obtain the output variance (i.e., the uncertainty of the output). Afterward, derive the standard deviation from the variance and add it to the mean output $\overline{C}(t)$ to derive the output's upper bound. Next, subtract it from the mean to obtain lower bounds. Thus, an uncertainty estimate associated with a model's prediction is obtained.

If $C(t)$ or $C(t, k, n)$ is a function of two parameters k and n, we need to expand the function into the Taylor series with two variables. Then, the above procedure could lead to the following expression of the variance of $C(t)$.

$$\sigma^2_{c'(t)} = \left(\left.\frac{\partial f}{\partial k}\right|_{\overline{k},\overline{n}}\right)^2 \sigma^2_{\overline{k}} + 2\left(\left.\frac{\partial f}{\partial k}\right|_{\overline{k},\overline{n}}\right)\left(\left.\frac{\partial f}{\partial n}\right|_{\overline{k},\overline{n}}\right)\mathrm{cov}(k, n) + \left(\left.\frac{\partial f}{\partial n}\right|_{\overline{k},\overline{n}}\right)^2 \sigma^2_{n} \tag{3.6.14}$$

If k and n are independent, the covariance term, $\mathrm{cov}(k, n)$, is zero, the variance becomes

$$\sigma^2_{c'(t)} = \left(J_{ck}|_{\bar{k},\bar{n}}\right)^2 \sigma^2_k + \left(J_{cn}|_{\bar{k},\bar{n}}\right)^2 \sigma^2_n. \tag{3.6.15}$$

Eqs. (3.6.13) and (3.6.15) manifest that the output variance depends on the sensitivity and variance of the input parameters. A small $C(t)$ sensitivity to a parameter (e.g., k) does not mean that the uncertainty of k to the output uncertainty is neglectable. The magnitude of the variance of k must also be considered. That is to say, sensitivity analysis alone does not tell the entire story of the output variance. This fact should be obvious but is often ignored by many.

3.7 Applications and Uncertainty Analysis

Anderson and Bowser (1986) presented a study of groundwater's role in lake acidification in Wisconsin. They used a two-dimensional finite difference numerical flow and solute transport model (spatially distributed model) for their investigation. This section uses their study to demonstrate that a well-mixed model is a well-suitable tool for this investigation. Fig. 3.17 shows the two-dimensional cross-section conceptual model in the study to investigate the interaction between precipitation, groundwater reservoir, and the lake.

In applying the well-mixed model to this problem, we conceptualize the groundwater and lake system interaction as two connected well-mixed models (one for the aquifer and the other is the lake), as illustrated in Fig. 3.18.

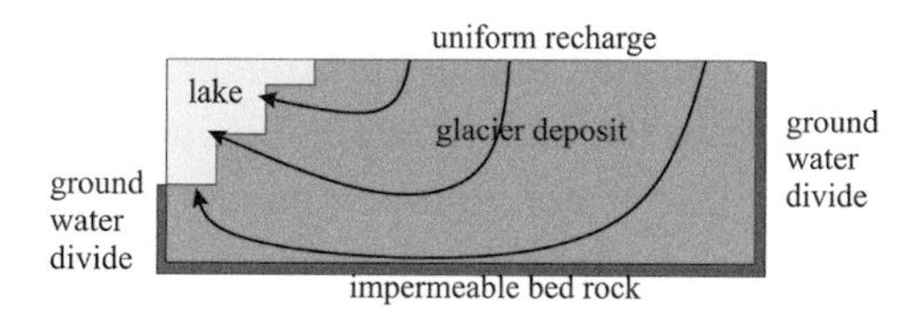

Figure 3.17 The conceptual model for the two-dimensional numerical for the study by Anderson and Bowser (1986).

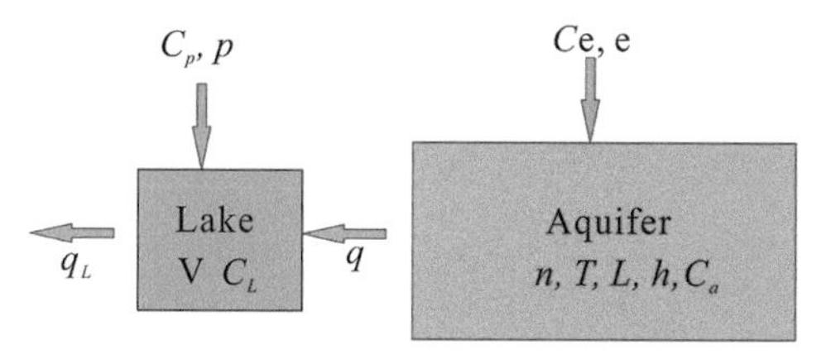

Figure 3.18 The conceptual model of the interaction between the aquifer and lake.

Suppose the discharge from the groundwater to the lake is at a steady state. The chemical reaction between the acid rain recharging to the aquifer and aquifer media follows the equilibrium reaction model (Section 3.5.4). Assuming the recharge and acid concentration are a step input, we have the following analytical solutions for flow,

$$q = \varepsilon = a(h - h_0) \quad a = 3T/L^2 \tag{3.7.1}$$

where q is the discharge from the groundwater to the lake, notice that a is not the partition coefficient and ε is the recharge to the groundwater. $T = Kh$ is the transmissivity, K is hydraulic conductivity, h is the aquifer's average water level, h_0 is the groundwater discharge datum or the steady-state lake level, and L is the aquifer length. The chemical mass balance equation for the aquifer system is

$$\frac{dC_a}{dt} + AC_a = AC_\varepsilon. \tag{3.7.2}$$

where C_a is the pH of aquifer water and C_ε is the pH of the rain. The solution to the equation for a step input of C_ε is

$$C_a(t) = C_a(0)\mathrm{e}^{(-At)} + C_\varepsilon\left[\left(1 - e^{(-At)}\right)\right]. \tag{3.7.3}$$

where $C_a(t)$ and $C_a(0)$ are the aquifer water pH values at time=*t* and time=*0*, respectively. The term $A = \varepsilon/(nhR)$ where n is the porosity, and R is the retardation factor, see Eq. (3.5.12). The first term on the right-hand side of Eq. (3.7.3) represents the diminishing influence of the initial aquifer pH. The second term represents the increase of the impact of recharge pH.

We next assume that the groundwater discharge pH is instantaneously mixed with the lake's. We further consider that the precipitation amount to the lake is p, its pH is C_p, evaporation from the lake is negligible, and lake's flow system has reached a steady state (i.e., lake water volume remains constant). The well-mixed chemical balance equation of the lake becomes

$$\frac{dC_L}{dt} + BC_L = IC_a \tag{3.7.4}$$

where C_L is the lake solute concentration, $B = (q+p)/V$ and $I = \left(\mathrm{C_a}q + \mathrm{C}_p p\right)/V$. The volume of lake water is V. Once the aquifer discharge concentration is obtained from Eq. (3.5.18), the term I can be determined, and the solution to Eq. (3.5.19) can be obtained:

$$\begin{aligned} C_L(t) &= E\left[1 - e^{-Bt}\right] - D\left[e^{-At} - e^{-Bt}\right] + C_L(0)e^{-Bt} \\ E &= \left(C_p p + C_\varepsilon q\right)/V; \quad D = q(C_a(0) - C_\varepsilon)/[V(B - A)] \end{aligned} \tag{3.7.5}$$

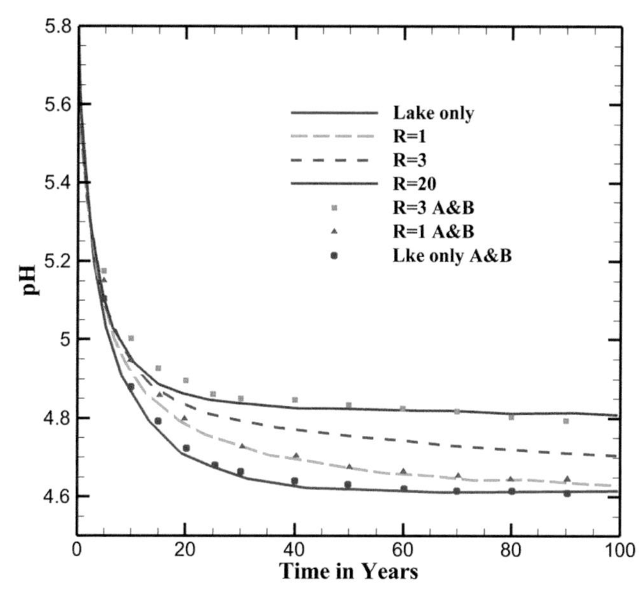

Figure 3.19 The comparison of the simulated PH values by the well-mixed and the two-dimensional distributed models. (Ritz and Yeh, 1986).

where $C_L(0)$ is the initial lake solute concentration. The comparison of the outputs from the well-mixed systems to the two-dimensional distributed model is shown in Fig. 3.19. The parameter values are listed in Rizi and Yeh (1988). The results are in close agreement due to the convergence (i.e., mixing) of the groundwater discharge at the outlet. Besides, the lake is assumed to be well-mixed in the distributed model. This figure also shows that the chemical response time increases as the retardation factor increases.

An analytical solution to the well-mixed model, such as Eq. (3.5.20), always has an advantage over the two-dimensional numerical model. It can explain a system's behavior without performing many simulations as a numerical model. The following first-order analysis (see Section (3.5)) Demonstrates This Advantage By examining the impacts of the uncertainty of parameters on the lake's pH value. As shown in Fig. 3.20, the vertical axis is the coefficient of variation squared (i.e., the output variance divided by its mean squared or the relative uncertainty), avoiding the effects of different mean values of the parameters.

3.8 Final Remarks

The well-mixed model is an ensemble mean model for the entire aquifer under investigation, avoiding complicated mathematics to describe spatially distributed processes. It is relatively straightforward and is a fundamental conceptual model for

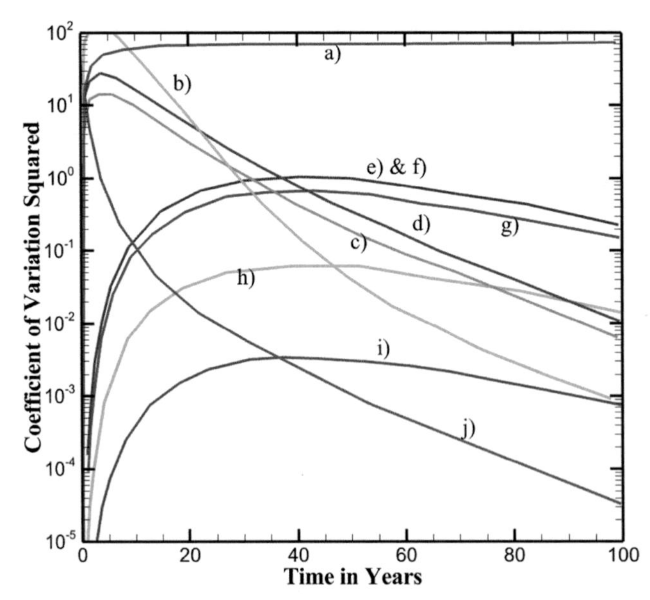

Figure 3.20 Plots of coefficient of variation squared of a well-mixed model for the study of the role of groundwater in delaying lake acidification. (Ritz and Yeh, 1986).

any complex natural system. It facilitates a better understanding of the principles behind complicated spatially distributed models in the later chapters.

Lastly, while the well-mixed model requires simple input information and parameters, the model's accuracy still demands detailed spatial and temporal data to correctly establish the system's total input (i.e., inflow and outflow). Its ensemble-mean output also demains detailed spatiotemporal data. Nevertheless, the model is a practical first-cut tool before comprehensive multidimensional numerical simulations.

3.9 Homework

(1) Use superposition and the solution from the step input to derive the solution for a slug input and plot the result as a function of time.

(2) During the last decade, intensive agricultural development has occurred in an area by introducing supplemental irrigation from wells using sprinkler irrigation. The irrigation water is pumped from a phreatic (water table) aquifer, which underlies the framing area and serves as a source of individual domestic water supply for the farms. The reliable water supply has made it possible to grow high-value crops using large amounts of nitrogen fertilizers.

Monitoring of some domestic water supply wells in the area has shown an increase of several ppm in the concentration of nitrate expressed as nitrogen (NO_3–N). This increase prompted a study by a state agricultural agency in which several hundred domestic and irrigation wells were sampled. The study showed that the average increase .in the concentration of NO_3–N was 4 ppm above the level 10 years ago (around 0.5 ppm). Although a few domestic wells showed concentrations greater than the U.S. Public Health Service standard of 10 ppm, the study attributed these to poor well construction and concluded **that nitrate contamination would not be a problem**. The study also showed from lysimeter tests on an experimental farm that only 15% of the applied fertilizer nitrogen is being leached below the root zone into the groundwater. Nitrate is usually considered to be a conservative tracer in groundwater.

Using the data given below, develop a simple model of this situation. Do the results from the model support the conclusion of the study?

Average annual precipitation	30 in/yr
Supplemental irrigation	20 in/yr
Evapotranspiration	25 in/yr
Average aquifer thickness	h = 100 ft
Average aquifer porosity	0.25

Average nitrogen fertilizer application 200 lb N/acre yr (1 acre = 43560 ft^2)

The soil in the area is very sandy, and surface runoff is insignificant. Several small streams drain the aquifer.

(3) A sewer construction program has been proposed to control eutrophication in a small lake surrounded by a suburban community. The lake receives practically all of its water from a phreatic groundwater system that surrounds the lake. Evaporation from the lake surface is equal to precipitation. Suburban development is distributed throughout the groundwater basin. Domestic wastewater is currently discharged into individual septic tanks, which drain into the groundwater system. Water supply is also obtained from the ground, and consumptive use is small.

The wastewater discharge into the aquifer has increased nitrate concentration in the aquifer and the lake. Biologists have concluded that nitrogen is the limiting nutrient in this lake. They suggest that nitrate-nitrogen concentration in the lake has to be reduced to less than 1 ppm to control excessive eutrophication.

For the conditions indicated below, estimate the time required after completing the sewer system for the eutrophication control to become effective. Assume that nitrate is conservative in the groundwater system and the lake. Natural nitrate levels in this area are less than 0.1 ppm NO_3–N.

Lake: Surface area 2 mi^2, average depth 25 ft, and current NO_3–N concentration 3 ppm.

Aquifer: Area 10 mi^2, average saturated thickness 75 ft; current NO_3–N concentration 4 ppm; average porosity 0.2; net recharge 12 in/yr.

(4) A large waste disposal basin provides a certain degree of treatment of a degradable contaminant before the water percolates down to an underlying aquifer. The basin receives an average of 2 ft/month of wastewater based on the area of the basin. The average depth of water in the basin is 5 feet. Evaporation from the pond is about 0.5 ft/month, and the remainder of the water seeps down to a phreatic aquifer, which has an area three times that of the basin. The average saturated thickness of the aquifer is 50 feet, and the porosity is 0.2.

The state regulations require the industry to use this basin for waste disposal to demonstrate that the basin will remove at least 90% of the contaminant. Several monitoring wells have been installed throughout the aquifer. After the basin has been operating for 6 months with an average input concentration of 20 ppm in the wastewater, the average concentration of the contaminant observed in the aquifer is 2 ppm greater than the initial concentration.

Will this basin system accomplish the required removal? Use appropriate equations and numerical results to support your conclusion. Assume that the basin seepage is the only significant recharge to the aquifer. It is also reasonable to assume a steady concentration in the basin because the basin is initially filled with contaminated water.

(5) Formulate a lumped parameter model for a chemical that undergoes a nonequilibrium first-order chemical reaction with porous media and derive a step input solution. Plot the normalized outflow concentration as a function of a dimensionless time, assuming the inflow is steady, and the initial concentration of the chemical in the solid phase is zero.

(6) Derive the first-order uncertainty analysis using the analytical solution to the chemical mass balance model with an equilibrium reaction and steady flow, considering uncertainty in porosity, recharge, and chemical reaction rate.

4
Molecular Diffusion

4.1 Introduction

Chapter 1 presents the ensemble average and stochastic concepts, demonstrated in Chapters 2 and 3 using the well-mixed models. The well-mixed models adopt a CV as large as the entire investigation domain (a lake or an aquifer) and quantify the volume-averaged processes, omitting spatial details. Mixing due to physical processes (such as diffusion, winds, turbulences in the lake, or variations in regional flow in aquifers) justifies the well-mixed approach. Even if this mixing does not prevail, the ensemble-average and ensemble REV concepts defend the approach. Nevertheless, the well-mixed model results omit details, failing to satisfy the scale of our interest. Consequently, the development of spatially distributed models becomes necessary.

Before proceeding into spatially distributed models, we introduce the molecular diffusion concept and Fick's Law to explain the mixing phenomena at a fluid-continuum-scale CV. For this purpose, this chapter describes diffusion phenomena first, formulates Fick's law, and develops the diffusion equation afterward. Examining the random velocity of Brownian particles and their dispersal in fluids, this chapter subsequently articulates the probabilistic nature of the molecular diffusion process and why Fick's Law is an ensemble mean law.

Next, this chapter introduces analytical solutions to the diffusion equation for various types of inputs. The advection–diffusion equation (ADE) formulation then follows, which couples the effect of fluid motion at the fluid continuum scale and random motion of fluid molecules at the molecular scale to quantify solute migration. Likewise, we present analytical solutions to the ADE for several input forms and discuss snapshots and breakthroughs for different input forms.

4.2 Molecular Diffusion and Fick's Law

Inspired by the earlier experiments of Thomas Graham, Adolf Fick conducted experiments that dealt with measuring the concentrations and fluxes of salt,

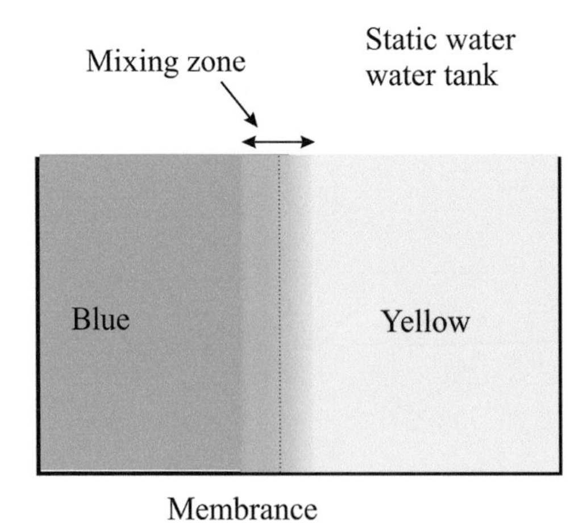

Figure 4.1 A mixing experiment in a static water tank.

diffusing between two reservoirs through tubes of water. In 1855, Fick reported his now well-known experimental principles for the transport of mass through diffusive means. This principle is analogous to those experimental relationships discovered at the same epoch by other scientists: Darcy's law (flow through porous media), Ohm's law (electric charge transport), and Fourier's Law (heat conduction).

Fick's law has become the core of our understanding of diffusion in solids, liquids, and gases. A diffusion process that follows Fick's law is referred to as Fickian and non-Fickian, otherwise.

To elucidate Fick's law and diffusion, consider a water tank where an impermeable membrane separates the tank into two portions: one contains blue and the other yellow water. After suddenly removing the membrane, we observe the formation of a greenish zone at the interface of the two portions (Fig. 4.1). This change in the color is mixing; the greenish zone (the area in which mixing of the blue and yellow particles occurs) is the mixing zone.

As time progresses, the average width of the mixed zone grows. We use the average width to overcome the irregular shape of the mixing zone. Suppose δ, is half the width, and we plot the square of δ as a function of time, and a trend appears, as shown in Fig. 4.2. The slope of the trend, $\Delta\delta^2/\Delta t$, is the rate of the growth of the mixing zone. A constant slope or a linear relation between δ^2 and t implies that the mixing zone grows at a constant rate. This rate generally varies with the fluid's temperature and molecular weight.

The zone's growth indicates a continuous increase in mixing between the two initially separated fluids. As the mixing zone expands and spreads out, it ultimately encounters the ends of the tank. Then, the entire tank becomes uniformly green. At this time, the blue and yellow particles in the tank are well mixed. The term "well-mixed"

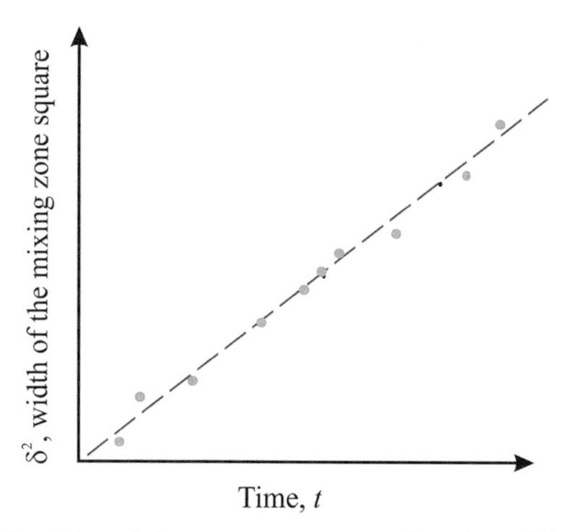

Figure 4.2 Growth of the mixing zone square as a function of time.

means that the yellow and blue particles distribute uniformly over the entire tank. Specifically, taking a small volume (control volume, CV) of the mixed fluid, one finds the number of blue and yellow particles in the CV (i.e., the concentration of blue or yellow particles) is the same at any tank location. Otherwise, the two fluids are poorly mixed (Chapter 1).

Since the water in the tank is static (zero velocity at the fluid-dynamics scale), the mixing is then attributed to the "random" motion of water molecules due to the unknown kinetic activity (internal energy) of molecules. We call this mixing process **Diffusion**, which describes the effects of molecule movements, not defined by the velocity at the fluid-dynamics scale. The use of the word "random motion" avoids the difficulty in describing or measuring each molecule's velocity and reflects our lack of interest in each molecule's behavior (i.e., microscopic behavior). Here, we like to remind readers that environmental fluids move at velocities of a multiplicity of scales (e.g., velocities at macroscopic and microscopic scales or the fluid-dynamics and molecular scales) (Fig. 4.3).

The water-tank experiment also points out that diffusion, mixing, or smearing is a macroscopic-scale phenomenon resulting from observing the number of blue and yellow particles over a control volume (CV) large than many fluid molecules. Accordingly, despite a well or poorly mixed solution, each blue molecule in the solute is still a blue molecule at the molecular level. Likewise, each yellow molecule remains a yellow molecule. The change of color (i.e., mixing) is the product of the volume (ensemble) averaging procedure, reflecting different numbers of blue and yellow molecules in a given volume larger than a molecule itself (i.e., the concentration of yellow or blue molecules in a volume). Since the volume average outcome depends on the size of the CV, mixing, dilution or

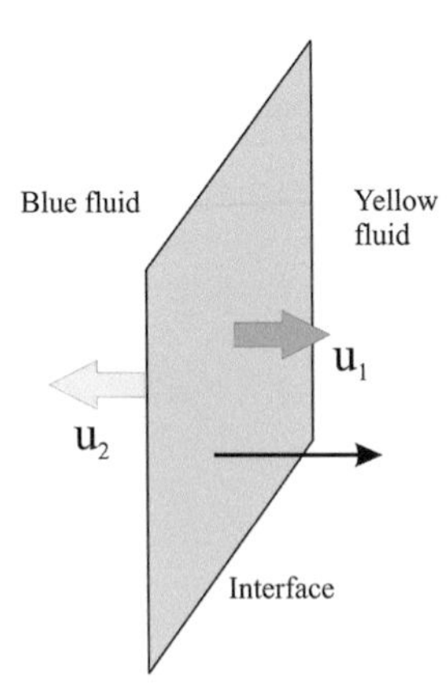

Figure 4.3 Illustration of the movement of yellow and blue fluid particles at the interface.

concentration is a scale-dependent observation, varying with our observation or averaging volume scale.

The water tank experiment further reveals that the solute with a high concentration of blue molecules moves into the fluid where the blue molecules are scarce and vice versa. Therefore, it is logical to speculate that diffusion occurs from a high concentration region to a low concentration region, depending on the concentration gradient. A mathematical analysis below based on a CV larger than many molecules quantifies this conjecture or observation.

Consider the movement of blue and yellow molecules at the interface in the water tank experiment. For the sake of simplicity, we assume that the movement of those molecules is one-dimensional, restricted in the x-axis. Let $C_1(x_1)$ and $C_2(x_2)$ be the concentrations (the number of the yellow molecules within two CVs) at the locations x_1 and x_2 separated by the interface at a given time t. Due to the "random" motion of the molecules, some molecules move from the yellow side to the blue side and vice versa. Thereby, the mass flux per area crossing the interface from blue solute to the yellow solute is $\rho C_1 u_1$ and from the yellow region to the blue region is $\rho C_2 u_2$. The terms u_1 and u_2 represent the velocity of yellow molecules crossing the interface from x_1 and that from x_2, respectively. The density of the fluid is ρ. Then, the net mass flux of yellow molecules per unit interfacial area, ρq [M/L^{2}T], at the interface is given by:

$$\rho q = \rho C_1 u_1 - \rho C_2 u_2. \tag{4.2.1}$$

Note that the concentration unit is mass per mass (ppm), and the internal energy of molecules determines the velocities, which are volume-averaged and temperature-dependent quantities.

If the fluid temperature is constant (i.e., isothermal conditions), yellow molecules collide with each other and also with other molecules due to variation in internal energy among molecules. After the collision, a molecule likely travels at a relatively constant velocity over some distance before colliding with others. The mean free path (ℓ) then denotes the average distance the molecule traverses before it experiences additional collisions and forgets its initial velocity. Within this distance, we may assume that the magnitudes of the velocities u_1 and u_2 are the same, although they have different signs (i.e., different directions)

$$u_m = |u_m| = |u_1| = |u_2| \tag{4.2.2}$$

where | | denotes the magnitude of the velocity without considering the direction. Then the net flux across the interface can be calculated by the formula:

$$\rho q = \rho(C_1 - C_2)u_m. \tag{4.2.3}$$

Assume that x_1 and x_2 are separated by a distance ℓ, i.e., $x_1 = x + \ell$ and $x_2 = x$. We can express net mass flux as

$$\rho q = \rho(C(x + \ell) - C(x))u_m. \tag{4.2.4}$$

Applying a first-order Taylor's series approximation (Chapter 3), the difference in concentration at $x + \ell$ and x divided by ℓ becomes the concentration gradient since the ℓ is small:

$$\frac{(C(x + \ell) - C(x))}{\ell} = -\frac{\partial C}{\partial x}. \tag{4.2.5}$$

The negative sign denotes the flux is from a high concentration to a low one. As a result, the net flux is given by

$$\rho q = -\rho u_m \ell \frac{\partial C}{\partial x}. \tag{4.2.6}$$

Let $D = u_m \ell$ and call it the diffusion coefficient with a dimension $[L^2/T]$. We then have a formula for diffusion along x.

$$q = -D\frac{\partial C}{\partial x}. \tag{4.2.7}$$

Eq. (4.2.7) is **Fick's law**. The above analysis implicitly assumes that all the molecules in the CV follow Eq. (4.2.7). However, as discussed in the later section, molecules move "randomly." For this reason, this equation does not necessarily apply to an individual molecule. Instead, it is suitable for the average of all the possible velocity, concentration, and flux values over many experiments – in the ensemble-average sense. The equation thereby is an ensemble mean equation, as is **Fick's Law** an ensemble mean law.

Generalization of Equation (4.2.7) to multi-dimensions leads to

$$\mathbf{q} = -D\nabla C \tag{4.2.8}$$

Or

$$q_x = -D\frac{\partial C}{\partial x},\ q_y = -D\frac{\partial C}{\partial y},\ q_z = -D\frac{\partial C}{\partial z}. \tag{4.2.9}$$

Fick's law, after all, states that in the ensemble mean sense, the flux of solute mass (the mass of a solute crossing a unit area per unit time in a given direction) is linearly proportional to the gradient of solute concentration in that direction. The proportionality constant is the diffusion coefficient, D [L^2/T], and it could be a tensor (different magnitudes in different directions) but is generally considered a scalar (the same magnitudes in all directions). The dimension of $\mathbf{q}$ is [L/T] if the concentration is mass/mass or ppm; It is [M/L^2T] if the dimension of the concentration is [M/L^3]. According to Eq. (4.2.8), the concentration gradient drives the mass flux, but molecules' random velocity (internal energy) is the real driving force. The concentration gradient is merely a proxy of random velocity effects rather than the driving force of diffusion. For example, suppose we freeze a water tank immediately after releasing a drop of ink in the water. Even though the concentration gradient persists, the ink does not spread since molecules' random motion is virtually zero.

4.3 Diffusion Equation

The above Fick's law recounts how molecules, on average, move at one location. A continuity equation then narrates their movement from one location to another. Consider the continuity equation or mass conservation of the tracer (i.e., the blue or yellow solute) over a CV. The continuity equation states that divergence ($\nabla\cdot$) of the mass flux equals the change in mass in a CV:

$$-\nabla\cdot\rho\mathbf{q} = \frac{\partial\rho C}{\partial t} \quad \text{or} \quad -\left(\frac{\partial\rho q_x}{\partial x} + \frac{\partial\rho q_y}{\partial y} + \frac{\partial\rho q_z}{\partial z}\right) = \frac{\partial\rho C}{\partial t} \tag{4.3.1}$$

In vector calculus, the divergence theorem relates the flux of a vector field through a closed surface to the divergence of the field in the enclosed volume (i.e., the term inside the bracket of the second equation).

Assuming that ρ is constant in time and space, we can rewrite the continuity equation as

$$-\nabla \cdot \mathbf{q} = \frac{\partial C}{\partial t} \text{ or } -\left(\frac{\partial q_x}{\partial x} + \frac{\partial q_y}{\partial y} + \frac{\partial q_z}{\partial z}\right) = \frac{\partial C}{\partial t} \tag{4.3.2}$$

Using Fick's Law for $\mathbf{q}$, we have a general diffusion equation that takes the following form.

$$\nabla \cdot (D\nabla C) = \frac{\partial C}{\partial t} \tag{4.3.3}$$

If D is constant, the diffusion equation becomes

$$D\nabla^2 C = \frac{\partial C}{\partial t} \tag{4.3.4}$$

The diffusion coefficient D depicts the rate of solute spreading due to random collisions of molecules. Theoretical analysis of Brownian motion (Section 4.4) has shown that the spreading rate will take time to reach a constant value (i.e., a constant D value). Once D stabilizes at a constant, Fick's Law becomes valid. In other words, molecules must travel a distance to experience many collisions with other molecules. Then the migrating molecules exercise independent random motion. As such, molecules spread out at a constant rate in the ensemble sense, and Fick's Law is applicable. Generally, reaching a constant spreading rate for most solutes takes less than 10^{-8} seconds. Because this period is much smaller than our observation time scale, we often consider the diffusion coefficient constant.

Similarly, while the diffusion coefficient theoretically is a tensor, we have little information about its tensor property, and we often assume it is a scalar. Also, the diffusion process generally is irreversible because it results from the random motion of molecules. However, accepting Fick's law for diffusion, we inevitably assume that diffusion is a reversible process in the ensemble sense. Likewise, the linear relationship between the mass flux and the concentration gradient in Fick's Law implies that a diffusing solute plume from a point source must be symmetrical around the point in space.

The diffusion equation written in the Cartesian coordinate system in a three-dimensional space is

$$D\left(\frac{\partial^2 C}{\partial x^2} + \frac{\partial^2 C}{\partial y^2} + \frac{\partial^2 C}{\partial z^2}\right) = \frac{\partial C}{\partial t}. \tag{4.3.5}$$

In a two-dimensional x–y space, it is

$$D\left(\frac{\partial^2 C}{\partial x^2} + \frac{\partial^2 C}{\partial y^2}\right) = \frac{\partial C}{\partial t}. \tag{4.3.6}$$

In a one-dimensional x coordinate system, we have

$$D\frac{\partial^2 C}{\partial x^2} = \frac{\partial C}{\partial t}. \tag{4.3.7}$$

Since Fick's law is an ensemble mean law, the diffusion equation, in turn, is an ensemble-mean equation, and it portrays either the ensemble-average behaviors of the diffusion process in a physical experiment or the averaged behaviors of many trials of the same experiment. Thus, the behaviors based on the diffusion equations could differ from our observation in one experiment unless the process lasted over a significant time. For example, they will agree with our observation when the entire water tank is well-mixed. Alternatively, our observation covers a large CV (e.g., observing the water tank experiment at far distances). At this time, the ergodicity requirement of the ensemble-mean equation is met.

Thinking Points

- How would gravity affect the spread of molecules? Should the diffusion equation include the gravity term since the concentration gradient is merely a surrogate of the energy?

4.4 Brownian Motion, Random Walk, Fickian Regime, and Ensemble Mean

The following discussion of Brownian motion explains Fick's law and the diffusion equation's stochastic ensemble mean nature. The botanist Robert Brown (1826) observed sustained irregular motion of particles of colloidal size immersed in a static fluid with a microscope's aid. Such an observed phenomenon remained a puzzle and a subject of controversy for a long time. In 1905, Einstein's paper explained the particle's motion as being moved by individual fluid molecules. The "Brownian motion" is maintained by constant collisions with "randomly" moving molecules of the surrounding fluid. Under normal conditions, a Brownian particle in a liquid suffers about 10^{21} collisions per second. The average magnitude of the impulse due to the collision varies and depends on the liquid's temperature. The direction of the force of atomic bombardment is also continually changing. The particle is hit more on one side than another at different times, leading to the motion's seemingly "random" nature. Such a bombardment causes a particle to wander around the liquid and particles to disperse in the liquid. This explanation thus unravels the myth of the observed Brownian motion. It further proves that microscopic-scale velocity exists beyond the velocity defined at the fluid dynamics scale.

The use of the word "random" rises from our lack of interest and cost-effective technology for quantifying each molecule's movement and the Brownian particle.

For this reason, the statistical theory of Brownian motion is appropriate and may shed some light on the solute transport in environmental fluids. The following brief analysis closely follows the presentation by Csanday (1973).

4.4.1 Fluid Dynamic Analysis

This analysis assumes that the Brownian particle is much larger than the surrounding fluid's molecules. Thus, once the particle moves, it experiences a viscous resistance, which is assumed to be linearly proportional to the velocity of the particle relative to that of fluid molecules (Stoke's law in fluid dynamics): The viscous force per unit mass of the particle is

$$\mathbf{F} = -\beta \mathbf{u} \tag{4.4.1}$$

where β is a constant [1/T]. The term $\mathbf{u}$ is the velocity vector [L/T] of the particle (relative to the fluid).

Besides, the particle is assumed to be small enough to be affected by the surrounding fluid molecules' bombardments. The bombardments exert forces on the particle, which is assumed to be "random" because of our inability to describe and our lack of interest in explicitly quantifying them. These forces cause the particle to accelerate randomly with time, $A(t)$ [L/T^2]

Ignoring the gravity force, the equation describing the movement of the Brownian particles is

$$\frac{d\mathbf{u}}{dt} = -\beta \mathbf{u} + A(t) \tag{4.4.2}$$

The solution to Eq. (4.4.2) is

$$\mathbf{u} = \mathbf{u}_0 e^{-\beta t} + e^{-\beta t} \int_0^t e^{\beta t'} \mathbf{A}(t') dt' \tag{4.4.3}$$

where $\mathbf{u}_0$ is the initial velocity of the particle and t' is the dummy variable for integration. Accordingly, the initial velocity and the impulse of the surrounding fluid molecules affect the particle's velocity at any time. The former, however, decays as time increases, and the latter (the random acceleration) dominates when t or βt is large (i.e., $\mathbf{u}_0 e^{-\beta t} \simeq 0$ and the second term in Eq. (4.4.3) approaches a constant). Physically, Eq. (4.4.3) states that the particle forgets its initial velocity after some time and moves according to the "random" impulses due to collision with the surrounding fluid molecules. The above analysis explains the "random" wondering the Brownian particle and many Brownian particles' spread, as observed by Robert Brown (1826).

Eq. (4.4.3) suggests that the time to reach "random" motion ranges from 10^{-9} to 10^{-3} seconds for a particle with a radius from 10^{-7} to 10^{-3} cm. For this reason, under our observation time scale (greater than 10^{-3} seconds), the particles conduct "random motions" free of the influence of their initial velocities. Again, the word "random" implies that scientists cannot describe the phenomenon with the current technology nor are interested in quantifying explicitly. Hence, a probabilistic analysis is most appropriate. In other words, instead of predicting exactly the particle's movement, we examine the probability of a Brownian particle where it may be after a given time of the random movement.

4.4.2 Stochastic Analysis

The above analysis assumes that we know the Brownian particle's velocity and the validity of using Stoke's law to derive the viscous force and force balance equation (Langevin's equation). In reality, these assumptions may be valid for limited cases. Consequently, a probabilistic (stochastic) analysis is desirable.

In a probabilistic approach, the velocity or displacement of a diffusing particle is a function of time, but its value for any given time is random, which may only be specified as a JPD (Chapter 1). Without loss of generality, we may assume the ensemble mean value of the velocity to be zero, and the movement of the particles is limited to x-direction (one-dimensional) for clarity. We thus write the mean and variance of the velocity as

$$\overline{u}(t) = 0. \text{ and } \overline{u^2}(t) = \text{constant} = \sigma_u^2 \tag{4.4.4}$$

Again, overbars denote the expected values (ensemble averages). Furthermore, a covariance is needed to specify any possible variance relationship between the velocity at t and $t + \tau$ (a requirement of JPD, Chapter 1):

$$R_{uu}(\tau) = \overline{u(t)u(t+\tau)} \tag{4.4.5}$$

τ is the time interval separating the two times. Then, after normalizing the covariance with the variance σ_u^2, we have an autocorrelation function $\rho(\tau)$, a characteristic time measuring the "persistence" of a given velocity value or the similarity of the velocity at different times. Precisely, it means that once the particle possesses a certainty velocity, it is likely to have a velocity of a similar magnitude and sign sometime later within this persistent time.

Of course, the displacement of a particle is related to its velocity by

$$x(t) = \int_0^t u(\eta)d\eta \tag{4.4.6}$$

Note that η is a dummy variable. Eq. (4.4.6) depicts the particle's distance from its original location at t. The distance could be either positive or negative (i.e., a different direction from the origin). After taking the ensemble averages on both sides of Eq. (4.4.6), we find that $\overline{x}(t) = 0$ since $\overline{u}(t) = 0$. This ensemble averaged position denotes a meandering particle's average position at time t, starting from the origin $x = 0$, after repeated experiments. Alternatively, it is the averaged positions of an infinite number of particles at time = t, after their random movements. In either case, particles spread around the origin, and the average of particles' displacements is zero.

Because x could be positive or negative, we square Eq. (4.4.6) to obtain $x^2(t)$ a measure of the particle's deviation from the origin. The rate of change in $x^2(t)$ becomes

$$\frac{d\left(x(t)^2\right)}{dt} = 2x(t)\frac{dx(t)}{dt} = 2\left[\int_0^t u(\eta)d\eta\right](u(t)) = 2\int_0^t u(t)u(\eta)d\eta \tag{4.4.7}$$

We can move $u(t)$ into the integral since it is the particle's velocity at time t and is a constant for the integration. After taking ensemble averages (overbar) of both sides of Eq. (4.4.7), we have.

$$\frac{d\left(\overline{x(t)^2}\right)}{dt} = \frac{d\sigma_x^2}{dt} = 2\int_0^t \overline{u(t)u(\eta)}\ d\eta = 2\sigma_u^2\int_0^t \rho(\tau)d\tau \tag{4.4.8}$$

The term σ_x^2 is the spatial variance, denoting all particles' average deviations from the origin or the average deviation from many trials. Again, the ensemble averaging procedure (expected values) is independent of the differentiation and integration operations. The velocity $u(t)$ is a constant of the integrand for a given t and η varies from 0 to t. For this reason, we can express $\overline{u(t)u(\eta)}$ as the autocorrelation multiplied by the velocity variance.

In statistics, spatial variance is defined as

$$\sigma_x^2 = \int_{-\infty}^{\infty} x^2 P(x, t)dx \tag{4.4.9}$$

The term $P(x, t)$ is the probability density function, describing the particle's probability at location x at time $= t$. Specifically, it is the number of particles at location x at a time t after releasing infinite numbers of particles at $x = 0$. Alternatively, it is the number of times a particle appears at the location x, after it is released at $x = 0$ over many experiments. Using the CV concept of the concentration (Chapter 1), we conclude that the $P(x, t)$ is the same as the concentration at x at time t, $C(x, t)$. As such,

$$\sigma_x^2 = \int_{-\infty}^{\infty} x^2 C(x,t)dx \tag{4.4.10}$$

Eq. (4.4.10) represents the concentration distribution's spatial variance or the spread of a tracer slug at t if ergodicity is met. As a result, Eq. (4.4.8) represents a tracer plume's spreading rate, which depends on the velocity's variance and autocorrelation function.

Suppose the velocity function for a Brownian particle Eq. (4.4.3) is valid. We then could evaluate the velocity autocorrelation function:

$$R_{uu}(\tau) = \overline{u(0)u(0+\tau)} = \overline{u_0\left(u_0 e^{-\beta\tau} + e^{-\beta\tau}\int_0^{\tau} e^{\beta t'} A(t')dt'\right)} = \sigma_{u_o}^2 e^{-\beta\tau} \tag{4.4.11}$$

Again, t' is the dummy variable for integration. If the velocity is a stationary stochastic process (or the velocity variance does not change with time), $\sigma_{u_o}^2 = \sigma_u^2$ (the initial velocity variance is the same at any t). The velocity autocorrelation function is $\rho_{uu}(\tau) = e^{-\beta\tau}$. After substituting the autocorrelation function into Eq. (4.4.8), we have

$$\frac{d\sigma_x^2}{dt} = 2\sigma_u^2\int_0^t e^{-\beta\tau}d\tau = 2\sigma_u^2\left(-\frac{e^{-\beta t}}{\beta}\Big|_0^t\right) = 2\sigma_u^2\left(-\frac{e^{-\beta t}}{\beta} + \frac{1}{\beta}\right) = \frac{2\sigma_u^2}{\beta} = 2D_m$$
$$\text{when } t \gg \frac{1}{\beta} \tag{4.4.12}$$

The term D_m is the diffusivity of the Brownian particles (or molecular diffusivity). In particular, Eq. (4.4.12) tells us that when time is greater than the persistent time ($1/\beta$), the tracer cloud's spreading rate is constant, or the cloud grows at a constant rate, twice the diffusivity value. Further, the particles forget the influence of the initial velocity and exercise purely random (uncorrelated) movements. As a reminder to readers, the conclusion rests on the ensemble-average concept: each particle's behavior may not necessarily follow this finding.

The next logical question is the tracer cloud's shape: is it a normal (or Gaussian) distribution? In probability theory, the central limit theorem (CLT) establishes that when independent random variables are added, their properly normalized sum tends toward a normal distribution even if the original variables do not follow normal distributions.

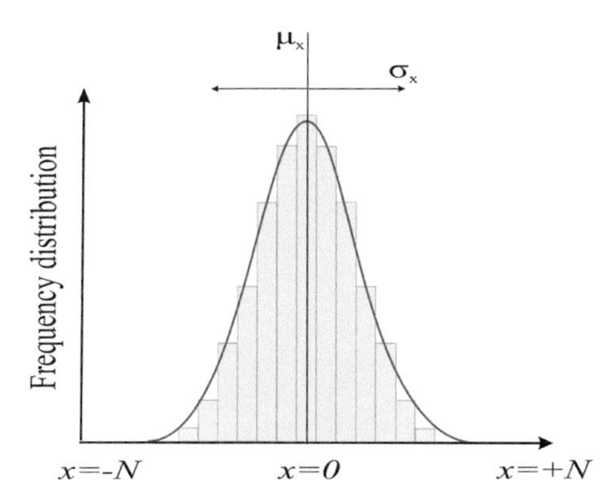

Figure 4.4 Histogram of the results of random walk from 1,000 experiments. μ_x and σ_x^2 are the mean and the standard deviation of the positions of the particle.

As remarked, the particle moves randomly after a short persistent time, varying the step and direction. Under this condition, the result of such movements can be illuminated most simply in an analysis of random walks along a straight line. The walk takes a unit step length with the probability of a forward or backward step being exactly ½. In plain English, one could conduct a random walk experiment along the straight line by tossing a coin to determine whether moving forward a step if the head occurs and backward a step otherwise. After flipping a coin N times, one could be at any of these locations (one starts the walk at the origin):

$$-N, -N+1, -N+2, \ldots -1, 0, 1, \ldots N-2, N-1, N.$$

What probability does the person reach a given location $-N < m < N$? Csanady (1973) answered this question using some elementary statistical analysis. Instead, we answer it using 1,000 random walk experiments, each involving N flipping a coin. If we record the location one reaches after N tosses during each experiment, we summarize the locations of the 1,000 experiments in a histogram, similar to Fig. 4.4.

The resulting histogram is a bell-shaped distribution with the maximum value at the origin and decaying away toward $-N$ and $+N$, approximately a normal probability distribution. It shows all possible locations where one may arrive after N tosses of the coin or those of a Brownian particle after N random movements. The continuous red curve in Fig. 4.4 denotes the theoretical normal probability distribution. In plain English, such a probability distribution states that since the precise movement of the particle is intangible, our best estimate is that the particle remains at the origin after N steps. The chance that the particle reaches $-N$ or N position is tiny.

Another way to interpret the probability distribution is to view the distribution as the spatial distribution of 1,000 particles after N random movements. This

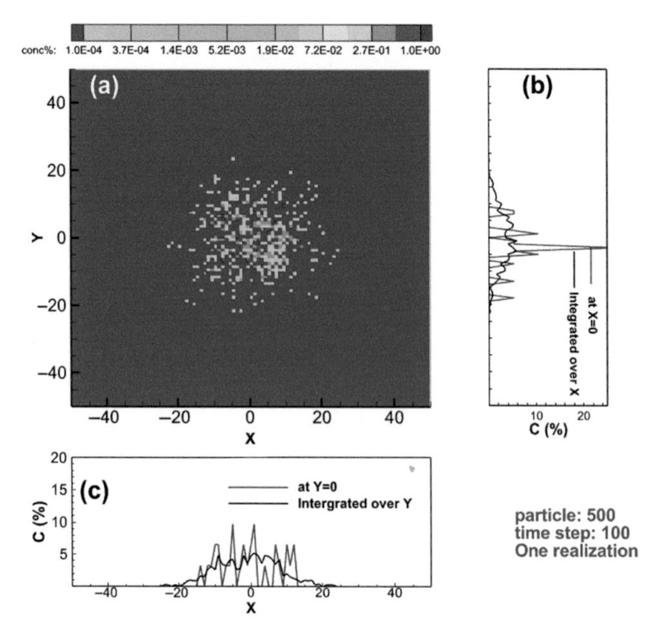

Figure 4.5 (a) two-dimensional distribution of 500 particles after 100 random walk steps. (b) The particle distribution along all ys at $x = 0$ (red line) and that for all xs (blue line).

distribution reveals that most 1,000 particles remain at the origin after N movements, and the rest spreads out.

To further elaborate this interpretation, we plot a possible distribution of 500 particles after 300 random steps in a two-dimensional x–y plane, as illustrated in Fig. 4.5a. The plot (the red line in Fig. 4.5b) is the number of particles along the y-axis at $x = 0$ (the marginal pdf in statistics). The blue line is the total number of particles along the y-axis at all xs. Fig. 4.5c shows a similar plot along the x-axis. Once more, due to the particles' intractable movement, the plots are a possible behavior of the 500 particles–one realization of all possible behaviors. Behaviors of another realization are portrayed in Fig. 4.6a, b, and c. They are similar to those in Fig. 4.5a, b, and c but different in detail and sparsely scatted; both figures only resemble a theoretical normal distribution in bulk behavior. Notice that the integrated number of particles over all the x or y directions exhibits smoother behavior than those along a single transect line, illustrating the spatial averaging effect.

The distribution from 1,000 realizations of 500 particles' random walk experiments after 300 steps, as shown in Fig. 4.7a, b, and c. It approaches the theoretical smooth normal distribution.

These figures explain the inherent discrete nature of the movement and dispersal of Brownian particles and the ensemble concept behind the normal distribution,

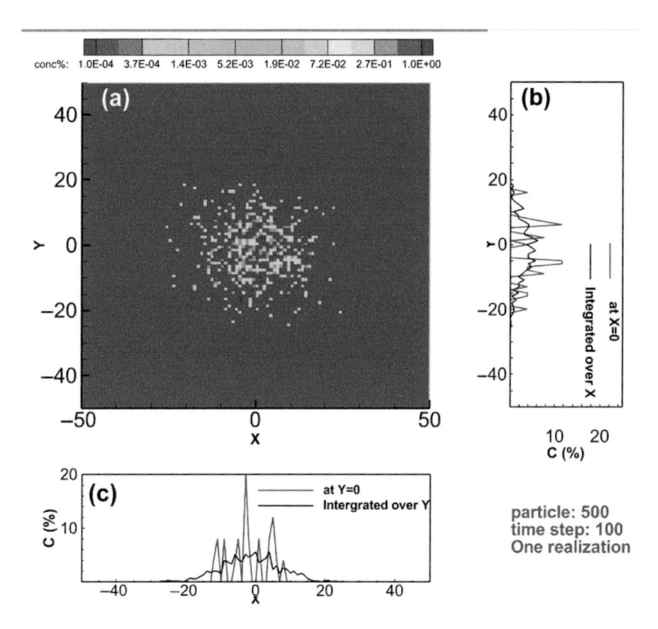

Figure 4.6 A similar plot as Fig. 4.5 for a different realization.

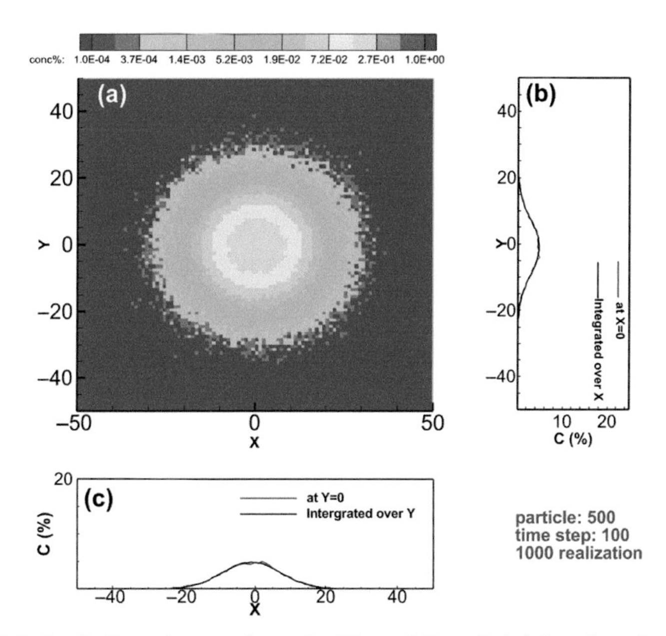

Figure 4.7 A similar plot as those in Figs. 4.5 and 4.6 but based on 1,000 realizations of random walk experiments.

embedded in Fick's law in the diffusion equation, as discussed in previous sections. Specifically, Fick's law does not describe exact but general behaviors (spatial or ensemble averages over a CV) of meandering molecules. This averaging concept is similar to the well-mixed modes in Chapters 2 and 3 but with different CV scales.

4.5 Analytical Solutions of the Diffusion Equation

To illustrate the behavior of the solution of the diffusion equation, we examine the solute to the one-dimensional diffusion equation. The analysis for the one-dimensional equation, in general, can be extended to multidimensional problems. Consider a one-dimensional diffusion process in an infinite domain. The governing equation is

$$\frac{\partial C}{\partial t} = D\frac{\partial^2 C}{\partial x^2} \tag{4.5.1}$$

Boundary conditions are $C(-\infty, t) = 0$, and $C(+\infty, t) = 0$, at all t, and an initial condition, $C(x, 0) = 0$ at all x. The solution to Eq. (4.5.1) varies with the form of input. Here we consider two cases: 1) spatial distribution of the input concentration is specified at $t = 0$, and 2) a concentration at a fixed point is specified as a function of time. Solution techniques for these two cases are available in many reference books, such as Conduction of Heat in Solids by Carslaw and Jaeger (1959, 1988) and Diffusion by Crank, J.C. (1956). In the following sections, we present the solution without discussing the solution techniques.

4.5.1 Specified Spatial Distribution of Input Concentration

This section examines the solution for three possible input forms: a) an impulse, b) an arbitrary, and c) a step input.

a) **An Impulse Input Or Slug Input**. At $t = 0$, a tracer with a mass, M, is suddenly released at $x = \xi$. For example, a drop of red ink is suddenly released into a thin sheet of static and clean water film, and we would like to quantify the concentration of the red ink in time and space. Mathematically, this input can be specified as

$$C(x, 0) = \frac{M}{\rho A}\delta(x - \xi)$$

Where A is the source area perpendicular to the x-axis (i.e., the thickness of the water film), δ is the Dirac delta function, which possesses the following properties:

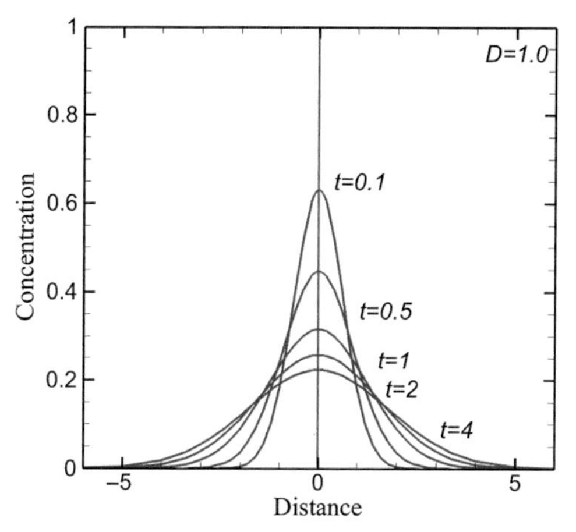

Figure 4.8 Spatial concentration distributions at different times due to the release of an impulse input. Any typical diffusion coefficient value normalizes *d*, which determines the unit of distance and time.

$$\int_{\xi^-}^{\xi^+} \delta(x-\xi)dx = 1 \text{ and } \delta(x-\xi) = 0 \; x \neq \xi \tag{4.5.2}$$

The delta function has a dimension [1/L]. The solution corresponding to this case is

$$C(x, t) = \frac{M}{\rho A\sqrt{4\pi Dt}} \exp\left[\frac{-(x-\xi)^2}{4Dt}\right] \tag{4.5.3}$$

The concentration in Eq. (4.5.3) has the dimension [mass/mass, ppm]. A graph of the solution as a function of x at different times is displayed in Fig. 4.8, showing that the concentration distribution in space is always symmetrical (a bell shape). This bell-shaped distribution is called Gaussian or normal distribution, the same as in statistics. Initially, the distribution is narrow and with a high peak concentration. As time progresses, the peak value decreases, and the distribution base broadens. However, the area under the distribution at different times must be the same – mass conservation.

b) **Arbitrary Input.** Suppose that the input concentration is specified as $C(x, 0) = f(x)/(\rho A)$ at $-\infty < x < \infty$, where f(x) is some arbitrary function, representing the mass of a solute at location x at $t = 0$. A real-world analogy could be a sudden spill of pesticides along a lakeshore. Despite the form of the arbitrary function, a f(x) function can be represented by a series of separated slugs (or impulse inputs, say, 1, 2, . . .n, Fig. 4.9). Each slug has a finite width dξ and a mass $f(\xi)\, d\xi$. That is,

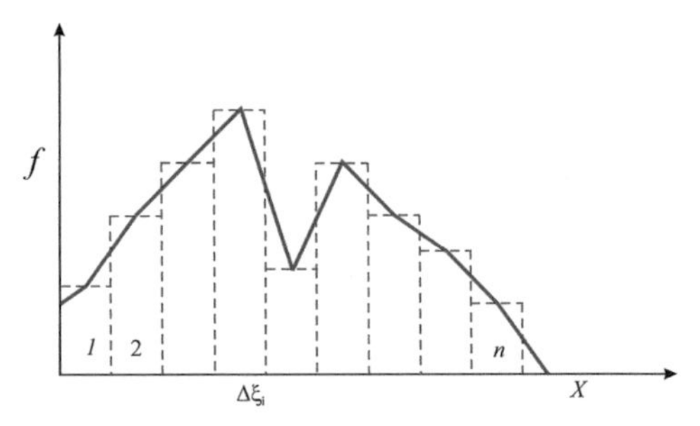

Figure 4.9 Schematic illustration of an arbitrary input concentration.

$$M(\xi) = f(\xi)d\xi = \left[\frac{M}{L}\right][L] \tag{4.5.4}$$

Notice that f is defined as mass per unit length. The solution for each slug i is given by

$$C_i(x, t) = \frac{f(\xi_i)d\xi_i}{\rho A\sqrt{4\pi Dt}} \exp\left[\frac{-(x-\xi_i)^2}{4Dt}\right], \tag{4.5.5}$$

representing the contribution to the solution due to the ith slug. Then, the total contribution at a given x and t from all slugs becomes

$$C(x, t) = \sum_{i=1}^{n} \frac{f(\xi_i)\Delta\xi_i}{\rho A\sqrt{4\pi Dt}} \exp\left[\frac{-(x-\xi_i)^2}{4Dt}\right] \tag{4.5.6}$$

If we allow the width of the slug to be infinitesimally small, a continuous solution for this case is

$$C(x, t) = \int_{-\infty}^{\infty} \frac{f(\xi)}{\rho A\sqrt{4\pi Dt}} \exp\left[\frac{-(x-\xi)^2}{4Dt}\right] d\xi \tag{4.5.7}$$

c) **Step Input.** This input describes a section of the one-dimensional domain being imposed with a given concentration everywhere at time equals zero (e.g., at $t = 0$, a concentration of C_0 is released over a segment of the domain $x < 0$. This situation is analogous to the sudden removal of the membrane in the water tank experiment. We then want to determine the spatial and temporal distributions of the concentration. First, mathematically, the step input can be prescribed by the following equation:

$$C(x, 0) = \left\{\begin{matrix} 0, x > 0 \\ C_0, x \le 0 \end{matrix}\right\} = \frac{f(x)}{\rho A} \tag{4.5.8}$$

Again, f is defined as mass per unit length. We can derive the solution to the diffusion with this step input by using the solution for arbitrary input Eq. (4.5.7) and setting the arbitrary input function as the step input. That is,

$$C(x, t) = \int_{-\infty}^{0} \frac{C_0}{\sqrt{4\pi Dt}} \exp\left[\frac{-(x-\xi)^2}{4Dt}\right] d\xi + \int_{0}^{\infty} 0 \exp\left[\frac{-(x-\xi)^2}{4Dt}\right] d\xi \tag{4.5.9}$$

note that $C_0 = M/\rho A$. Eq. (4.5.9) leads to the following solution

$$C(x, t) = \frac{C_0}{\sqrt{\pi}} \int_{-\infty}^{0} \exp\left[-\left(\frac{(x-\xi)}{\sqrt{4Dt}}\right)^2\right] \frac{d\xi}{\sqrt{4Dt}} \tag{4.5.10}$$

After transformation using $u = (x-\xi)/\sqrt{4Dt}$, we have

$$\begin{aligned} C(x, t) &= -\frac{C_0}{\sqrt{\pi}} \int_{\infty}^{x/\sqrt{4Dt}} \exp\left[-u^2\right] du \\ &= \frac{C_0}{\sqrt{\pi}} \left(\int_{-\infty}^{0} \exp\left[-u^2\right] du - \int_{0}^{x/\sqrt{4Dt}} \exp\left[-u^2\right] du \right) \\ &= \frac{C_0}{\sqrt{\pi}} \left[\frac{\sqrt{\pi}}{2} - \int_{0}^{x/\sqrt{4Dt}} \exp\left[-u^2\right] du \right] \\ &= \frac{C_0}{2} \left[1 - erf\left[\frac{x}{\sqrt{4Dt}}\right] \right] = \frac{C_0}{2} erfc\left[\frac{x}{\sqrt{4Dt}}\right] \end{aligned} \tag{4.5.11}$$

where *erf* is the error function, which is

$$erf(z) = \frac{2}{\sqrt{\pi}} \int_{0}^{z} \exp\left(-\xi^2\right) d\xi \text{ and } erfc(z) = 1 - erf(z), \tag{4.5.12}$$

and it is called the complementary error function. The behavior of the function is shown below.

As illustrated in Fig. 4.10, the error function has these properties:

$$erf(-z) = -erf(z), \quad erf(0) = 0, \quad erf(\infty) = +1.0, \text{ and } erf(-\infty) = -1.0.$$

Using Eq. (4.5.10), we show the diffusion process solutions at $t = 0.02$, 0.1, and 0.2 for the step input, $C_0 = 1.0$, and $D = 1$ (any consistent unit), in Fig. 4.11. Using the water tank as an example, a sharp interface between the yellow and blue fluids exists at time zero. As time progresses, the sharp front is smeared, becoming greenish, due to some yellow particles moving to the blue side and vice versa. At

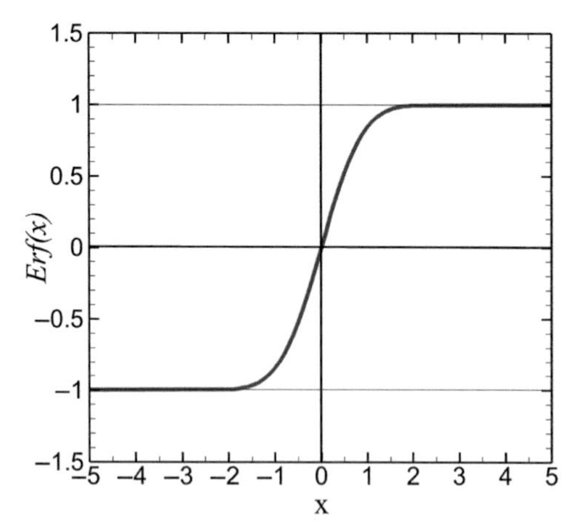

Figure 4.10 The behavior of the error function erf.

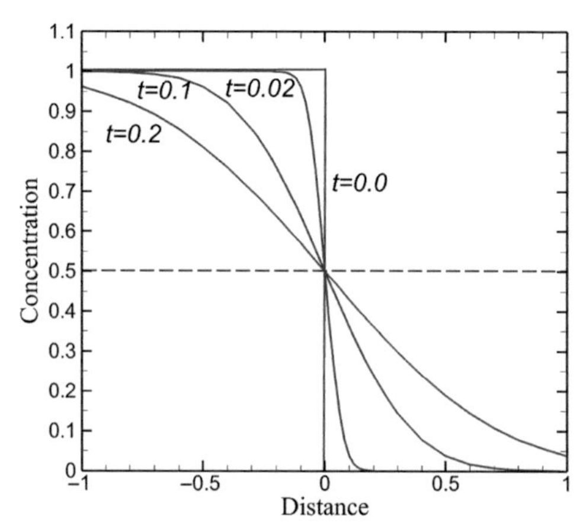

Figure 4.11 Spatial yellow particle concentration distribution on either side of the original position of the membrane at different times.

time approaches infinite, the finial concentration approaches 0.5 everywhere in the solution domain. That is, the water tank is well-mixed, the concentration is homogeneous. If one takes a CV larger than many molecules, one finds the numbers of the yellow and blue particles are the same everywhere in the tank.

4.5.2 Input Concentration Specified as a Function of Time at a Fixed Location

Instead of specifying input spatial concentration distribution in a domain, we now show the solution for the case where input concentration is specified as a function

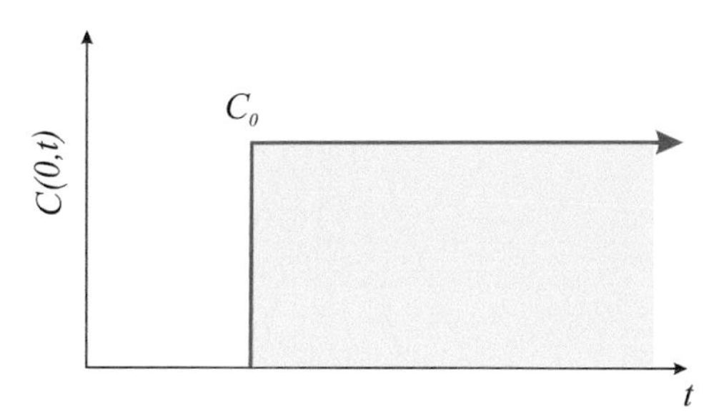

Figure 4.12 An illustration of a step input function in time at $x = 0$.

of time at a fixed location. This situation is analogous to a continuous release of a tracer at a point in the initial clean water tank. Without loss of generality, we assume the domain is infinite and unbounded, and at initial time $t = 0$, the concentration is zero everywhere along the x-axis.

Suppose the concentration at the point x = 0 is suddenly raised to C_0 over an infinite period. We like to determine the concentration distribution along the x-axis at all times, $C(x,t)$. This problem is a step input in time at a location x = 0, as in Fig. 4.12. It has an analytical solution

$$C(x, t) = C_0\left[1 - erf\left(\frac{x}{\sqrt{4Dt}}\right)\right] = C_0 \; erfc\left[\frac{x}{\sqrt{4Dt}}\right] \text{ for } x > 0, \tag{4.5.13}$$

and for $x < 0$,

$$C(x, t) = C_0\left[1 - erf\left(\frac{x}{\sqrt{4Dt}}\right)\right] \tag{4.5.14}$$

(Note: Eq. 4.5.13 and Eq. 4.5.14 are IDENTICAL). The behavior of the solution is graphed in Fig. 4.13 for $C_0 = 1.0$, $t = 0$, 0.02, 0.1, and 0.2 at various distances. Notice that when time approaches infinity, the concentration should approach unity everywhere.

The diffusion equation is mathematically linear (i.e., the output is linearly proportional to the input), allowing the superposition of the above solutions to derive the analytical solutions for different situations (e.g., different input forms or boundary conditions). Please see Fisher et al. (1979).

4.6 Advection–Diffusion Process

The previous section examines tracer diffusion in a static fluid. This section investigates the diffusion process in a fluid moving at a steady-state uniform fluid-dynamic-scale velocity, u (constant in time). Suppose the diffusion is independent

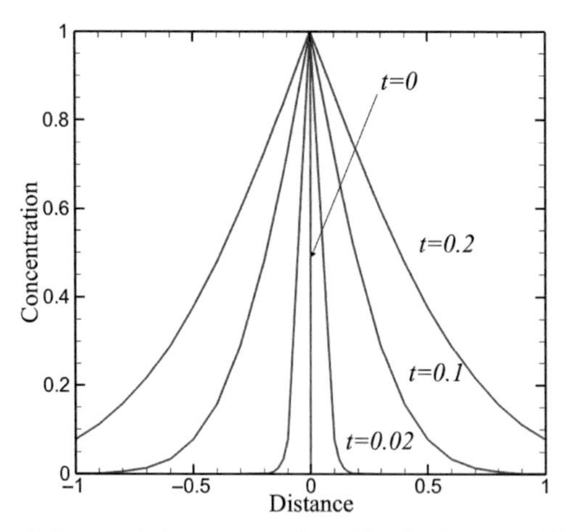

Figure 4.13 Plot of the spatial concentration distribution at different times for a step input at location $x = 0$.

of the fluid-dynamic-scale velocity. Observing simultaneous molecular diffusion and fluid movement at the fixed location becomes analogous to observing a diffusion experiment in a static water tank at a fixed location; meanwhile, the tank moves at u toward or away from the observation location. Accordingly, the total tracer mass flux (q_T) crossing the observation location would consist of two processes. One is the mass flux due to the tank's movement (convective or advective flux, uC). The other is diffusion (diffusive flux, q_D) in the tank, representing the random microscopic motion of the tank fluid molecules. That is, we see fluid movements at fluid-dynamic and molecular scales.

Mathematically, the total mass flux crossing the fixed location is

$$\mathbf{q}_T = \mathbf{q}_D + \mathbf{u}C \tag{4.6.1}$$

Again, the continuity equation for the solute mass states that

$$-\nabla \cdot \mathbf{q}_T = \frac{\partial C}{\partial t} \tag{4.6.2}$$

Using the expression for $\mathbf{q_T}$ in Eq. (4.61), we have

$$-\nabla \cdot (\mathbf{q}_D + \mathbf{u}C) = \frac{\partial C}{\partial t} \tag{4.6.3}$$

Expanding the divergence of the product of velocity and concentration $\nabla \cdot (\mathbf{u}C)$, Eq. (4.6.3) becomes

$$-\nabla \cdot \mathbf{q}_D - \mathbf{u}\nabla C - C\nabla \mathbf{u} = \frac{\partial C}{\partial t} \tag{4.6.4}$$

Recall, the mass balance of water states that

$$-\nabla \cdot \rho \mathbf{u} = \frac{\partial \rho}{\partial t} \tag{4.6.5}$$

If the fluid is homogeneous (ρ is constant in space) and incompressible (ρ is constant in time), we then have

$$\nabla \cdot \mathbf{u} = 0 \tag{4.6.6}$$

for either steady uniform (constant **u** in space) or non-uniform (variable **u** in space) flow. As a result, the term, $C\nabla \vec{u}$, is zero, and Eq. (4.6.4) becomes

$$-\nabla \cdot \mathbf{q}_D - \mathbf{u}\nabla C = \frac{\partial C}{\partial t} \tag{4.6.7}$$

Notice that the advective velocity is outside the concentration gradient term in Eq. (4.6.7), but the steady velocity may vary with **x**. Substituting Fick's Law for the diffusive flux, we have the advection–diffusion equation,

$$D\nabla^2 C - \mathbf{u}\nabla C = \frac{\partial C}{\partial t} \tag{4.6.8}$$

In Eq. (4.6.8) the first term represents the diffusion flux per volume of the fluid, and the second term denotes the advective flux per volume of the fluid. The advection–diffusion equation in Cartesian coordinates is

$$D\left(\frac{\partial^2 C}{\partial x^2} + \frac{\partial^2 C}{\partial y^2} + \frac{\partial^2 C}{\partial z^2}\right) - \left(u_x \frac{\partial C}{\partial x} + u_y \frac{\partial C}{\partial y} + u_z \frac{\partial C}{\partial z}\right) = \frac{\partial C}{\partial t} \tag{4.6.9}$$

where u_x, u_y, and u_z are the fluid-dynamic-scale velocity components in the x, y, and z-direction, respectively. The diffusion term depicts the random movement of molecules. In other words, the advection–diffusion equation describes both molecular and fluid-dynamic-scale movements of the fluid molecules.

4.6.1 Temporal and Spatial Evolutions

The following two scenarios illustrate the combined effects of advection and diffusion on a solute plume: an impulse (slug) input and a step input at the location $x = \xi$. The corresponding solution can be easily derived if we recognize it is the sum of advection and diffusion processes. Specifically, the advection–diffusion equation becomes a diffusion equation after the fixed (i.e., Eulerian) coordinate system is transformed into a moving (i.e., Lagrangian) coordinate system. In other words, instead of observing the advection–diffusion process at a fixed location, one observes the process on a platform moving at the same velocity as the advective velocity. As such, one on this moving platform observes the diffusion process only, and the solutions for diffusion in the previous section are readily available.

4.6.1.1 Impulse (Slug) Input

Based on the above reasoning, one can convert the solution of diffusion (Eq. 4.5.3) for a slug input to that of the advection–diffusion equation, which takes the following form:

$$C(x, t) = \frac{M}{\rho A\sqrt{4\pi Dt}} \exp\left[\frac{-[(x-ut)-\xi]^2}{4Dt}\right] \tag{4.6.10}$$

Note that $(x-ut)$ is the moving coordinate or the concentration observation location x relative to the distance the advective velocity has traveled, ut. The term ξ is the location where a slug of a tracer is released. Although this solution here is valid for an unbounded domain, it can be used to derive the solutions that consider boundary effects by using the image method in hydraulics and other fields.

Utilizing Eq. (4.6.10), we consider the one-dimensional problem, where a slug of tracer is leased from $\xi = 0$, migrating from left to right of Fig. 4.14 with $u = 1$ and $D = 0.1$ (any consistent unit). Suppose the concentration everywhere in the solution domain is zero at the time $t = 0$. We present this solution in two ways: (1) spatial distribution, C-x plot, at a given time (called **snapshots**), and (2) temporal distribution C-t (**breakthroughs**) at a given location.

Snapshots. As illustrated in Fig. 4.14, the snapshot is a bell-shaped curve over the distance at a given time. The shape of the distribution is symmetric around the

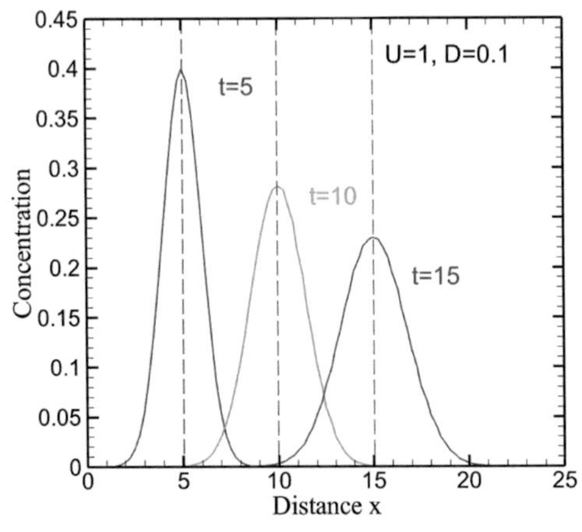

Figure 4.14 Plots of spatial concentration distribution, $C(x)$, snapshots at different times.

peak concentration. The advection velocity and time determine peak concentration location ($x_{peak} = ut$), while the diffusion coefficient determines the lateral spread of the concentration distribution. In other words, the distance of the peak concentration from the source at different times reveals the velocity of the advection–diffusion equation. A plot of the square of the concentration distribution's width at some specified concentration vs. time (e.g., Fig. 4.2) can yield the rate of spreading, which is twice the diffusion coefficient in the advection–diffusion equation (see section 8.3, Chapter 8).

Scrutinizing without solving Eq. (4.6.10), one could guess the solution's behaviors. A snapshot is a plot of the solution by varying the value of x with fixed t, u and D values. Since the solution is the product of $M/\left(\rho A\sqrt{4\pi Dt}\right)$ and $\exp\left[-[(x-ut)-\xi]^2/4Dt\right]$, the square term $[(x-ut)-\xi]^2$ makes the exponential function an even function in terms of $(x-ut)-\xi$, decaying symmetrically from $x = ut + \xi$. At this x value, Eq. (4.6.12) is reduced to

$$C(x, t) = \frac{M}{\rho A\sqrt{4\pi Dt}} = C_{\max},$$

which is the peak (or maximum) concentration, independent of x. Therefore, the solution of Eq. (4.6.10) , plotted as a function of x at different times , maintains the symmetrical shape.

Breakthroughs. The temporal solute distribution at a given location, C vs. t plot, is a breakthrough curve (BTC). Fig. 4.15 shows the BTCs at x = 5, 10, and 15 under the same flow field as in Fig. 4.14. The concentration BTC at early times is always skewed and becomes almost symmetric at late times. The peak arrival time is slightly early than the travel time of the advection velocity to the observation location (i.e., $t = x/u$).

BTC is an evaluation of Eq. (4.6.2) by varying t values with fixed x, u, and D values. In this case, the product of $M/\left(\rho A\sqrt{4\pi Dt}\right)$ and $\exp\left[-[(x-ut)-\xi]^2/4Dt\right]$ is a function of t, not x. As a result, the solution is an odd function at $x = ut + \xi$. These rudimentary mathematic facts explain the skewed distribution of the BTCs, although the derivative of the solution with respect to t allows determining the peak time and value. The resultant derivative is a complex nonlinear function, difficult to simplify.

Physically, the skew distribution phenomenon is analogous to the Doppler effect of sound propagation in elementary physics, although sound propagation differs from solute transport. For example, one hears a high pitch horn sound from a car moving toward and a low pitch horn sound when it moves away. The combination of the car's and horn sound wave's velocity explains this phenomenon.

To elucidate the similarity, consider the one-dimensional ADE problem in which advection–diffusion occurs from left to right of Fig. 4.15 with $u = 1$ and $D = 1$.

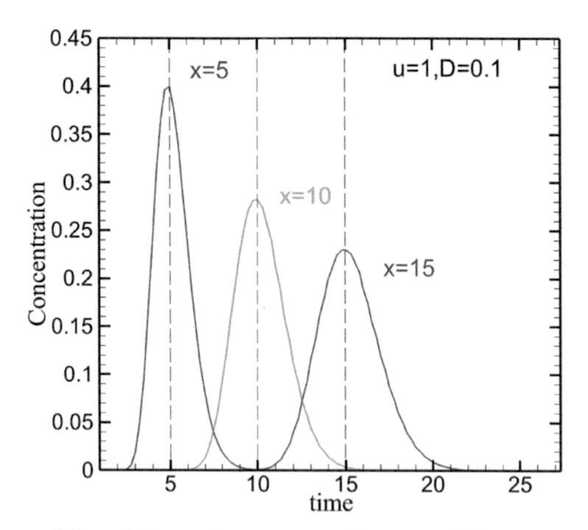

Figure 4.15 Plots of breakthrough curves, $C(t)$, at different distances from the injection point.

Suppose the concentration everywhere in the solution domain is zero at the initial time, $t = 0$. Then, an impulse input C_0 is suddenly released at $x = 0$ at time $t = 0$. Find a breakthrough curve (C vs. t) observed at $x = 6$.

Instead of solving an advection–diffusion equation for the BTC at a fixed observation location, one first solves the diffusion equation for the snapshot due to the impulse input and then moves the observation location successively toward the diffusing plume. For example, consider concentration distributions resulting from diffusion at eleven different times ($t = 1, 2, 3, \ldots 11$) in Fig. 4.16. We move the observation location, $x = 6$, toward the tracer release point according to the velocity, u, and the time after the release to reproduce the advection. At $t = 1$, we move the observation location from $x = 6$ to $x = 5$; at the new location, we determine the concentration value resulting from the diffusion at that location and time (the circle shown at x = 5). At $t = 2$, we again move the observation location from $x = 5$ to $x = 4$ by 1 and then record the concentration value resulting from the diffusion equation corresponding to $t = 2$ (the circle shown at $x = 4$). By moving the observation point successively toward the diffusing plume, we obtain a temporal concentration distribution (the thick solid line shown in the figure), which is, in essence, the BTC at $x = 5$.

This resulting breakthrough curve (the red curve in Fig. 4.16) is skewed, characterized by sharply rising concentrations and a long-tailing recession limb. As the center of the concentration plume moves toward the observation location, we record concentration due to the advection and diffusion effects moving toward us. Thus, we see a rapidly rising concentration profile. On the other hand, when the plume center passes the observation point, we observe the concentration diffusing backward from the plume moving away. We consequently observe the long-tailing

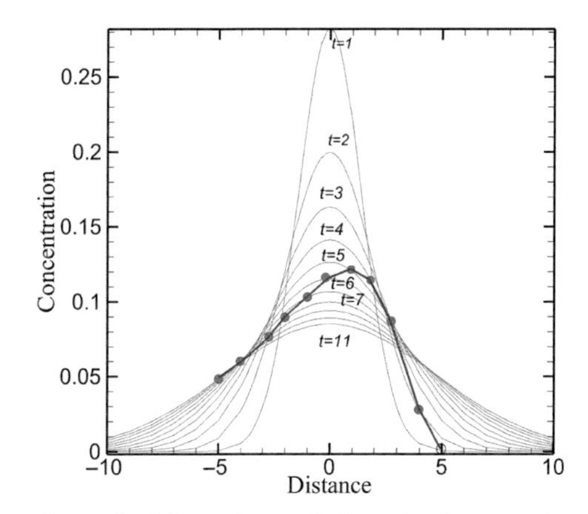

Figure 4.16 A schematic illustration of the physics causing skewness of a breakthrough curve.

effect of the plume. Therefore, we can conclude that a breakthrough curve resulting from the advection–diffusion equation is always skewed. This elementary explanation is necessary since the breakthrough curve's long tailing has frequently been misunderstood as the effects of reactive tracers or heterogeneity.

4.6.1.2 Step Input

Consider the problem of an ideal, frictionless pipe filled with a fluid displaced by another fluid with a tracer in concentration C_o with a velocity u. At time $t = 0$, a sharp front exists so that,

$$C(x,0) = \left\{ \begin{array}{ll} 0, & x > 0 \\ C_0, & x \leq 0 \end{array} \right\} \tag{4.6.11}$$

Again, if we transform the fixed coordinate system of the problem into the moving coordinate system, the solution is already available as presented in the diffusion equation solution (Eq. 4.5.11). We merely have to adjust for the moving coordinates, and we have,

$$C(x,t) = \frac{C_0}{2}\left[1 - erf\left[\frac{(x-ut)}{\sqrt{4Dt}}\right]\right] \tag{4.6.12}$$

This solution is graphed as snapshots at t = 1, 5, 10, and 15 in Fig. 4.17. $u = 4, D = 1$ $C_0 = 1$. The vertical dashed lines cross the 0.5 concentration line indicate the distances the solute has traveled by advection only at the corresponding times. The location with a concentration equal to 0.5 is the advective front. The area bounded by the concentration curve from 1 to 0.5, the concentration 1 line, and the vertical dashed line of the advective front (yellow area) is the solute diffused to the front of

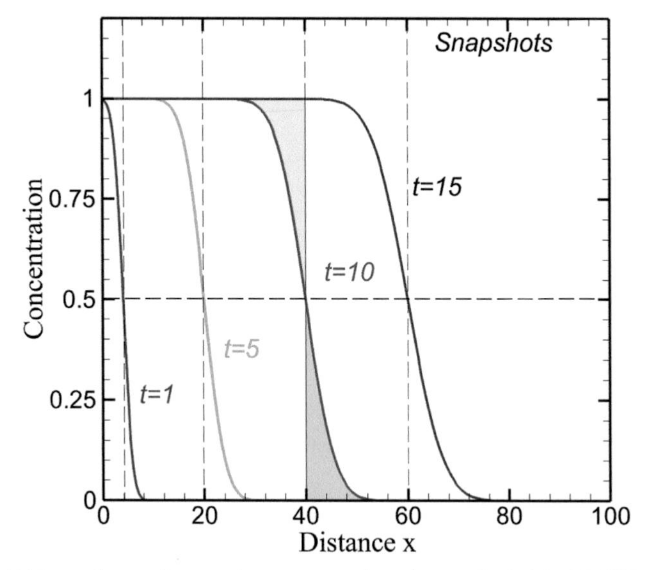

Figure 4.17 Snapshots of a step input at $x = 0$ at time = 1, 5, 10, 15. HWR516 ade step input snapshot.lpk. The vertical dashed lines indicate the locations of the advective front at different times.

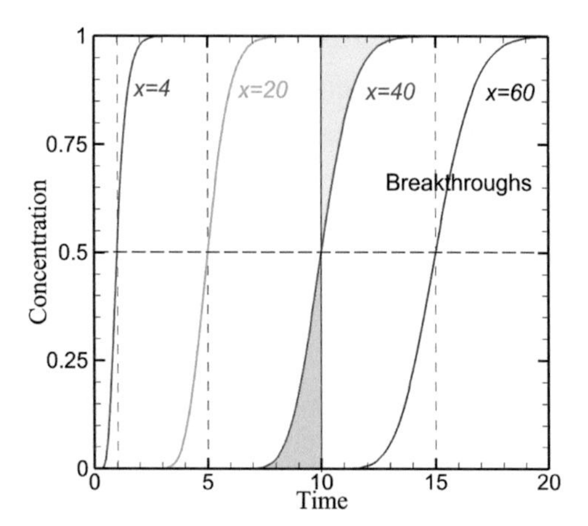

Figure 4.18 Concentration breakthrough curves at different observation distances.

the advective front (blue area). These two areas must be equal. Like the snapshots of a slot input, the $C(x,t)$ distributions are symmetrical about the advection front.

If we plot the concentrations at $x = 4$, 20, 40, and 60 as a function of time, we have breakthrough curves, as shown in Fig. 4.18. The vertical lines denote the tracer advection fronts; the blue area corresponds to the amount of solute arriving at the observation location earlier than the advection front, while those arriving later than the front is colored yellow. Again, similar to the breakthrough curves in a slug input case, the breakthrough curves are asymmetrical about the advecting front (0.5 concentration). In other words, the yellow area is larger than the blue area. The areas become close, or the breakthrough curves approach symmetrical distributions at large times.

A slightly different solution has been derived (e.g., Otaga and Banks, 1961) for a constant input concentration at a point. Physically, this solution could represent a solute migration in a pipe for which the water flow is steady, and a constant input concentration C_0 is introduced at $x = 0$ at time $t = 0$ and continued. The problem can be formulated mathematically using the ADE with the following boundary and initial conditions.

$$C(0, t) = C_0 \quad 0 < t < \infty,$$
$$C(x, 0) = 0 \quad 0 < x < \infty$$

The solution for $t < \infty$ becomes

$$C(x, t) = \frac{C_0}{2}\left[erfc\left(\frac{x - ut}{\sqrt{4Dt}}\right) + erfc\left(\frac{x + ut}{\sqrt{4Dt}}\right)\exp\left(\frac{ux}{D}\right)\right] \tag{4.6.13}$$

This solution differs from Eq. (4.6.12) by adding the second term. Intuitively, the effects of this additional term (boundary effects) exist near the boundary. However, the effects are minor and diminish as the solute moves away from the boundary. We will leave the proof of this statement for the reader to investigate in the exercise.

4.7 Remarks

The diffusion and advection–diffusion concepts and their analytical solutions have played an essential role in understanding and analyzing solution transport in environments. Widely being adopted in all science and engineering fields as they are, the philosophy, assumptions, ensemble mean concept, scale issues, and limitations have frequently been overlooked.

For example, the advection–diffusion equation (ADE) concept is identical to the well-mixed models in Chapters 2 and 3. In the well-mixed model, the advection is the inflow and outflow, and diffusion is the mixing agent. The well-mixed model assumes that the diffusion coefficient's magnitude is infinite. As a result, the solute mixing due to diffusion reaches a steady state immediately once the solute enters the reservoir. That is, the solute is instantaneously well-mixed in the reservoir. The CV in the well-mixed model is the entire reservoir. The concentration is either a volume average or an ensemble average over the whole reservoir.

In contrast, the CV in the ADE is much smaller than the reservoir (i.e., scale difference) but sufficiently large to contain many molecules. Consequently, the concentration from the ADE represents the volume-averaged or ensemble-averaged concentration over many molecules. For this reason, ADE is a spatially distributed model, while the well-mixed model is a lumped model.

Lastly, the analytical solutions presented in this chapter are based on the resident concentration instead of flux concentration. Differences between the solution based on the two types of concentration concepts have been explored by Kreft and Zuber (1978). However, they are beyond the scope of this introductory book.

4.8 Homework

(1) Compare the solution of a well-mixed model and the advection–diffusion equation for a continuous step input of a tracer under a steady-state flow condition and discuss the difference.
(2) What is the distribution of the slope of the concentration breakthrough curve resulting from a continuous step input? Prove it mathematically.
(3) Is it valid to assume that molecular diffusion is independent of the fluid-dynamic-scale velocity? Discuss the dependence on solute migration.
(4) Compare the solutions (Eqs. 4.6.12 and 4.6.13) and discuss their differences and relevance in real-world scenarios.

5

Numerical Methods for Advection–Diffusion Equations

5.1 Introduction

Chapter 5 presents the advection–diffusion equation (ADE) formulation, which couples the effect of fluid motion at the fluid continuum scale and random motion of fluid molecules at the molecular scale to quantify solute migration. It then provides analytical solutions to the ADE for several input forms and discusses snapshots and breakthroughs for different input forms. Closed-form analytical solutions are beneficial for many purposes but are restricted to simplified conditions (e.g., uniform velocity, initial and boundary conditions). This chapter introduces numerical methods for solving the ADE applicable to multidimensional, variable velocity, irregular boundary, and initial conditions. However, the introduction focuses on one- and two-dimension examples for convenience. Once the numerical methods' algorithms are understood, one can quickly expand them to other complicated situations.

Specifically, we focus on one-dimensional ADE.

$$D\frac{\partial^2 C}{\partial x^2} - u\frac{\partial C}{\partial x} = \frac{\partial C}{\partial t} \tag{5.1.1}$$

where D is diffusion coefficient, u is the fluid-dynamic-scale velocity. We present three numerical methods for ADE, including (1) the Finite Difference Approach, (2) the Methods of characteristics (Eulerian-Lagrangian), and (3) the Finite Element Approach.

5.2 Finite Difference Approximation

The finite difference method is the most common numerical method for solving the ADE Eq. (5.1.1). It uses a first-order approximation of the spatial and time derivatives of the continuous variables in the equation. It discretizes the one-dimensional domain along the x-axis into many blocks, as illustrated in Fig. 5.1.

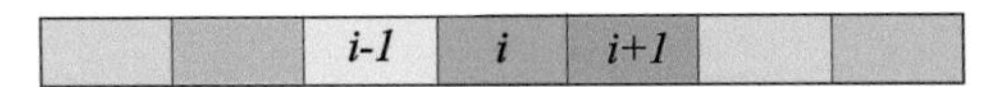

Figure 5.1 The block-centered finite difference's spatial discretization of a one-dimensional solution domain.

The center of each block is denoted by index i, ranging from 1, 2, ... N, the total number of blocks.

This discretization is called block-center finite-difference, while the nodal finite-difference designates the index i to the nodes of the blocks. Their formulations are identical, but each has its convenience in some situations.

5.2.1 Spatial Derivatives

The finite difference approximation of the spatial derivatives could use several schemes, including the forward, backward, central, and upwind differences, based on the Taylor series expansion of a continuous function.

5.2.1.1 Forward Difference

The forward difference evolves from the expansion of the continuous concentration function $C(x+\Delta x, t)$, where Δx is the distance between two nodes, into an infinite Taylor series:

$$C(x+\Delta x) = C(x) + \Delta x\left(\left.\frac{dC}{dx}\right|_x\right) + \frac{(\Delta x)^2}{2!}\left.\frac{d^2C}{dx^2}\right|_x + \frac{(\Delta x)^3}{3!}\left.\frac{d^3C}{dx^3}\right|_x + HOT. \quad (5.2.1)$$

Notice that the time argument of $C(x, t)$ is omitted for convenience, and d is the derivative with respect to x at a given time t. The first term on the right side of Eq. (5.2.1) is the zero-order, the second is the first-order, the third is the second-order, and so on. HOT denotes the higher-order terms (i.e., those with power higher than cubic and those with third-order and higher-order derivatives). Dropping terms higher than the first-order yields the first-order approximation of the derivative:

$$\frac{dC}{dx} \simeq \frac{C(x+\Delta x) - C(x)}{\Delta x} = \frac{C_{i+1} - C_i}{\Delta x} \quad O[\Delta x]. \quad (5.2.2)$$

The subscript i represents the concentration at position x, and $i+1$ denotes position $x + \Delta x$. The forward difference has the first-order accuracy $O[\Delta x]$.

5.2.1.2 Backward Finite Difference

The backward finite difference approach expands the concentration function $C(x - \Delta x, t)$ into an infinite Taylor series:

$$C(x - \Delta x) = C(x) - \Delta x \left(\frac{dC}{dx}\bigg|_x \right) + \frac{(\Delta x)^2}{2!} \frac{d^2C}{dx^2}\bigg|_x - \frac{(\Delta x)^3}{3!} \frac{d^3C}{dx^3}\bigg|_x + HOT. \quad (5.2.3)$$

Rearrange the above equation leads to

$$\frac{dC}{dx} = \frac{C_i - C_{i-1}}{\Delta x} + \frac{(\Delta x)}{2!} \frac{d^2C}{dx^2} - \frac{(\Delta x)^2}{3!} \frac{d^3C}{dx^3} + \cdots. \quad (5.2.4)$$

Dropping terms higher than the first-order, we have the first-order approximation of the derivative

$$\frac{dC}{dx} \approx \frac{C_i - C_{i-1}}{\Delta x} \quad O[\Delta x]. \quad (5.2.5)$$

That is, the slope between C(x) and C(x – Δx) (i.e., a backward finite difference) approximates the derivative.

5.2.1.3 Central Finite Difference

The central finite difference approach provides a second-order accuracy for the spatial derivative. It can be formulated by subtracting Eq. (5.2.3) from Eq. (5.2.1) to obtain

$$C_{i+1} - C_{i-1} = 2\Delta x \frac{dC}{dx} + \frac{2(\Delta x)^3}{3!} \frac{d^3C}{dx^3} + \cdots \quad (5.2.6)$$

And then

$$\frac{dC}{dx} \approx \frac{C_{i+1} - C_{i-1}}{2\Delta x} \quad O\left[(\Delta x)^2\right]. \quad (5.2.7)$$

This approximation is more accurate than the previous two approximations if $\Delta x < 1$.

5.2.1.4 Upwind Difference

Another finite difference scheme often used in solving the ADE is the upwind finite difference, written as

$$\frac{\partial C}{\partial x} \approx (1 - \alpha) \frac{(C_{i+1} - C_{i-1})}{2\Delta x} + \alpha \frac{(C_{i+1} - C_i)}{\Delta x}. \quad (5.2.8)$$

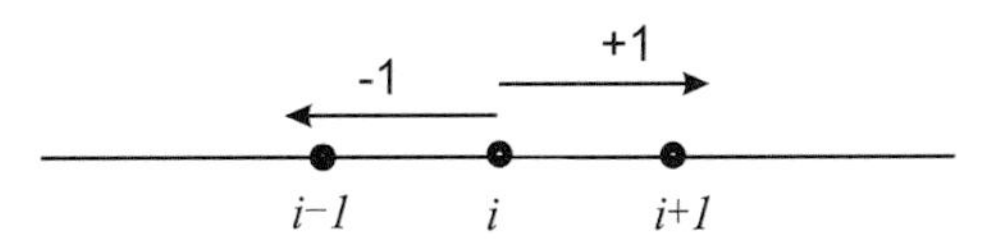

Figure 5.2 A schematic illustration of the upwind difference.

The term $\alpha = \pm 1$ is a weigh, depending on the direction of the velocity (Fig. 5.2),

$$\begin{aligned} \alpha = -1 \quad & \frac{\partial C}{\partial x} \approx \frac{C_i - C_{i-1}}{\Delta x} \\ \alpha = +1 \quad & \frac{\partial C}{\partial x} \approx \frac{C_{i+1} - C_i}{\Delta x}. \end{aligned}$$

The negative sign is for flow opposite to the direction of the increment of the block index and otherwise.

5.2.1.5 Finite Difference Analog for the Second-Order Derivative

We can also derive the finite difference approximation of the second derivative in the ADE based on Taylor series expansion. Specifically, by adding Eq. (5.2.1) to Eq. (5.2.3), we have

$$C_{i+1} + C_{i-1} = 2C_i + 2\frac{(\Delta x)^2}{2!}\frac{d^2C}{dx^2} + 2\frac{(\Delta x)^4}{4!}\frac{d^4C}{dx^4} + \cdots. \tag{5.2.9}$$

Rearranging Eq. (5.2.9) leads to the finite difference approximation of the second-order spatial derivative:

$$\left.\frac{d^2C}{dx^2}\right|_i \approx \frac{C_{i+1} - 2C_i + C_{i-1}}{(\Delta x)^2} \quad o\left((\Delta x)^2\right). \tag{5.2.10}$$

The vertical line after the second-order derivative indicates that the derivative is evaluated at the node i.

5.2.2 Temporal Discretization

After approximating the spatial derivatives, we next proceed to approximate the temporal derivative using explicit, implicit, and weighted average schemes discussed below.

5.2.2.1 Explicit Scheme

The ADE at the current time t is written as

$$\left(D\frac{\partial^2 C}{\partial x^2} - u\frac{\partial C}{\partial x}\right)^t = \left(\frac{\partial C}{\partial t}\right)^t. \tag{5.2.11}$$

Approximating the time derivative on the right-hand side of the equation becomes the next issue. The explicit scheme takes the forward difference approach for this approximation. That is,

$$\left(D\frac{\partial^2 C}{\partial x^2}\right)^t - \left(u\frac{\partial C}{\partial x}\right)^t = \frac{C^{t+1} - C^t}{\Delta t} \tag{5.2.12}$$

The C^t represents the concentration of all blocks at current time *t,* and C^{t+1} is the concentration of one block at the next time *t+1*. Δt is the time step between $t + 1$ and *t*.

C^{t+1} at the given block is unknown and can be derived from

$$C^{t+1} = \Delta t\left[D\frac{\partial^2 C}{\partial x^2} - u\frac{\partial C}{\partial x}\right]^t + C^t. \tag{5.2.13}$$

In other words, the explicit scheme linearly extrapolates the known concentration at each block at the current time to the unknown concentrations at the next time. As such, this scheme is explicit and easy to comprehend and implement. However, the explicit scheme is conditionally stable. The time step size must be small to avoid instability and oscillation of the concentration solution. For this reason, it is most useful for simple one-dimensional cases.

5.2.2.2. Implicit Scheme

As indicated below, the implicit scheme treats the concentration variable at time $t +$ 1 at all blocks in ADE as unknowns.

$$\left(D\frac{\partial^2 C}{\partial x^2} - u\frac{\partial C}{\partial x}\right)_i^{t+1} = \left(\frac{\partial C}{\partial t}\right)_i^{t+1}. \tag{5.2.14}$$

Replacing the forward difference for the time derivative yields

$$\left(D\frac{\partial^2 C}{\partial x^2} - u\frac{\partial C}{\partial x}\right)_i^{t+1} = \frac{C_i^{t+1} - C_i^t}{\Delta t}. \tag{5.2.15}$$

Moving the unknowns in the above equation to the left-hand side produces

$$\left(D\frac{\partial^2 C}{\partial x^2} - u\frac{\partial C}{\partial x}\right)_i^{t+1} - \frac{C_i^{t+1}}{\Delta t} = \frac{-C_i^t}{\Delta t}. \tag{5.2.16}$$

With known C^t at each block on the right-hand side and the specified *D* and *u* for each block on the left side of the equation, one solves a system of equations for C^{t+1} at each finite-difference block. This implicit scheme is unconditionally stable but first-order accurate in time.

5.2.2.3 Weighted Average Scheme

A flexible and general method for the temporal approximation of the ADE is the weighted average scheme:

$$\left(D\frac{\partial^2 C}{\partial x^2} - u\frac{\partial C}{\partial x}\right)_i^{t+\theta} = \left(\frac{\partial C}{\partial t}\right)_i^t. \quad (5.2.17)$$

It assumes that the temporal change in concentration at a block (i.e., $(\partial C/\partial t)_i^{t+1}$) is the weighted average of the diffusion and advection fluxes at $t + 1$ and t

$$\theta\left(D\frac{\partial^2 C}{\partial x^2} - u\frac{\partial C}{\partial x}\right)_i^{t+1} + (1-\theta)\left(D\frac{\partial^2 C}{\partial x^2} - u\frac{\partial C}{\partial x}\right)_i^t = \frac{C_i^{t+1} - C_i^t}{\Delta t}. \quad (5.2.18)$$

The weight is $\theta (0 \leq \theta \leq 1)$. After moving the unknown terms at $t + 1$ to the left-hand side and the known terms to the right-hand side, we have

$$\theta\left(D\frac{\partial^2 C}{\partial x^2} - u\frac{\partial C}{\partial x}\right)_i^{t+1} - \frac{C_i^{t+1}}{\Delta t} = -(1-\theta)\left(D\frac{\partial^2 C}{\partial x^2} - u\frac{\partial C}{\partial x}\right)_i^t - \frac{C_i^t}{\Delta t}. \quad (5.2.19)$$

If $\theta = 0$, the equation is an explicit scheme; when $\theta = 1/2$, it is the Crank–Nicolson scheme and $\theta = 1$, the formula is an implicit scheme. The Crank–Nicolson scheme is unconditionally stable with a higher order of accuracy. As a result, the numerical formulation of ADE adopts such a general weighted scheme.

5.2.3 Finite Difference Analog for the Advection–Diffusion Equation

Now, we apply the following approximations for spatial derivatives of ADE:

$$\frac{\partial^2 C^t}{\partial x^2{}_i} = \frac{C_{i+1}^t - 2C_i^t + C_{i-1}^t}{\Delta x^2} \quad (5.2.20)$$

$$\frac{\partial^2 C^{t+1}}{\partial x^2{}_i} = \frac{C_{i+1}^{t+1} - 2C_i^{t+1} + C_{i-1}^{t+1}}{\Delta x^2} \quad (5.2.21)$$

$$\frac{\partial C^t}{\partial x_i} = \frac{(1-\alpha)\left(C_{i+1}^t - C_{i-1}^t\right)}{2\Delta x} + \alpha\frac{\left(C_i^t - C_{i-1}^t\right)}{\Delta x} \quad (5.2.22)$$

$$\frac{\partial C^{t+1}}{\partial x_i} = \frac{(1-\alpha)\left(C_{i+1}^{t+1} - C_{i-1}^{t+1}\right)}{2\Delta x} + \alpha\frac{\left(C_i^{t+1} - C_{i-1}^{t+1}\right)}{\Delta x} \quad (5.2.23)$$

Where α is the upwind weighting factor, we then use the weighted scheme for the temporal derivative:

$$\theta\left(D\frac{\partial^2 C}{\partial x^2} - u\frac{\partial C}{\partial x}\right)_i^{t+1} - \frac{C_i^{t+1}}{\Delta t} = -(1-\theta)\left(D\frac{\partial^2 C}{\partial x^2} - u\frac{\partial C}{\partial x}\right)_i^t - \frac{C_i^t}{\Delta t}. \quad (5.2.24)$$

Substituting Eqs. (5.2.20) through (5.2.23) into Eq. (5.2.24) and collecting terms with the same subscripts, we have derived the following finite difference analog approximating the ADE:

$$\theta\left[\frac{D}{\Delta x^2}+\frac{(1-\alpha)}{2\Delta x}u+\frac{\alpha}{\Delta x}u\right]C_{i-1}^{t+1}+\left[\theta\left(\frac{-2D}{\Delta x^2}-\frac{\alpha}{\Delta x}u\right)-\frac{1}{\Delta t}\right]C_i^{t+1}+\theta\left[\frac{D}{\Delta x^2}-\frac{(1-\alpha)}{2\Delta x}u\right]C_{i+1}^{t+1}$$
$$=-(1-\theta)\left[\frac{D}{\Delta x^2}+\frac{(1-\alpha)}{2\Delta x}u+\frac{\alpha}{\Delta x}u\right]C_{i-1}^{t}+\left[-(1-\theta)\left(\frac{-2D}{\Delta x^2}-\frac{\alpha}{\Delta x}u\right)-\frac{1}{\Delta t}\right]C_i^{t}$$
$$-(1-\theta)\left[\frac{D}{\Delta x^2}-\frac{(1-\alpha)}{2\Delta x}u\right]C_{i+1}^{t}$$
$$(5.2.25)$$

The above finite difference analog can be expressed in the form:

$$AC_{i-1}^{t+1}+BC_i^{t+1}+EC_{i+1}^{t+1}=H_i^t \quad (5.2.26)$$

where

$$A=\theta\left[\frac{D}{\Delta x^2}+\frac{(1-\alpha)}{2\Delta x}u+\frac{\alpha}{\Delta x}u\right], \quad (5.2.27)$$

$$B=\left[\theta\left(\frac{-2D}{\Delta x^2}-\frac{\alpha}{\Delta x}u\right)-\frac{1}{\Delta t}\right], \quad (5.2.28)$$

$$E=\theta\left[\frac{D}{\Delta x^2}-\frac{(1-\alpha)}{2\Delta x}u\right], \text{ and} \quad (5.2.29)$$

$$H_i=-(1-\theta)\left[\frac{D}{(\Delta x)^2}+\frac{(1-\alpha)}{2\Delta x}u+\frac{\alpha}{\Delta x}u\right]C_{i-1}^{t}+\left[-(1-\theta)\left(\frac{-2}{\Delta x^2}D-\frac{\alpha}{\Delta x}u\right)-\frac{1}{\Delta t}\right]C_i^{t}$$
$$+\left[-(1-\theta)\left[\frac{D}{\Delta x^2}-\frac{(1-\alpha)}{2\Delta x}u\right]\right]C_{i+1}^{t}$$
$$(5.2.30)$$

Next, we apply the above formulation to a one-dimensional domain consisting of seven blocks, as shown in Fig. 5.3. Each block has a length of Δx and an index, i.

Expanding Eq. (5.2.26) using the spatial index, i = 2, 3,... 7 leads to the following system of equations:

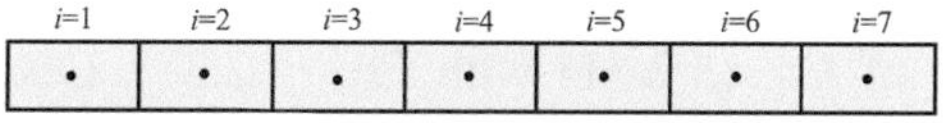

i=1	i=2	i=3	i=4	i=5	i=6	i=7
•	•	•	•	•	•	•

Figure 5.3 Finite different analog to a one-dimensional domain.

$$
\begin{array}{llllllll}
AC_1^{t+1} & +BC_2^{t+1} & +EC_3^{t+1} & & & & = & H_2^t \\
& AC_2^{t+1} & +BC_3^{t+1} & +EC_4^{t+1} & & & = & H_3^t \\
& & AC_3^{t+1} & +BC_4^{t+1} & +EC_5^{t+1} & & = & H_4^t. \\
& & & AC_4^{t+1} & +BC_5^{t+1} & +EC_6^{t+1} & = & H_5^t \\
& & & & AC_5^{t+1} & +BC_6^{t+1} \quad +EC_7^{t+1} & = & H_6^t
\end{array}
\tag{5.2.31}
$$

where A, B, E, and H are known, based on Eqs. (5.2.27)–(5.2.30). The system of equations is then ready to be solved for unknown C^{t+1} after the initial and boundary conditions are specified.

Incorporation of Initial and Boundary Conditions The next step is to incorporate initial and boundary conditions. The initial condition $c(x, 0) = 0$ ($c_i^0 = 0$, where i = 1, 2, 3, 4, 5, 6, and 7) is implemented by substituting it to H_i^0 (Eq. 5.2.30).

Afterward, we include the flux boundary condition at the first block and the prescribed concentration at the end of the domain, i.e., x = L. That is,

$$q_1(t) = uC - D\frac{\partial C}{\partial x} \tag{5.2.32}$$

$$C(L, t) = C_7^t = 0. \tag{5.2.33}$$

The flux boundary condition, Eq. (5.2.32), can be discretized as

$$q_1 = u_1 C_1 - D\left(\frac{(1-\alpha)(C_2 - C_0)}{2\Delta x} + \alpha\frac{(C_1 - C_0)}{\Delta x}\right)$$

If $\alpha = 1$, it comes

$$q = u_1 C_1 - D\frac{C_2 - C_1}{\Delta x} \tag{5.2.34}$$

or

$$C_1 = \frac{q_1 \Delta x}{(u_1 \Delta x + D)} + \frac{D}{(u_1 \Delta x + D)} C_2. \tag{5.2.35}$$

Two approaches are available to implement the boundary conditions.

Approach 1: Replace C_1^{t+1} in the first line of Eq. (5.2.31) with Eq. (5.2.35) and C_7^{t+1} of the last line of Eq. (5.2.31) with Eq. (5.2.33). Once the two boundary conditions are included, Eq. (5.2.31) becomes

$$
\begin{aligned}
A\left(\frac{q_1 \Delta x}{(u_1 \Delta x+D)}\right)+\left(B+\frac{AD}{(u_1 \Delta x+D)}\right)C_2^{t+1}+EC_3^{t+1} &= H_2^t \\
AC_2^{t+1}+BC_3^{t+1}+EC_4^{t+1} &= H_3^t \\
AC_3^{t+1}+BC_4^{t+1}+EC_5^{t+1} &= H_4^t \\
AC_4^{t+1}+BC_5^{t+1}+EC_6^{t+1} &= H_5^t \\
AC_5^{t+1}+BC_6^{t+1}+EC_7^t &= H_6^t
\end{aligned} \tag{5.2.36}
$$

The first term of the first line in Eq. (5.2.36) is known because the flux, D, and A are specified. Similarly, the last term of the last line is known because of the prescribed concentration boundary condition. These terms can be moved to the right-hand side of the equations. Now, we have a new system of equations (5.2.37) as follows:

$$
\begin{aligned}
0+\left(B+\mathrm{A}\frac{D}{(u_1 \Delta x+D)}\right)C_2^{t+1}+EC_3^{t+1} &= H_2^t-A\left(\frac{q_1 \Delta x}{(u_1 \Delta x+D)}\right) \\
Ac_2^{t+1}+BC_3^{t+1}+EC_4^{t+1} &= H_3^t \\
AC_3^{t+1}+BC_4^{t+1}+EC_5^{t+1} &= H_4^t \\
AC_4^{t+1}+BC_5^{t+1}+EC_6^{t+1} &= H_5^t \\
AC_5^{t+1}+BC_6^{t+1}+0 &= H_6^t-EC_7^t
\end{aligned} \tag{5.2.37}
$$

The above system of equations can be written in matrix forms

$$
\begin{bmatrix}
\left(B+\mathrm{A}\frac{D}{(u_1 \Delta x+D)}\right) & E & & & \\
A & B & E & & \\
 & A & B & E & \\
 & & A & B & E \\
 & & & A & B
\end{bmatrix}
\begin{Bmatrix} C_2^{t+1} \\ C_3^{t+1} \\ C_4^{t+1} \\ C_5^{t+1} \\ C_6^{t+1} \end{Bmatrix}
=
\begin{Bmatrix} H_2^t-A\left(\frac{q_1 \Delta x}{(u_1 \Delta x+D)}\right) \\ H_3^t \\ H_4^t \\ H_5^t \\ H_6^t-EC_7^t \end{Bmatrix} \tag{5.2.38}
$$

After solving the above matrix equation, we can determine C_1^{t+1} using Eq. (5.2.35).

Approach 2. This approach includes the boundary flux Eq. (5.2.34) as the first equation and the constant concentration at the other boundary as the last equation of the system of equations.

$$
\begin{array}{lllllll}
\left(\frac{D}{\Delta x}+u\right)C_1^{t+1} & \left(\frac{D}{\Delta x}\right)C_2^{t+1} & & & & & =q \\
 & AC_2^{t+1} & +BC_3^{t+1} & +EC_4^{t+1} & & & =H_2^t \\
 & & AC_3^{t+1} & +BC_4^{t+1} & +EC_5^{t+1} & & =H_3^t \\
 & & & AC_4^{t+1} & +BC_5^{t+1} & +EC_6^{t+1} & =H_4^t \\
 & & & & AC_5^{t+1} & +BC_6^{t+1} \quad +EC_7^{t+1} & =H_5^t \\
 & & & & & AC_6^{t+1} \quad +BC_7^{t+1} & =H_6^t \\
 & & & & & 1 & =C(L)
\end{array}
\tag{5.2.39}
$$

The matrix form of the equations is

$$
\begin{bmatrix}
\left(\frac{D}{\Delta x}+u\right) & \left(\frac{D}{\Delta x}\right) & & & & & =H_6^t \\
 & A & B & E & & & \\
 & & A & B & E & & \\
 & & & A & B & E & \\
 & & & & A & B & E \\
 & & & & & A & B \\
 & & & & & & 1
\end{bmatrix}
\begin{Bmatrix} C_1^{t+1} \\ C_2^{t+1} \\ C_3^{t+1} \\ C_4^{t+1} \\ C_5^{t+1} \\ C_6^{t+1} \\ C_7^{t+1} \end{Bmatrix}
=
\begin{Bmatrix} q \\ H_2^t \\ H_3^t \\ H_4^t \\ H_5^t \\ H_6^t \\ C(L) \end{Bmatrix}.
\tag{5.2.40}
$$

Setting the last element in the last row of the coefficient matrix to be one forces C_7^{t+1} equal to the prescribed $C(L)$ at the block and time $t + 1$.

5.2.4 Solution to the Matrices

We express the above matrices as the coefficient matrix [S] multiplying the unknown solution vector {C}, resulting in the known forcing vector $\{T\}$:

$$[S]\{C\} = \{T\}.$$

The solution of the linear system of equations is

$$\{C\} = [S]^{-1}T$$

where $[S]^{-1}$ is the inverse of $[S]$. Select an empty column in the Excel sheet with the number of rows equal to the number of unknowns to be solve. At the first element of the column, type "=MMULT(MINVERS(S),T) and CTL+SHIFT+ENTER" to obtain the solution vector {C}.

Given values of C at all blocks at the initial time or the previous time step and the boundary conditions, all terms on the right-hand side of Eq. (5.2.33) are determined. Moreover, the equations' coefficients on the left-hand side are also known. Thus, we can solve the system of equations (Eq. 5.2.33) for C at all blocks at time level = $t + 1$. Repeat the above procedure to advance the solution to the next time step. This rudimentary example elucidates the algorithm of the finite difference method. With more involved programming, the above procedure could expedite the calculation for a large domain and many time steps. Notice that the system of equations has a tri-diagonal characteristic (i.e., a band matrix with nonzero elements only on the main diagonal). An efficient method, The Thomas Algorithm is recommended below.

The Thomas algorithm is illustrated using the following tri-diagonal matrix:

$$\begin{array}{lllllll} b_1C_1 & +d_1C_2 & & & & = & h_1 \\ a_2C_1 & +b_2C_2 & +d_2C_3 & & & = & h_2 \\ & a_3C_2 & +b_3C_3 & +d_3C_4 & & = & h_3\,. \\ & & a_4C_3 & +b_4C_4 & +d_4C_5 & = & h_4 \\ & & & a_5C_4 & +b_5C_5 & = & h_5 \end{array} \tag{5.2.41}$$

Procedure:

Step (1), divide the first equation of Eq. (5.2.38) by b_1, and we have

$$C_1 + \left(\frac{d_1}{b_1}\right)C_2 = \left(\frac{h_1}{b_1}\right) \tag{5.2.42}$$

which can be expressed as

$$C_1 + \beta_1 C_2 = y_1,\ \beta_1 = \left(\frac{d_1}{b_1}\right),\ y_1 = \left(\frac{h_1}{b_1}\right). \tag{5.2.43}$$

Step (2), multiplying Eq. (5.2.40) with $-a_2$ yields

$$-a_2C_1 - a_2\beta_1C_2 = -a_2y_1 \tag{5.2.44}$$

Adding Eq. (5.2.41) to the second equation of Eq. (5.2.38) results in

$$(b_2 - a_2\beta_1)C_2 + d_2C_3 = h_2 - a_2y_1 \tag{5.2.45}$$

which can be expressed as

$$\alpha_2C_2 + d_2C_3 = h_2 - a_2y_1 \quad \alpha_2 = (b_2 - a_2\beta_1). \tag{5.3.46}$$

Dividing Eq. (5.2.43) by α_2 yields

$$C_2 + \beta_2 C_3 = y_2, \ \beta_2 = \left(\frac{d_2}{\alpha_2}\right), \ y_2 = \left(\frac{h_2 - a_2 y_1}{\alpha_2}\right) \tag{5.2.47}$$

Step (3). We multiply Eq. (5.2.44) with $-a_3$ to yield

$$-a_3 C_2 - a_3 \beta_2 C_3 = -a_3 y_2 \tag{5.2.48}$$

Adding Eq. (5.2.45) to the third equation in Eq. (5.2.38), we have

$$(b_3 - a_3 \beta_2) C_3 + d_3 C_4 = h_3 - a_3 y_2 \tag{5.2.49}$$

which can also be expressed as

$$\alpha_3 C_3 + d_3 C_4 = h_3 - a_3 y_2 \quad \alpha_3 = (b_3 - a_3 \beta_2).$$

Dividing Eq. (5.2.47) by α_3 yields

$$C_3 + \beta_3 C_4 = y_3, \quad \beta_3 = \left(\frac{d_3}{\alpha_3}\right), \quad y_3 = \left(\frac{h_3 - a_3 y_2}{\alpha_3}\right) \tag{5.2.50}$$

Repeating the above procedure for the rest equations in Eq. (5.2.38) shows that the procedure has a particular pattern:

$$\begin{aligned} \alpha_i &= b_i - a_i \beta_{i-1} \\ \beta_i &= d_i / \alpha_i \\ y_i &= (h_i - a_i y_{i-1}) / \alpha_i \end{aligned} \tag{5.2.51}$$

where i = 1,2, . . . n, n is the total number of equations. The above procedure is the forward substitution. After the forward substitution, we have

$$\begin{array}{cccccccccc} C_1 & +\beta_1 C_2 & & & & & & & = & y_1 \\ & C_2 & & +\beta_2 C_3 & & & & & = & y_2 \\ & & & C_3 & & +\beta_3 C_4 & & & = & y_3 \,. \\ & & & & & C_4 & & +\beta_4 C_5 & = & y_4 \\ & & & & & & & C_5 & = & y_5 \end{array} \tag{5.2.52}$$

Now, C_5 is known explicitly from the last line of Eq. (5.2.49), and it can be substituted back to the next line to solve C_4. Subsequently, the backward substitution procedure solves for all Cs. The backward substitution procedure can be expressed in a general formula:

$$C_i = y_i - \beta_i C_{i+1}. \tag{5.2.53}$$

The regular pattern of the forward and backward substitution promotes the development of a computer program to solve the tri-diagonal matrix resulting from the finite difference discretization of the advection–diffusion equation. Again, this algorithm can be programmed in Excel with relative ease.

5.2.4.1 Homework

Use Excel to implement the finite difference formulation of the problem in Fig. 5.3 and solve the matrices using the Excel function described earlier for two time steps from the initial condition. Compare the solution with those solved by the Thomas algorithm.

5.2.5 Numerical Dispersion

While the numerical formulation of the advection–diffusion equation is straightforward, obtaining a stable and accurate solution confronts many issues. We will examine some of the issues.

Consider the 1-D advection–dispersion Equation

$$\frac{\partial C}{\partial t} + u\frac{\partial C}{\partial x} = D\frac{\partial^2 C}{\partial x^2}. \tag{5.2.54}$$

Supposedly, we use the upwind finite difference approach to discretize the governing equation's spatial derivatives.

$$\frac{C_i^{t+1} - C_i^t}{\Delta t} + u\frac{C_i^t - C_{i-1}^t}{\Delta x} = D\frac{C_{i-1}^t - 2C_i^t + C_{i+1}^t}{(\Delta x)^2} \tag{5.2.55}$$

The stability criterion for the solution of the above finite difference equation is

$$\Delta t \leq \left[\frac{u}{\Delta x} + \frac{2D}{\Delta x^2}\right]^{-1} \tag{5.2.56}$$

(see Roache, 1976).

Notice that $\Delta x/u$ is the time for the advection to travel over a block while $\Delta x^2/2D$ represents the time for diffusion to cover the entire block. The criterion suggests that the time step size should be smaller than the time for advection and diffusion travel over a block. That is, the time step depends on the values of u and D.

Next, we expand C_i^{t+1} in Eq. (5.2.51) in a Taylor series gives

$$C_i^{t+1} = C_i^t + \Delta t\left(\frac{\partial C}{\partial t}\right) + \frac{\Delta t^2}{2}\frac{\partial^2 C}{\partial t^2} + O(\Delta t^3). \tag{5.2.57}$$

Since the first time derivative term can be related to ADE:

$$\frac{\partial C}{\partial t} = D\frac{\partial^2 C}{\partial x^2} - u\frac{\partial C}{\partial x} \tag{5.2.58}$$

Substitution of the right-hand side of Eq. (5.2.55) as Cs in ADE determines the second-time derivative term on the right-hand side of Eq. (5.2.54). That is,

$$\begin{aligned}\frac{\partial^2 C}{\partial t^2} &= \frac{\partial}{\partial t}\left(\frac{\partial C}{\partial t}\right) = D\frac{\partial^2}{\partial x^2}\left[D\left(\frac{\partial^2 C}{\partial x^2}\right) - u\frac{\partial C}{\partial x}\right] - u\frac{\partial}{\partial x}\left[D\left(\frac{\partial^2 C}{\partial x^2}\right) - u\left(\frac{\partial C}{\partial x}\right)\right]\\ &= D^2\frac{\partial^2}{\partial x^2}\left(\frac{\partial^2 C}{\partial x^2}\right) - 2D\left[u\frac{\partial}{\partial x}\left(\frac{\partial^2 C}{\partial x^2}\right)\right] + u^2\frac{\partial^2 C}{\partial x^2}\end{aligned} \tag{5.2.59}$$

Substituting (5.2.56) into (5.2.54), we have

$$C_i^{t+1} = C_i^t + \Delta t\left(\frac{\partial C}{\partial t}\right) + \frac{\Delta t^2}{2}\left[u^2\frac{\partial^2 C}{\partial x^2}\right] + \text{HOT} + \text{HOD} \tag{5.2.60}$$

where HOT and HOD stand for higher-order terms and derivatives.

Likewise, the spatial terms in the finite-difference analogs can be expressed in a Taylor series:

$$C_{i-1}^t = C_i^t - \Delta x\frac{\partial C}{\partial x} + \Delta x^2\frac{\partial^2 C}{2!\partial x^2} + \cdots \tag{5.2.61}$$

$$C_{i+1}^t = C_i^t + \Delta x\frac{\partial C}{\partial x} + \Delta x^2\frac{\partial^2 C}{2!\partial x^2} + \cdots\cdot \tag{5.2.62}$$

Substituting Eqs. (5.2.57), (5.2.58), (5.2.59) into Eq. (5.2.52), we have

$$\frac{\partial C}{\partial t} + u\frac{\partial C}{\partial x} = D\frac{\partial^2 C}{\partial x^2} + \frac{1}{2}\left[\left(u\Delta x - u^2\Delta t\right)\frac{\partial^2 C}{\partial x^2}\right] + \cdots\cdot \tag{5.2.63}$$

Eq. (5.2.60) is the continuous partial differential equation of our finite difference analog of the ADE (i.e., Eq. 5.2.52). This equation shows that our finite-difference analog is not equal to the ADE we intend to solve but an ADE with additional terms. Since the second term of the right-hand side is related to the second derivative of concentration, Eq. (5.2.59) can also be rewritten as

$$\frac{\partial C}{\partial t} + u\frac{\partial C}{\partial x} = \left[D + \frac{1}{2}\left(u\Delta x - u^2\Delta t\right)\right]\frac{\partial^2 C}{\partial x^2} + \cdots \tag{5.2.64}$$

As a result, the coefficient of the second partial derivative of concentration, representing the effect of diffusion, contains not only the physical diffusion coefficient, D, but also a numerical error:

$$D_{numerical} = \frac{1}{2}\left(u\Delta x - u^2\Delta t\right) \tag{5.2.65}$$

The numerical error acts like diffusion and is thus called numerical dispersion (or diffusion). The numerical dispersion term results from spatial and time (temporal) approximation of advection transport (u term). Eq. (5.2.61) manifests that we can control the numerical dispersion if we let

$$u\Delta x = u^2\Delta t. \tag{5.2.66}$$

That is, we can eliminate the numerical dispersion if we restrict the ratio below equal to unity:

$$\frac{u\Delta t}{\Delta x} = 1 \tag{5.2.67}$$

This ratio is the **Courant number**. Physically, the Courant number criterion states that for a given u, the simulation time step should be chosen so that the advection front precisely reaches the center of the next finite-difference block. However, the stability criterion discussed previously,

$$\Delta t \leq \left[\frac{u}{\Delta x} + \frac{2D}{\Delta x^2}\right]^{-1} \tag{5.2.68}$$

leads to the fact that

$$\left(u\Delta x - u^2\Delta t\right) \neq 0. \tag{5.2.69}$$

In other words, while we can control the numerical dispersion, we encounter the problem of stability (under or overshooting). See Fig. 5.4. On the other hand, numerical dispersion occurs when the stability is controlled.

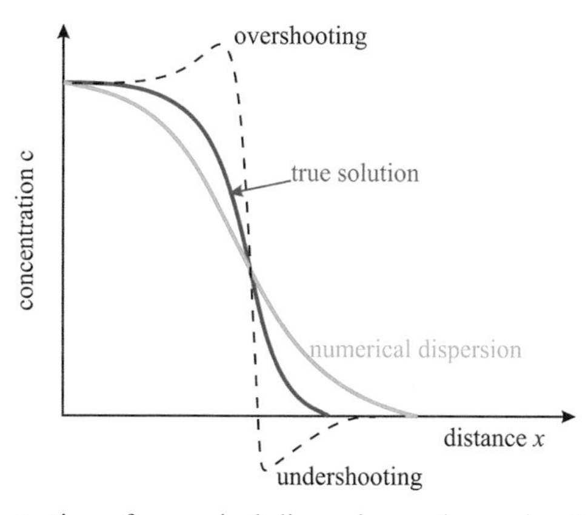

Figure 5.4 An illustration of numerical dispersion and overshooting/undershooting.

While these problems are active research topics, the general rule of thumb is that the time step and grid size of the discretization should satisfy the following conditions:

$$\text{Mesh Peclet Number} = \frac{u\Delta x}{D} < 2$$

$$\text{Courant Number} = \frac{u\Delta t}{\Delta x} < 1.$$

Specifically, the **mesh peclet number** criterion dictates that the grid size in the numerical analysis should be small such that the diffusion overtakes the advection in a grid cell (i.e., the spread of solute due to diffusion must be larger than that due to advection). Notice that these analyses are based on one-dimensional and uniform velocity. Implementing the above criteria to simulate multidimensional solution transport in heterogeneous domains (Chapter 9), which involves divergent and convergent flow and velocity-dependent dispersion coefficient, could be complex. Nevertheless, a small grid size and time step could yield solutions, less susceptible to these errors.

5.3 Method of Characteristics (MOC)

One of the problems associated with the finite difference approach for the advection–diffusion equation is the overshooting and undershooting when the diffusion coefficient is small or equal to zero (i.e., a violation of the mesh Peclet number criterion). To alleviate this problem, one can use the method of characteristics approach (Eulerian–Lagrangian Solution, Chen et al., 1984), discussed as follows.

Consider a 2-D solute transport equation:

$$\frac{\partial C}{\partial t} = D_m \frac{\partial^2 C}{\partial x^2} + D_m \frac{\partial^2 C}{\partial y^2} - u_x \frac{\partial C}{\partial y} - u_y \frac{\partial C}{\partial y}. \tag{5.3.1}$$

Again, we will use the total (or substantial, material) derivative of *c(x, y, t)*,

$$\frac{dC}{dt} = \frac{\partial C}{\partial t} + \frac{\partial C}{\partial x}\frac{dx}{dt} + \frac{\partial C}{\partial y}\frac{dy}{dt} \tag{5.3.2}$$

where

$$\frac{dx}{dt} = u_x \quad \frac{dy}{dt} = u_y. \tag{5.3.3}$$

Applying Eq. (5.3.2) and (5.3.3) to Eq. (5.3.1), we have

$$\frac{dC}{dt} = D_m \frac{\partial^2 C}{\partial x^2} + D_m \frac{\partial^2 C}{\partial y^2}. \tag{5.3.4}$$

Equation (5.3.4) thus describes the change in concentration perceived by an observer moving at the same velocity as the fluid (or at a moving framework, or Lagrangian coordinates).

Equations (5.3.3) and (5.3.4) are the characteristics of Eq. (5.3.1). Accordingly, we solve the ADE in two steps: (1) Solving Eq. (5.3.3) for convective term first, and (2) then solving Eq. (5.3.4) for the diffusive term. In other words, instead of solving the ADE directly, we solve the equivalent equations of the ADE. The equivalent equations are the characteristics of ADE, i.e.,

$$\frac{dx}{dt} = u_x \quad \frac{dy}{dt} = u_y \tag{5.3.5}$$

$$\frac{dC}{dt} = D_m \frac{\partial^2 C}{\partial x^2} + D_m \frac{\partial^2 C}{\partial y^2}. \tag{5.3.6}$$

They are the equations of the Eulerian–Lagrangian approach: Eq. (5.3.5) corresponds to the Eulerian (advection) part, and Eq. (5.3.6) is the Lagrangian part (viewing the advection–diffusion process on the moving coordinate system, described by Eq. (5.3.5).

Implementation of the Solution involves the following steps.

Step 1. Discretization of the two-dimensional solution domain using a block-centered grid system. Variables and parameters are specified at the center of the block with the dimension, Δx and Δy. The finite difference block is identified by the indices, i, and j. Next, assign boundary and initial conditions to the boundaries and all blocks, respectively (Fig. 5.5).

Step. 2. Particle Tracking. Introduce a set of moving points distributed uniformly in the solution domain with initial coordinates, $x_{p,o}$ $y_{p,o}$. and initial concentration $C_{p,0}$ where p is an index identifying the moving points, and 0 denotes the initial concentration.

Step. 3. Assume that we know the heads at all stationary points within the grid system from the solution of a two-dimensional flow equation, and we have calculated the velocities $u_x(i, j)$ and $u_y(i, j)$ at each block.

Use these velocities at the center of blocks, u_x and u_y, to determine the velocities for the moving points at any location (x, y) within the block by

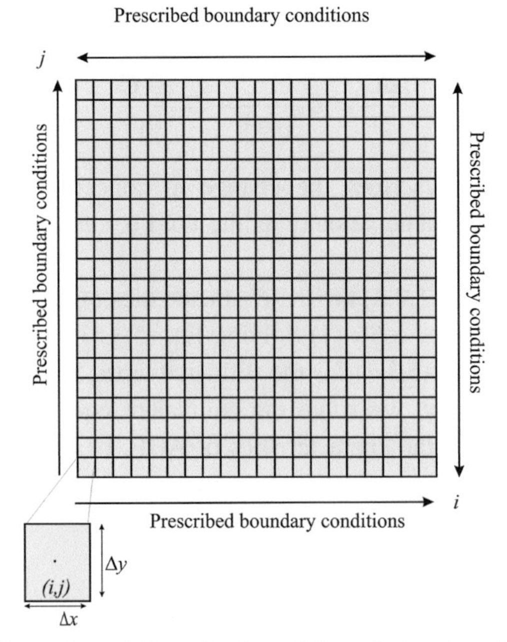

Figure 5.5 An illustration of discretization of the solution domain using a block-centered grid system.

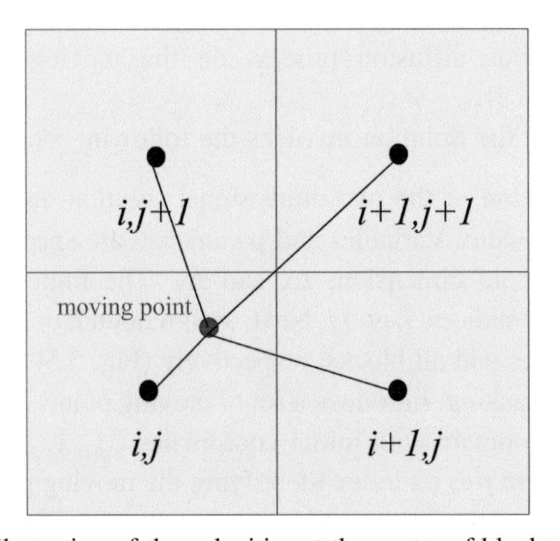

Figure 5.6 An illustration of the velocities at the center of blocks.

a linear interpolation scheme based on velocity values at adjacent blocks (Fig. 5.6).

Step. 4. Determine the new positions of the moving points according to the moving point velocities. That is,

$$\left(\frac{dx}{dt}\right)_p^{n+1} = u_x\left(x_p^n, y_p^n\right), \qquad \left(\frac{dy}{dt}\right)_p^{n+1} = u_y\left(x_p^n, y_p^n\right) \tag{5.3.7}$$

$$\frac{x_p^{n+1} - x_p^n}{\Delta t} = u_x \qquad \Delta t = t^{n+1} - t^n \tag{5.3.8}$$

where $n + 1$ is a new time level, and n is the current time level. The new coordinates of a moving point are evaluated by

$$x_p^{n+1} = x_p^n + \Delta t \cdot u_x\left(x_p^n, y_p^n\right) \quad \text{and} \quad y_p^{n+1} = y_p^n + \Delta t \cdot u_y\left(x_p^n, y_p^n\right). \tag{5.2.9}$$

This is an Euler formula with a first-order accuracy in time $O(\Delta t)$. Higher-order approximation (such as the Runge–Kutta type formula) can also be used.

Step 5. After moving the points, coordinates of the new position of the moving point $\left(x_p^{n+1}, y_p^{n+1}\right)$ are tested to see which block the moving point lies in. Now a temporary concentration in each block C_{ij}^{*n} is determined by averaging all moving points' concentration C_p^n within the block.

Step 6. We next calculate the change in concentration due to diffusion by solving the finite difference analog form of Eq. (5.3.4). That is,

$$\frac{dC}{dt} = \frac{\Delta C_{ij}^n}{\Delta t} = \left[D_m \frac{\partial^2 C^*}{\partial x^2} + D_m \frac{\partial^2 C^*}{\partial y^2}\right] \tag{5.3.10}$$

where

$$\frac{\partial^2 C^*}{\partial x^2} = \left(C_{i-1,j}^{*n} - 2C_{ij}^{*n} + C_{i+1,j}^{*n}\right)/\Delta x^2 \tag{5.3.11}$$

$$\frac{\partial^2 C^*}{\partial y^2} = \left(C_{i,j-1}^{*n} - 2C_{ij}^{*n} + C_{i,j+1}^{*n}\right)/\Delta y^2. \tag{5.3.12}$$

Step 7. Update the concentration of moving points by

$$C_p^{n+1} = C_p^n + \Delta C_{ij}^n \tag{5.3.13}$$

This approach assumes that the moving points within a block undergo the exact change in concentration due to diffusion.

Step 8. The concentrations at the stationary grid points are updated for the new time step by:

$$C_{ij}^{n+1} = C^{*n}_{ij} + \Delta C_{i,j}^{n}. \tag{5.3.14}$$

Eq. (5.3.14) completes the step from t^n to t^{n+1}. The above eight steps are repeated for each subsequent time step. Fig. 5.7 shows the flow chart of these steps.

The advantage of the method of characteristics is its ability to simulate solute transport under purely convection conditions (i.e., the diffusion coefficient is zero). However, even if the Courant number criterion discussed earlier constrains the time step, MOC still suffers from numerical diffusion or dispersion since the moving point concentration is an average over the block unless the block is reduced to an infinitesimally small point. Nevertheless, this numerical diffusion could represent the effects of molecular diffusion. The criteria for stability and numerical dispersion are for constant velocity. When the velocity varies spatially and temporally, difficulties in applying these criteria are expected. Reducing the block size would involve a new setup of the solution domain. Reducing the time step is likely the most convenient remedy. Details of the approach are available in Geode (1990). Despite these weaknesses, simulating solute transport in highly heterogeneous aquifers prefers the MOC approach (Srivastava and Yeh, 1992;

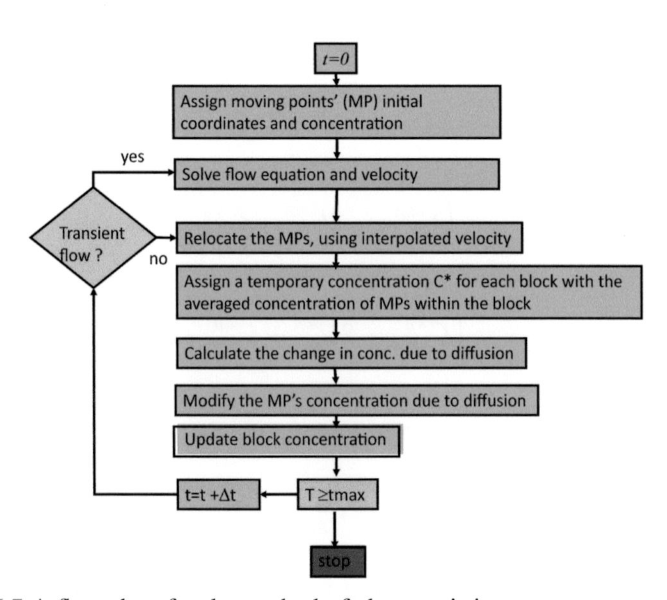

Figure 5.7 A flow chart for the method of characteristics.

Yeh et al., 1993; Ni et al., 2009; Su et al., 2020) since advection dominates the transport process at the scale of our interests.

5.3.1 Homework

Formulate the finite difference analog of Eq. (5.3.10) and matrices for a two-dimensional domain (4 × 4 blocks), assuming upper and lower boundaries are no diffusion flux and the left and the right-hand side boundaries are constant concentrations.

5.4 Finite Element Approach

The finite element method is more versatile than the finite-difference method because of its flexibility for treating irregular shapes of the domain. However, it also approximates the solution of a partial differential equation. The following segments introduce the method's general concept, and a more in-depth discussion of the method is available in many finite element textbooks (e.g., Istok, 1989).

First, consider steady flow in a one-dimensional confined aquifer with a uniform hydraulic conductivity K, subject to a constant recharge, Q, and constant head boundaries on both ends (Fig. 5.8).

The governing partial differential equation for the flow is given by

$$K\frac{d^2h}{dx^2} + Q = 0. \tag{5.4.1}$$

Suppose $\widehat{h}(x)$ is an approximate solution to the partial differential equation. Substituting it into Eq. (5.4.1), we expect the following result:

$$K\frac{d^2\widehat{h}(x)}{dx^2} + Q = R(x) \neq 0 \tag{5.4.2}$$

Since $\widehat{h}(x)$ it is an approximate solution, it will not satisfy the partial differential equation and leads to R(x) residual for any x.

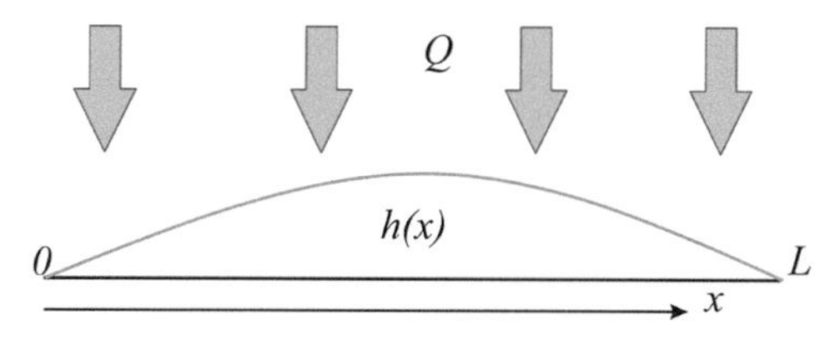

Figure 5.8 A schematic illustration of a one-dimensional confined aquifer subject to a constant recharge.

One finite element approach to account for the approximate nature of the solution uses the weighted residual method, i.e.,

$$\int_0^L w_i(x)R(x)dx = 0 \tag{5.4.3}$$

where w_i is the weighting function, i is the index for each unknown to be sought in the approximate solution. Several choices for the weighting functions are available, including (1) Collocation Method, (2) Subdomain Method, (3) Galerkin's Method, and (4) Least Squares Method. Here we will focus on the last two methods, which are wildly used.

5.4.1 Least Squares Method

This method uses the error (residual) as the weighting function.

$$Error = E = \int_0^L [R(x)]^2 dx. \tag{5.4.4}$$

The approach aims to find a solution that yields a minimum error over the entire solution domain. It uses optimization to accomplish this goal. That is, the error is minimized to the unknown coefficients in the approximate solution.

EXAMPLE: Suppose the approximate solution is

$$\widehat{h}(x) = A \sin \frac{\pi x}{L} \tag{5.4.5}$$

where A is unknown (i.e., a regression coefficient) and $\sin(\pi x/L)$ is the shape of the solution (i.e., shape function). Substitution of the approximate Solution into Eq. (5.4.2) and Eq. (5.4.4) leads to

$$E = \int_0^L \left[-\frac{K\pi^2}{L^2} A \sin \frac{\pi x}{L} + Q \right]^2 dx = \frac{A^2 L}{2} \left[\frac{K\pi^2}{L^2} \right]^2 - \frac{4QK\pi}{L} A + Q^2 L. \tag{5.4.6}$$

Next, we minimize E by taking its derivative to the unknown A and set the resulting equation to zero:

$$\frac{\partial E}{\partial A} = AL \left[\frac{K\pi^2}{L^2} \right]^2 - \frac{4QK\pi}{L} = 0. \tag{5.4.7}$$

Thus, the unknown A is

$$A = \frac{4QL^2}{\pi^3 K}. \tag{5.4.8}$$

Therefore, the approximate solution for the problem is

$$\widehat{h}(x) = \frac{4QL^2}{\pi^3 K} \sin \frac{\pi x}{L}. \tag{5.4.9}$$

This example illustrates the concept of the least square approach in the finite element method.

5.4.2 Galerkin's Method.

Instead of using the least-squares criterion, this method uses a weighted residual criterion:

$$\int_0^L w_i(x) R(x) dx = 0 \tag{5.4.10}$$

in which the shape function of the approximate solution is used as $w_i(x)$. Again, we use

$$\widehat{h}(x) = A \sin \frac{\pi x}{L} \tag{5.4.11}$$

as our approximate solution. The weighted residual criterion becomes

$$\int_0^L \sin \frac{\pi x}{L} \left[-K \frac{A\pi^2}{L^2} \sin \frac{\pi x}{L} + Q \right] dx = 0. \tag{5.4.12}$$

Note that $\sin(\pi x/L)$ is the shape function. Integration of the criterion yields

$$\frac{K\pi^2 A}{2L} - \frac{2QL}{\pi} = 0. \tag{5.4.13}$$

Then, the unknown is derived:

$$A = \frac{4QL^2}{\pi^3 K}. \tag{5.4.14}$$

Subsequently, the approximate solution is given as

$$\widehat{h}(x) = \frac{4QL^2}{\pi^3 K} \sin \frac{\pi x}{L}. \tag{5.4.15}$$

These two methods essentially result in the same approximate solution.

5.4.2.1 Selection of Shape Functions

The shape functions sometimes are called interpolation functions, basis functions, or interpolating polynomial. The finite element method assigns values of a field variable at finite element nodes. It then uses a shape function to represent the variation of the field variable over the element.

Example: Consider the hydraulic head h(x) in a one-dimensional aquifer as a continuous field variable, as shown in Fig. 5.9. The aquifer is discretized into four finite elements (#1, #2, #3, and #4). Each element has two nodes (1 and 2) and. These nodes are numbered as 1, 2, 3, 4, and 5 globally (over the entire aquifer).

If we choose a shape function, N_i^e, for each element e (where i = 1,2 nodes) such that

$$N_1^e = \frac{1}{L}(X_2 - x) \quad \text{and} \quad N_2^e = \frac{1}{L}(x - X_1) \tag{5.4.16}$$

Where L is the length of element #1. X_1 and X_2 are the coordinates of nodes 1 and 2 of this element, and x is any distance from X_1. The head variation between the two nodes thus is fully described by

$$h(x) = h_1 N_1^e + h_2 N_2^e. \tag{5.4.17}$$

For example, examine element #1 and assume that we know head values at x_1 and x_2. We want to determine the head value at x = $L/3$.

$$h\left(\frac{L}{3}\right) = \frac{1}{L}(x_2 - x)h_1 + \frac{1}{L}(x - x_1)h_2 = \frac{1}{L}\left(L - \frac{L}{3}\right)h_1 + \frac{1}{L}\left(\frac{L}{3} - 0\right)h_2 = \frac{2}{3}h_1 + \frac{1}{3}h_2 \tag{5.4.18}$$

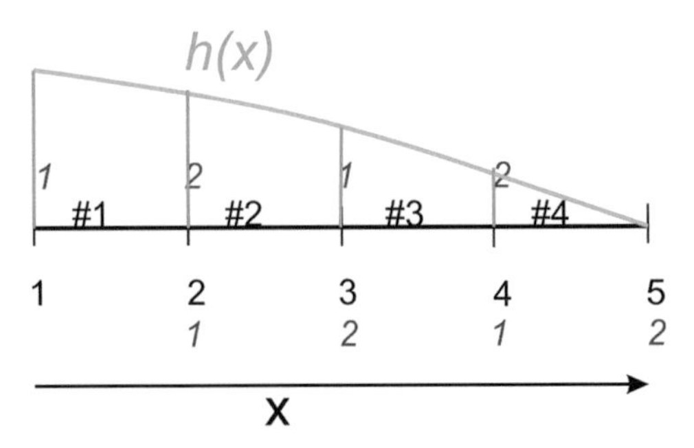

Figure 5.9 A schematic illustration of hydraulic head in a one-dimensional aquifer.

Therefore, the head field within an element can be expressed in general as

$$h(x) = \sum_{i=1}^{n_e} h_i N_i^e(x) \tag{5.4.19}$$

where n_e is the total number of nodes in the element; h_i is the head at node i; N_i^e is the shape function associated with node i.

In general, polynomials could be used as shape functions. The degree (order) of the polynomial depends on: (1) the number of nodes assigned to the element, (2) the nature of the unknown at each node, and (3) continuity requirements imposed at the nodes. A linear shape function is the simplest function commonly used.

5.4.2.2 Properties of the Linear Shape Function

Consider a one-dimensional element with two nodes i and j, separated by a distance of L. The coordinates of node i is X_i, and node 2 is X_j. The linear shape function of for node i is $N_i(x)$ and is $N_j(x)$ for node j, and they can be expressed as

$$N_i(x) = \frac{X_j - x}{L} \text{ and } N_j(x) = \frac{x - X_i}{L}. \tag{5.4.20}$$

The two shape functions must have the following properties illustrated in Fig. 5.10:

$$\begin{aligned} &at \quad x = X_i \quad N_i = \frac{X_j - X_i}{L} = 1 \\ &at \quad x = X_j \quad N_i = \frac{X_j - X_j}{L} = 0 \end{aligned} \tag{5.4.21}$$

and

$$\text{at } x = X_j \quad N_i = 1 \text{ and at } x = X_j \quad N_j = 0. \tag{5.4.22}$$

Further,

$$N_i(x) + N_j(x) = 1 \quad \text{at any x.}$$

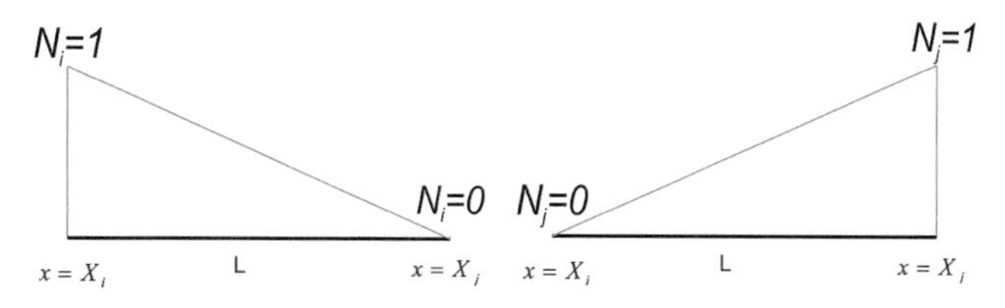

Figure 5.10 Illustration of essential properties of linear shape function.

5.4.2.3 Development of Linear Shape Functions

We illustrate how the linear shape function is derived for the one-dimensional element. Once the principle is understood, one can expand this concept and procedure to derive the linear shape functions for two- or three-dimensional elements.

Consider 1-D flow through a soil block. If we know head values, Φ_i and Φ_j, at nodes *i* and *j*, respectively (Fig. 5.11), we ask how to determine a shape function to represent the head variation ϕ between Φ_i and Φ_j?

According to Fig. 5.11, $\phi = \alpha_1 + \alpha_2 x$ and we know $\phi = \Phi_i$ at $x = X_i$ and $\phi = \Phi_j$ at $x = X_j$. The question is that what are the coefficients α_1 and α_2 that will satisfy the above two facts:

$$\Phi_i = \alpha_1 + \alpha_2 X_i \tag{5.4.23}$$

$$\Phi_j = \alpha_1 + \alpha_2 X_j \tag{5.4.24}$$

Solving the two equations for α_1 and α_2, we have

$$\alpha_1 = \frac{\Phi_i X_i - \Phi_j X_j}{L} \quad \text{and} \quad \alpha_1 = \frac{\Phi_j - \Phi_i}{L} \tag{5.4.25}$$

where $L = X_j - X_i$. Substituting α_1 and α_2 into the linear polynomial produces

$$\phi = \left(\frac{\Phi_i X_j - \Phi_j X_i}{L}\right) + \left(\frac{\Phi_j - \Phi_i}{L}\right) x = \left(\frac{X_j - x}{L}\right)\Phi_i + \left(\frac{x - X_i}{L}\right)\Phi_j \tag{5.4.26}$$

That is,

$$\phi = N_i \Phi_i + N_j \Phi_j \text{ or } \phi = [N]\{\Phi\} \tag{5.4.27}$$

In other words, using the linear shape function, we say that the head variation between the two known values is described by a linearly weighted sum of Φ_i and Φ_j .

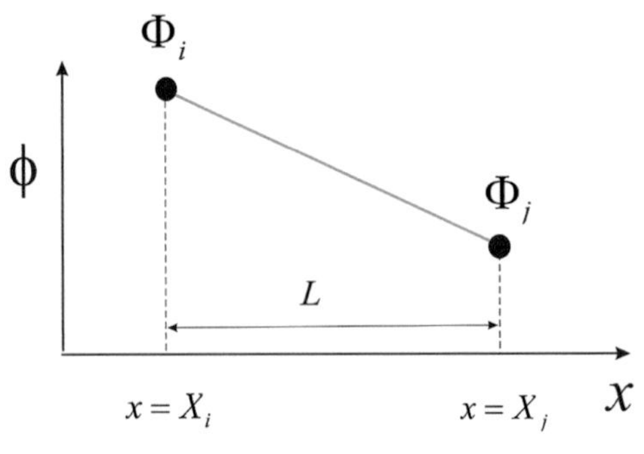

Figure 5.11 Illustration of linear shape function.

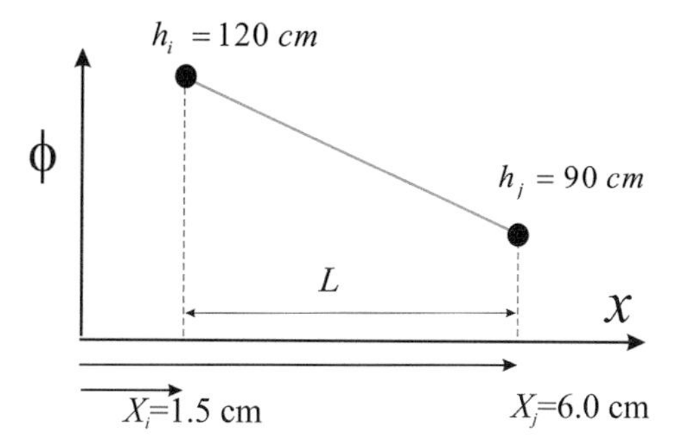

Figure 5.12 Illustration of the use of the shape function in 1-D flow.

Now, we can determine the head value and its gradient at any x in the element using Eq. (5.4.27). Consider a head field in one element in a one-dimensional aquifer domain (Fig. 5.12).

According to the linear shape function, the head value, $h(x)$, within the element is given by:

$$h(x) = \left(\frac{X_j - x}{L}\right)h_i + \left(\frac{x - X_i}{L}\right)h_j \tag{5.4.28}$$

where $L = X_j - X_i = 4.5cm$. We can use Eq. (5.4.2*) to determine $h(x = 4.0\text{ cm})$.

$$h(4.0) = \left(\frac{6.0 - 4.0}{4.5}\right) \times 120 + \left(\frac{4 - 1.5}{4.5}\right) \times 90 = 103.33\text{ cm}$$

Meanwhile, we can determine the hydraulic gradient over the element:

$$\frac{dh}{dl} = -\frac{1}{L}h_i + \frac{1}{L}h_j = \frac{1}{L}\left(h_j - h_i\right) = -6.67\text{ cm/cm}.$$

5.4.3 Galerkin's Method

The above basic knowledge of the finite element method allows us to formulate the finite element model for one-dimensional ADE. Consider solute transport in a semi-infinite sand column and assume that the following ADE can describe the process

$$\frac{\partial C}{\partial t} + u\frac{\partial C}{\partial x} - D\frac{\partial^2 C}{\partial x^2} = 0 \tag{5.4.29}$$

with boundary conditions

$$C(0, t) = C_0 \qquad t \geq 0 \tag{5.4.30}$$

$$C(\infty, t) = 0 \qquad t \geq 0 \tag{5.4.31}$$

and initial conditions

$$C(x, 0) = 0 \qquad x > 0\,. \tag{5.4.32}$$

5.4.3.1 Formulation of Element Matrices

We will use the Galerkin Criterion at the element level,

$$\int_0^L \ell\left(\widehat{C}\right) N_i dx = 0 \tag{5.4.33}$$

where $i = 1$ and 2, N_i is a linear element shape function, and let

$$\ell(C) = \frac{\partial C}{\partial t} + u\frac{\partial C}{\partial x} - D\frac{\partial^2 C}{\partial x^2} = 0. \tag{5.4.34}$$

The Galerkin criterion becomes

$$\int_0^L \left(\frac{\partial C}{\partial t} + u\frac{\partial C}{\partial x} - D\frac{\partial^2 C}{\partial x^2}\right) N_i dx = 0. \tag{5.4.35}$$

Integration of the third term in the bracket by parts yields

$$\int_0^L D\frac{\partial^2 C}{\partial x^2} N_i dx = N_i DC|_{x=0}^{x=L} - \int_0^L D\frac{\partial C}{\partial x}\frac{\partial N_i}{dx} dx = q_L - q_0 - \int_0^L D\frac{\partial C}{\partial x}\frac{\partial N_i}{dx} dx. \tag{5.4.36}$$

The first term on the right-hand side of Eq. (5.4.36) yields the diffusive mass flux at $x = L$ and $x = 0$. Subsequently, the Galerkin criterion for the entire aquifer becomes

$$\int_0^L \left(\frac{\partial C}{\partial t} + u\frac{\partial C}{\partial x}\right) N_i dx + \int_0^L D\frac{\partial C}{\partial x}\frac{\partial N_i}{dx} dx = q_L - q_0. \tag{5.4.37}$$

Now, we use the following as the approximation solution

$$C(x, t) \simeq \widehat{C}(x, t) = \sum_{j=1}^{2} N_j(x)\widehat{C}_j(t) \tag{5.4.38}$$

where N_j is the shape function for the element and $\widehat{C}_j(t)$ is the approximated solution of concentration at node j of the element. Substitution of this approximated Solution into Eq. (5.4.37) results in

$$\sum_{j=1}^{2}\int_{0}^{L}\left[N_iN_j\frac{d\widehat{C}_j}{dt}+uN_i\frac{\partial N_j}{\partial x}\widehat{C}_j+D\frac{\partial N_i}{\partial x}\frac{\partial N_j}{\partial x}\widehat{C}_j\right]dx=q_L-q_0. \qquad (5.4.39)$$

This formula is called the consistent formulation of the ADE. Notice that we change $\partial C/\partial t$ to dC/dt because the derivative applies to C at that particular node.

Another approach is called the lumped approach, which is given as follows

$$\sum_{j=1}^{2}\int_{0}^{L}\left[N_i\frac{d\widehat{C}_j}{dt}+uN_i\frac{\partial N_j}{\partial x}\frac{\partial N_i}{\partial x}\widehat{C}_j+D\frac{\partial N_i}{\partial x}\frac{\partial N_j}{\partial x}\widehat{C}_j\right]dx=q_L-q_0. \qquad (5.4.40)$$

The difference between the lump and the consistent approach is illustrated below.

Next step is to form the element matrix by summing Eq. (5.4.40) over the two nodes in an element. Specifically, writing Eq. (5.4.39) or (5.4.40) for each element (two nodes: 1 and 2) in a matrix form leads to

$$\underset{2\times 2}{[A]}\ \underset{2\times 1}{\left\{\widehat{C}\right\}}+\underset{2\times 2}{[B]}\ \underset{2\times 1}{\left\{\frac{d\widehat{C}}{dt}\right\}}=q_L-q_0. \qquad (5.4.41)$$

Elements in matrices [A] and [B] are

$$a_{ij}=\int_{0}^{L}\left[D\frac{\partial N_i}{\partial x}\frac{\partial N_j}{\partial x}+uN_i\frac{\partial N_j}{\partial x}\right]dx$$

$$b_{ij}=\int_{0}^{L}N_iN_jdx \text{ for the consistent formulation}$$

$$b_{ij}=\int_{0}^{L}N_idx \text{ for the lumped formulation}$$

Rewrite these elements in matrix form, knowing i and j range from 1 to 2.

$$[A] = \begin{bmatrix} a_{11} & a_{12} \\ a_{21} & a_{22} \end{bmatrix} = D\int_0^L \begin{bmatrix} \frac{\partial N_1}{\partial x}\frac{\partial N_1}{\partial x} & \frac{\partial N_1}{\partial x}\frac{\partial N_2}{\partial x} \\ \frac{\partial N_2}{\partial x}\frac{\partial N_1}{\partial x} & \frac{\partial N_2}{\partial x}\frac{\partial N_2}{\partial x} \end{bmatrix} dx + u\int_0^L \begin{bmatrix} N_1\frac{\partial N_1}{\partial x} & N_1\frac{\partial N_2}{\partial x} \\ N_2\frac{\partial N_1}{\partial x} & N_2\frac{\partial N_2}{\partial x} \end{bmatrix} dx. \tag{5.4.42}$$

The subscript 1 or 2 denotes Node 1 or Node 2 in each element. Next, applying the linear shape function

$$N_i = 1 - \frac{x}{L}, N_j = \frac{x}{L} \tag{5.4.43}$$

to Eq. (5.4.42), we can express the integrals within the A matrix as

$$\int_0^L \frac{\partial N_1}{\partial x}\frac{\partial N_1}{\partial x} dx = \int_0^L \left(\frac{-1}{L}\right)\left(\frac{-1}{L}\right) dx = \frac{1}{L} \tag{5.4.44}$$

$$\int_0^L \frac{\partial N_1}{\partial x} dx = \int_0^L \left(1 - \frac{x}{L}\right)\left(\frac{-1}{L}\right) dx = -\frac{1}{2}. \tag{5.4.45}$$

Therefore,

$$[A] = \frac{D}{L}\begin{bmatrix} 1 & -1 \\ -1 & 1 \end{bmatrix} + \frac{u}{2}\begin{bmatrix} -1 & 1 \\ -1 & 1 \end{bmatrix}. \tag{5.4.46}$$

Similarly,

$$[B] = \int_0^L \begin{bmatrix} N_1N_1 & N_1N_2 \\ N_2N_1 & N_2N_2 \end{bmatrix} dx = L\begin{bmatrix} 1/3 & 1/6 \\ 1/6 & 1/3 \end{bmatrix} \tag{5.4.47}$$

for the consistent formulation and

$$[B] = \int_0^L \begin{bmatrix} N_1 & 0 \\ 0 & N_2 \end{bmatrix} dx = \begin{bmatrix} 1/2 & 0 \\ 0 & 1/2 \end{bmatrix} \tag{5.4.48}$$

for the lumped formulation.

The lumped formulation differs from the consistent formulation in the weights. Specifically, the lumped approach weights dC/dt each node equally, forming a diagonal matrix, which fosters a stable solution for the transient problem.

Now, we will employ a finite difference approximation of time derivative term:

$$[A]\left\{\varepsilon\left\{\widehat{C}\right\}^{t+\Delta t}+(1-\varepsilon)\left\{\widehat{C}\right\}^{t}\right\}+\frac{1}{\Delta t}[B]\left(\left\{\widehat{C}\right\}^{t+\Delta t}-\left\{\widehat{C}\right\}^{t}\right)=\{Q\}. \quad (5.4.49)$$

The term ε is the time weighting factor:

$\varepsilon = 1$ Implicit scheme
$\varepsilon = 1/2$ Crank–Nicolson
$\varepsilon = 0$ the time approximation is explicit

Using the consistent formulation, we have the finite element formulation for each element.

$$\left[\varepsilon\frac{D}{L^e}\begin{bmatrix}1 & -1\\ -1 & 1\end{bmatrix}+\varepsilon\frac{u}{2}\begin{bmatrix}-1 & 1\\ -1 & 1\end{bmatrix}+\frac{L^e}{\Delta t}\begin{bmatrix}1/3 & 1/6\\ 1/6 & 1/3\end{bmatrix}\right]\begin{Bmatrix}\widehat{C}_1\\ \widehat{C}_2\end{Bmatrix}^{t+\Delta t}=$$

$$\left[(\varepsilon-1)\frac{D}{L^e}\begin{bmatrix}1 & -1\\ -1 & 1\end{bmatrix}+(\varepsilon-1)\frac{u}{2}\begin{bmatrix}-1 & 1\\ -1 & 1\end{bmatrix}+\frac{L^e}{\Delta t}\begin{bmatrix}1/3 & 1/6\\ 1/6 & 1/3\end{bmatrix}\right]\begin{Bmatrix}\widehat{C}_1\\ \widehat{C}_2\end{Bmatrix}^{t}. \quad (5.4.50)$$

Eq. (5.4.50) is the element (or local) matrix of the ADE. Next, we will assemble the element matrices of each element to form a global matrix and then incorporate the given boundary and initial conditions. Afterward, we solve the global matrix equation for each node's concentration at the next time step.

5.4.4 Formulation of Global Matrices

The following example illustrates how the element matrices are assembled to form a global matrix. Suppose the solution domain consists of three elements and four global nodes. Each element has two nodes (see Fig. 5.13).

Notice that there are four nodes over the solution domain. The matrices of the global equation will have 4×4 matrices and 1×4 vectors containing the unknown Cs at $t + \Delta t$ and the known Cs at t (see Eq. 5.4.51). We first fill the global matrices with the information of element 1 (yellow rectangles).

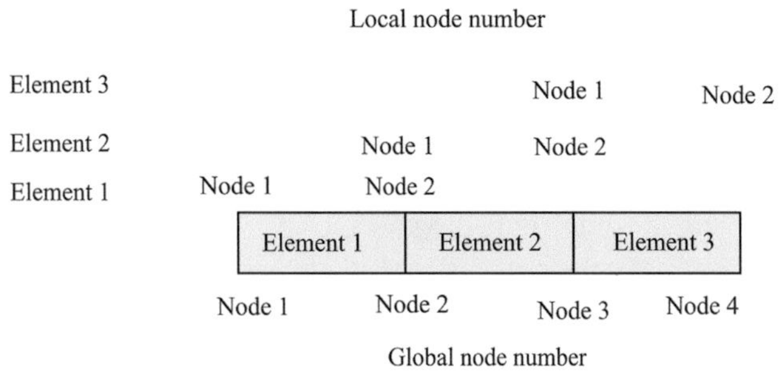

Local node number

Element 3 — Node 1, Node 2

Element 2 — Node 1, Node 2

Element 1 — Node 1, Node 2

Element 1	Element 2	Element 3

Node 1 Node 2 Node 3 Node 4

Global node number

Figure 5.13 Illustration of the example for the assemblage of the elements and nodes.

$$\left[\frac{\varepsilon D}{L^e}\begin{bmatrix}1 & -1 & 0 & 0\\ -1 & 1 & 0 & 0\\ 0 & 0 & 0 & 0\\ 0 & 0 & 0 & 0\end{bmatrix}+\frac{L^e}{\Delta t}\begin{bmatrix}1/3 & 1/6 & 0 & 0\\ 1/6 & 1/3 & 0 & 0\\ 0 & 0 & 0 & 0\\ 0 & 0 & 0 & 0\end{bmatrix}+\frac{u\varepsilon}{2}\begin{bmatrix}-1 & 1 & 0 & 0\\ -1 & 1 & 0 & 0\\ 0 & 0 & 0 & 0\\ 0 & 0 & 0 & 0\end{bmatrix}\right]\begin{Bmatrix}\hat{c}_1\\ \hat{c}_2\\ \hat{c}_3\\ \hat{c}_4\end{Bmatrix}^{t+\Delta t}=$$

$$\left[\frac{(\varepsilon-1)D}{L^e}\begin{bmatrix}1 & -1 & 0 & 0\\ -1 & 1 & 0 & 0\\ 0 & 0 & 0 & 0\\ 0 & 0 & 0 & 0\end{bmatrix}+\frac{L^e}{\Delta t}\begin{bmatrix}1/3 & 1/6 & 0 & 0\\ 1/6 & 1/3 & 0 & 0\\ 0 & 0 & 0 & 0\\ 0 & 0 & 0 & 0\end{bmatrix}+\frac{u(\varepsilon-1)}{2}\begin{bmatrix}-1 & 1 & 0 & 0\\ -1 & 1 & 0 & 0\\ 0 & 0 & 0 & 0\\ 0 & 0 & 0 & 0\end{bmatrix}\right]\begin{Bmatrix}\hat{c}_1\\ \hat{c}_2\\ \hat{c}_3\\ \hat{c}_4\end{Bmatrix}^{t} \tag{5.4.51}$$

Afterward, we add the matrices of element 2 (the blue rectangles) to the matrices of element 1 in the global matrices. The components of elements 1 and 2 in the overlapping area are added together to obtain a new value.

$$\left[\frac{\varepsilon D}{L^e}\begin{bmatrix}1 & -1 & 0 & 0\\ -1 & 2 & -1 & 0\\ 0 & -1 & 1 & 0\\ 0 & 0 & 0 & 0\end{bmatrix}+\frac{L^e}{\Delta t}\begin{bmatrix}1/3 & 1/6 & 0 & 0\\ 1/6 & 2/3 & 1/6 & 0\\ 0 & 1/6 & 1/3 & 0\\ 0 & 0 & 0 & 0\end{bmatrix}+\frac{u\varepsilon}{2}\begin{bmatrix}-1 & 1 & 0 & 0\\ -1 & 0 & 1 & 0\\ 0 & -1 & 1 & 0\\ 0 & 0 & 0 & 0\end{bmatrix}\right]\begin{Bmatrix}\hat{c}_1\\ \hat{c}_2\\ \hat{c}_3\\ \hat{c}_4\end{Bmatrix}^{t+\Delta t}=$$

$$\left[\frac{(\varepsilon-1)D}{L^e}\begin{bmatrix}1 & -1 & 0 & 0\\ -1 & 2 & -1 & 0\\ 0 & -1 & 1 & 0\\ 0 & 0 & 0 & 0\end{bmatrix}+\frac{L^e}{\Delta t}\begin{bmatrix}1/3 & 1/6 & 0 & 0\\ 1/6 & 2/3 & 1/6 & 0\\ 0 & 1/6 & 1/3 & 0\\ 0 & 0 & 0 & 0\end{bmatrix}+\frac{u(\varepsilon-1)}{2}\begin{bmatrix}-1 & 1 & 0 & 0\\ -1 & 0 & 1 & 0\\ 0 & -1 & 1 & 0\\ 0 & 0 & 0 & 0\end{bmatrix}\right]\begin{Bmatrix}\hat{c}_1\\ \hat{c}_2\\ \hat{c}_3\\ \hat{c}_4\end{Bmatrix}^{t} \tag{5.4.52}$$

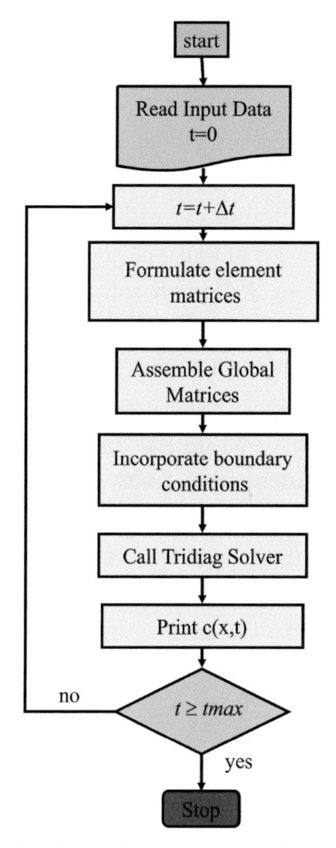

Figure 5.14 A flow chart for finite element approach.

After assembling matrices of elements 2 and 1 in the global matrices, assembly of the matrices of element 3 (green rectangles) to the global matrices comes after following the above procedure.

$$\left[\frac{\varepsilon D}{L^e}\begin{bmatrix}1 & -1 & 0 & 0\\ -1 & 2 & -1 & 0\\ 0 & -1 & 2 & -1\\ 0 & 0 & -1 & 1\end{bmatrix}+\frac{L^e}{\Delta t}\begin{bmatrix}1/3 & 1/6 & 0 & 0\\ 1/6 & 2/3 & 1/6 & 0\\ 0 & 1/6 & 2/3 & 1/6\\ 0 & 0 & 1/6 & 1/3\end{bmatrix}+\frac{u\varepsilon}{2}\begin{bmatrix}-1 & 1 & 0 & 0\\ -1 & 0 & 1 & 0\\ 0 & -1 & 0 & 1\\ 0 & 0 & -1 & 1\end{bmatrix}\right]\begin{Bmatrix}\hat{c}_1\\ \hat{c}_2\\ \hat{c}_3\\ \hat{c}_4\end{Bmatrix}^{t+\Delta t}=$$

$$\left[\frac{(\varepsilon-1)D}{L^e}\begin{bmatrix}1 & -1 & 0 & 0\\ -1 & 2 & -1 & 0\\ 0 & -1 & 2 & -1\\ 0 & 0 & -1 & 1\end{bmatrix}+\frac{L^e}{\Delta t}\begin{bmatrix}1/3 & 1/6 & 0 & 0\\ 1/6 & 2/3 & 1/6 & 0\\ 0 & 1/6 & 2/3 & 1/6\\ 0 & 0 & 1/6 & 1/3\end{bmatrix}+\frac{u(\varepsilon-1)}{2}\begin{bmatrix}-1 & 1 & 0 & 0\\ -1 & 0 & 1 & 0\\ 0 & -1 & 0 & 1\\ 0 & 0 & -1 & 1\end{bmatrix}\right]\begin{Bmatrix}\hat{c}_1\\ \hat{c}_2\\ \hat{c}_3\\ \hat{c}_4\end{Bmatrix}^{t} \tag{5.4.53}$$

After all element matrices are assembled, creating the global matrices for the solution domain, ε, D, u, and L values and boundary conditions are given and incorporated into the global matrix equation, we thus have the ADE's final global finite element equations. Solving this global matrix equation for the concentration is next. Since this problem is a one-dimensional solute transport problem, it has a tridiagonal matrix, which the Thomas algorithm can efficiently solve, as discussed previously. Excel can solve these global matrices conveniently for small domains with a small number of elements. A flowchart for a computer program based on the finite element formulation to solve the ADE is illustrated in Fig. 5.14.

The above explanation and examples present some basic concepts and methods for solving ADE in one- or two-dimensional problems. As we demonstrated, programming with Excel for one-dimensional problems permits readers to organize thoughts to implement the finite difference and element theories efficiently and logically. The theories and Excel exercises should provide a sufficient basis for understanding and developing more complex two- or three-dimensional ADE with variable velocity and diffusion coefficient. For example, two- or three-dimensional ADE will involve four or six adjacent finite-difference blocks or finite-element nodes. Finite element methods for two- or three-dimensional problems require more complex shape functions. The assemblage procedure is similar. However, these expansions are beyond the scope of this book. Istok (1989) provides detailed formulations.

6

Shear Flow Dispersion

6.1 Introduction

This chapter is the most crucial part of the book. It narrates the fundamental building block of the concept of dispersion in porous media and macrodispersion in field-scale aquifers. A comprehensive understanding of this chapter is essential to unravel the myth of macrodispersion, anomalous dispersion, scale-dependent dispersion, dual-domain, and other recently developed dispersion models for solute transport in aquifers (see Chapters 9 and 10).

This chapter explains how the molecular diffusion concept and Fick's law for molecules' random motion in Chapter 4 had been upscaled to the hydrodynamic dispersion concept and Fick's law that accounts for the effects of fluid–dynamics–scale velocity variation on solute transport in pipes. It also explores the relationships between diffusion, dispersion, cross-sectionally averaged concentration and Fick's law for dispersion validity. Of the most importance, it discusses the limitations of extending Fick's law for molecular-scale velocity variations to describe the effects of large-scale velocity variations.

6.2 Concept of Hydrodynamic Dispersion

Like molecular diffusion, the hydrodynamic dispersion (or dispersion for short) concept arises from our attempt to overlook the velocity variation smaller than the CV of our conceptual flow model. To illustrate this progression, we first examine the solute movement in idealized frictionless and frictional pipes. A mathematical model comes after describing solute movement with a fully known fluid–dynamics–scale velocity field in the frictional pipe. Volume averaging the models over the pipe's cross-sectional area comes next, leading to the hydrodynamic dispersion concept for our pragmatic interest and observation scales (i.e., upscale). The similarities and differences between the molecular diffusion and hydrodynamic dispersion concepts are summarized in terms of our observation, model, and interest scale afterward. We subsequently present the critical analysis by G. I. Taylor that validates the concept of hydrodynamic dispersion, and finally, we discuss the concept's limitations.

6.2.1 Movement of Ideal Non-viscous Fluid in Frictionless Pipes

Consider injecting a slug of red tracer uniformly over a cross-section of frictionless pipe in an ideal non-viscous fluid flowing through under a steady laminar flow condition. Suppose the pipe is perfectly uniform, circular, and straight. The slug moves uniformly along the pipe as a piston-type of displacement, maintaining its shape with some smearing at its front and back, as described by the advection–diffusion Equation (ADE) in the previous chapter. Fig. 6.1a shows that the slug maintains its shape even after some time, but its width broadens, and concentration is reduced (or diluted). The snapshot of the cross-section averaged concentration (Fig. 6.1b) exhibits a narrow normal distribution at early times. It maintains a bell-shaped curve with a reduced peak and broadened base at later times (similar to the snapshots in Fig. 4.14). We, therefore, conclude that the piston-type displacement corresponds to the advection by the uniform fluid-dynamics-scale velocity. The smearing is owing to molecular diffusion or random motion of fluid molecules, which we are unable and have little interest in describing exhaustively because of its microscopic scale nature. This smearing could be pictured as the effect of the volume average (more precisely, the ensemble average since the velocity variation is unknown).

The above discussion appears to deduce that a three-dimensional ADE can adequately describe the movement of the tracer slug's interface and its spread in the three-dimensional space of the pipe. Furthermore, cross-section area averaging the three-dimensional concentration distribution could yield the one-dimensional tracer snapshots that satisfy our interest and can be conveniently predicted by the one-dimensional ADE. The following mathematical analysis may support this conjecture.

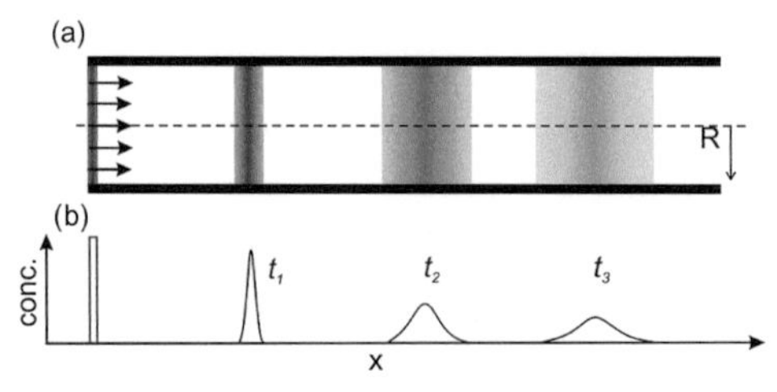

Figure 6.1 Illustrations of a tracer slug moving with an ideal, non-viscous fluid flowing through a frictionless pipe; (a) two-dimensional snapshots, and (b) snapshots of the cross-section averaged concentration at three times. The lines with arrows on the left denote the fluid-dynamics-scale velocities, while the dashed line is the center of the pipe, and R is its radius.

The analysis first defines the three-dimensional space using the x-coordinate along the pipe length and the r-coordinate (radial distance from the pipe center), assuming that the flow in the pipe is radially symmetric around the center of the pipe. Suppose $C(x,r,t)$ represents the point concentration of the displacing fluid at location x and r at time t. The three-dimensional advection–diffusion equation for the solute movement in the pipe is

$$D_m \frac{\partial^2 C(x, r, t)}{\partial x^2} + D_m \frac{\partial^2 C(x, r, t)}{\partial r^2} - u \frac{\partial C(x, r, t)}{\partial x} = \frac{\partial C(x, r, t)}{\partial t} \tag{6.2.1}$$

The first and second terms on the left-hand side of Eq. (6.2.1) denote the effects of molecular diffusion. The term D_m is the molecular diffusion coefficient, accounting for random motion of the fluid molecules at the microscopic scale, and is assumed isotropic. The third term represents the advection term, where u denotes the uniform fluid-dynamics-scale velocity driving the piston-like movement of the fluid.

Now, define the average concentration over any cross-sectional area as

$$\overline{C}(x, t) = \frac{1}{\pi R^2} \int_0^R C(x, r, t) 2\pi r dr. \tag{6.2.2}$$

Notice that the cross-sectional average concentration $\overline{C}(x, t)$ is only a function of x and t. Further, we express the point concentration within the pipe as the mean (the cross-sectional area average) and perturbations (the deviations from the average):

$$C(x, r, t) = \overline{C}(x, t) + C'(x, r, t). \tag{6.2.3}$$

Note that the average of the perturbation terms must be zero. That is,

$$\frac{1}{\pi R^2} \int_0^R C'(x, r, t) 2\pi r dr = 0. \tag{6.2.4}$$

Substituting Eq. (6.2.3) into Eq.(6.2.1) leads to

$$\begin{aligned} &D_m \frac{\partial^2 \left(\overline{C}(x, t) + C'(x, r, t)\right)}{\partial x^2} + D_m \frac{\partial^2 \left(\overline{C}(x, t) + C'(x, r, t)\right)}{\partial r^2} - \\ &u \frac{\partial \left(\overline{C}(x, t) + C'(x, r, t)\right)}{\partial x} = \frac{\partial \left(\overline{C}(x, t) + C'(x, r, t)\right)}{\partial t} \end{aligned} \tag{6.2.5}$$

in which we assume that the molecular diffusion coefficient D_m is a scalar and constant, implying the molecular diffusion process is isotropic: independent of directions. We also assume that Fick's law for molecular diffusion is valid at this scale (no time-dependent diffusion, Chapter 4). Integrating each term in Eq. (6.2.5) over the cross-sectional area leads to

$$\frac{1}{\pi R^2}\int_0^R \left(\begin{array}{l} D_m \dfrac{\partial^2\left(\overline{C}(x,\ t)+C'(x,\ r,\ t)\right)}{\partial x^2} + D_m \dfrac{\partial^2\left(\overline{C}(x,\ t)+C'(x,\ r,\ t)\right)}{\partial r^2} - \\ u \dfrac{\partial\left(\overline{C}(x,\ t)+C'(x,\ r,\ t)\right)}{\partial x} \end{array} \right) 2\pi r dr =$$

$$\frac{1}{\pi R^2}\int_0^R \left(\frac{\partial\left(\overline{C}(x,\ t)+C'(x,\ r,\ t)\right)}{\partial t} \right) 2\pi r dr \tag{6.2.6}$$

where the radius of the pipe is R. After the integration, terms containing C' will be zero, the contribution from the second derivatives of C' in the above equation is small, and $\overline{C}(x, t)$ is independent of r, Eq. (6.2.6) become

$$D_m \frac{\partial^2 \overline{C}(x, t)}{\partial x^2} - u \frac{\partial \overline{C}(x, t)}{\partial x} = \frac{\partial \overline{C}(x, t)}{\partial t}. \tag{6.2.7}$$

Notice that the integration is carried over r, independent from x and t, so the differentiation and integration are exchangeable. This resulting equation is identical to the ADE in Chapter 4. This analysis thus confirms that a simple one-dimensional ADE can describe the evolution of the cross-sectional area-averaged concentration in this idealized frictionless pipe.

The above analysis appears complex, but the result is foreseeable. Since the flow is laminar, steady, and uniform, all the streamlines' velocities must be the same, except for the Brownian-type random motion of tracer molecules, which distributes the tracer particles irregularly over the cross-section at the microscopic scale. For this reason, Fick's law for diffusion in Eq. (6.2.7) approximates the distribution as continuous and smeared in the ensemble mean sense. This approximation is further reinforced by averaging the three-dimensional point concentrations over the cross-sectional area, facilitating the need to satisfy the ergodicity embedded in Fick's law if the cross-section is sufficiently large compared to molecules. Thus, the ADE is valid for the tracer movement under this idealized situation.

Question: How would a spatially varying diffusion coefficient affect the results?

6.2.2 Movement of a Viscous Fluid in a Frictional Pipe

Albeit the above conjecture is valid, fluids always are viscous, and pipes are always frictional. As we conduct a tracer experiment in a realistic pipe, we observe that the tracer profile becomes parabolic after a tracer is injected into the flowing

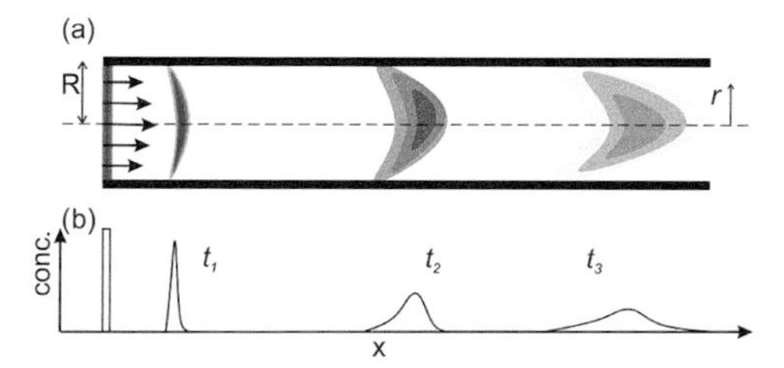

Figure 6.2 *Schematic illustrations of a slug of tracer movement in a viscous fluid and frictional pipe. (a) two-dimensional snapshots and (b) snapshots of the cross-section averaged concentrations at three times. R is the radius of the pipe, and r is the radial coordinate.*

fluid uniformly across the pipe (Fig. 6.2a). Such a profile indicates a non-uniform velocity field: the maximal velocity at the center and the minimal at the wall. As time progresses, the convex profile becomes more profound (the one on the fast lane moves further ahead) and becomes more smeared. The concentration distribution becomes bean-shaped narrow widths near the pipe wall, and broad widths in the center of the pipe. Noticeably, wall friction and fluid viscosity induce a non-uniform fluid-dynamics-scale velocity field, and the random motion of molecules at the scale below these velocities smears the profile. These effects become profound as the solute migrates further down the pipe.

Figure 6.2b shows snapshots of the cross-section averaged concentrations at different times (t_1, t_2, and t_3). The averaged concentration initially appears as a slug and becomes a non-symmetric bell-shaped distribution (a sharp rise before the peak and a slow decay after the peak). The peak concentration decreases, the distribution's base broadens, and more tracers lag behind the peak as the tracer migrates. These snapshots are similar to but different from those in Section 6.2.1 and the solution of the one-dimensional ADE (Fig. 4.14). Taylor's analysis in Section 6.3 provides explanations for this difference.

The above discussion suggests that if the velocity at every location is known and the diffusion coefficient is constant and specified, a three-dimensional axis-symmetrical advection–diffusion equation should adequately quantify the three-dimensional concentration distribution in the pipe (Eq. 6.2.1). To implement this approach in a real-world experiment, one needs to determine the velocity along the pipe's radius and assume the pipe's roughness is the same throughout the pipe. However, we neither measure the velocity nor are we interested in the concentration at a point within the pipe generally. Instead, we measure the average velocity using the inflow or outflow rate over the pipe's cross-sectional area A and collect the averaged concentration over the cross-sectional area at the end of the pipe. That is,

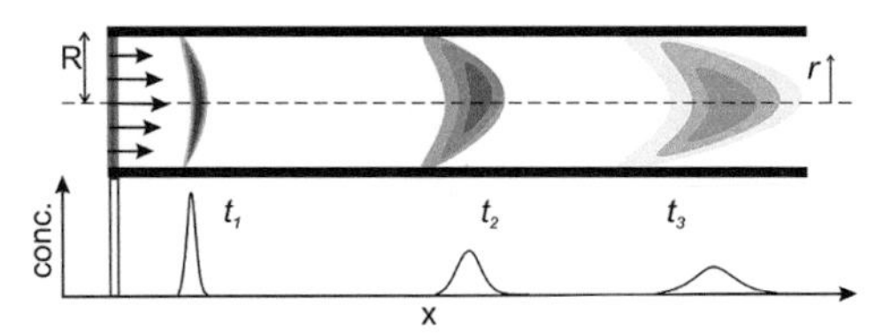

Figure 6.3 The parabolic velocity profile in a pipe and the average velocity. The centerline of the pipe is the dashed line, r is the radial coordinate from the centerline, and R is the radius of the pipe.

$$\overline{u} = \frac{Q}{A} = \frac{1}{A}\int u(r)dA = \frac{1}{\pi R^2}\int_0^R u(r)2\pi r dr \quad \text{and} \quad \overline{C}(x, t) = \frac{1}{\pi R^2}\int_0^R C(x, r, t)2\pi r dr \tag{6.2.8}$$

where R is the pipe radius and $\overline{u}$ is the mean or average velocity (see Fig. 6.3), constant over the length of the pipe, and likewise, $\overline{C}(x, t)$ is the concentration averaged over the cross-sectional area, and is a function of x and t only. Thus, for the sake of practical purpose, one may assume that the following one-dimensional advection–diffusion equation may be adequate for quantifying the concentration BTC at the end of the pipe:

$$D_m \frac{\partial^2 \overline{C}(x, t)}{\partial x^2} - \overline{u}\frac{\partial \overline{C}(x, t)}{\partial x} = \frac{\partial \overline{C}(x, t)}{\partial t}. \tag{6.2.9}$$

Eq. (6.2.9) notably neglects the variation in fluid-dynamics-scale velocity and concentration deviations around their means in the r direction. The former leads to the solute's earlier arrival at the pipe center and later arrival near the wall compared to the average velocity. The latter (i.e., concentration deviations) represents additional solute mass omitted by the average concentration. Eq. (6.2.9) alone, thus, cannot reproduce observed BTC at the end of the pipe as such.

The above shortcomings coerce us to postulate that these effects on the cross-sectional average concentration may be analogous to the smearing effects due to the random motion of molecules (molecular velocity variation below the fluid-dynamics scale) in the analysis of the advection–diffusion process in Chapter 4. If this is true, Fick's law should describe the effects of fluid-dynamics-scale velocity variation by using a coefficient D_h (hydrodynamic or mechanic dispersion coefficient) similar to the molecular diffusion coefficient to account for the effects of deviations of the velocity and concentration from the averaged ones. As a result, a new coefficient, D, the sum of molecular diffusion D_m and the hydrodynamic dispersion D_h must replace D_m in Eq. (6.2.9). In other words, the total mass flux crossing a cross-sectional area, unaccounted for by the average velocity, must be

$$q = -D\frac{\partial \overline{C}}{\partial x} = -(D_h + D_m)\frac{\partial \overline{C}}{\partial x}. \tag{6.2.10}$$

Suppose this conjecture is true. The following advection–dispersion equation describes the spatiotemporal evolution of the average concentration

$$D\frac{\partial^2 \overline{C}(x,t)}{\partial x^2} - \overline{u}\frac{\partial \overline{C}(x,t)}{\partial x} = \frac{\partial \overline{C}(x,t)}{\partial t} \tag{6.2.11}$$

The coefficient D_h represents the effect of the fluid–dynamics–scale velocity variation around its mean or average over the pipe's cross-sectional area. On the other hand, the term D_m accounts for the effects of random velocity at the molecular scale, omitted by the fluid-dynamics-scale velocity. Specifically, Eq. (6.2.11) considers the effects of two-scale velocity variations (i.e., molecular and fluid-dynamic scales). Since the magnitude of the mechanic dispersion coefficient is much higher than the coefficient of molecular diffusion, we often neglect molecular diffusion or lump it into the hydrodynamic dispersion.

The above reasoning implies that the dispersion is just a correction term for the effects of velocity and concentration variations around their means neglected in Eq. (6.2.9). Now, logical questions arise: Is this parsimonious correction term valid? How can we prove this? An analysis of shear flow dispersion in the next section answers these questions.

Before proceeding into the next section, we emphasize that a three-dimensional (or the axis-symmetric) advection–diffusion equation should satisfactorily reproduce the tracer concentration over a CV that encompasses many molecules and defines the fluid-dynamic scale velocity at any time and location in the pipe. The dispersion concept arises merely from our averaging approach for practical purposes (the principle of parsimony), avoiding complex solute migration in the pipe. Table 6.1 summarizes the philosophic differences between molecular diffusion and the dispersion concept based on the discussion above.

6.3 G. I. Taylor's Analysis of Shear Flow Dispersion

G. I. Taylor (1953) investigated the validity of using Fick's law for the dispersion discussed above. He examined the displacement of a tracer slug in a steady flowing fluid in a circular and straight pipe with a radius of R, as shown in Fig. 6.3. His analysis has been elucidated in a textbook by Fisher et al. (1979). Below is an excerpt from the textbook.

Table 6.1 *A summary of the differences in philosophy and rationales between the molecular diffusion and dispersion concepts.*

	Molecular Diffusion	Dispersion
The scale of velocity variation	Molecular (microscopic scale)	Macroscopic scale (fluid-dynamics scale) greater than many molecules,
Mechanism	Random motion of molecules	Fluid–dynamics–scale velocity variations due to friction and viscosity
Rationale	The randomness of molecule velocity is too small to measure; our instrument size or scale is often much greater than the size of a molecule.	Point fluid-dynamics-scale velocities in a pipe may be measurable but impractical (i.e., the problem's scale is much greater than our instrument scale).
The scale of our interest	Concentration over a volume that is larger than many molecules	Concentration over a cross-section of a pipe.

First, the average concentration of the cross-sectional area is written as

$$\overline{C}(x, t) = \frac{1}{\pi R^2}\int_0^R C(x, r, t)2\pi r dr \tag{6.3.1}$$

where $C(x,r,t)$ is the point concentration in the pipe at any time t. The cross-sectional average concentration is denoted by $\overline{C}(x, t)$ as a function of x and t only. Since the flow is assumed to be steady and uniform, the point velocity variation exists only in the r direction, i.e., $u(r)$. We can express the average (or mean) velocity over the cross-sectional area as

$$\overline{u} = \frac{1}{\pi R^2}\int_0^R u(r)2\pi r dr. \tag{6.3.2}$$

Notice that both averaged concentration and velocity are independent of any velocity distributions.

Now, we express the point concentration and velocity in terms of the volume-average mean and perturbations:

$$u(r) = \overline{u} + u'(r) \quad \text{and} \quad C(x, r, t) = \overline{C}(x, t) + C'(x, r, t). \tag{6.3.3}$$

Note that the average of the perturbation terms must be zero. That is,

$$\frac{1}{\pi R^2}\int_0^R C'(x, r, t)2\pi r dr = 0 \quad \text{and} \quad \frac{1}{\pi R^2}\int_0^R u'(r)2\pi r dr = 0. \tag{6.3.4}$$

According to our previous discussion, it is reasonable to assume that an axis-symmetrical advection–diffusion equation, Eq. (6.2.1), provides a good description of the tracer displacement along the frictionless pipe. Therefore, we substitute Eq. (6.3.3) for $u(r)$ and $C(x, r, t)$ in Eq. (6.2.1) to acount for the effects of velocity variation due to friction of the pipe, and we have

$$D_m \frac{\partial^2 \left(\overline{C}(x,\ t) + C'(x, r, t)\right)}{\partial x^2} + D_m \frac{\partial^2 \left(\overline{C}(x,\ t) + C'(x, r, t)\right)}{\partial r^2} - \\ \left(\overline{u} + u'(r)\right) \frac{\partial \left(\overline{C}(x,\ t) + C'(x, r, t)\right)}{\partial x} = \frac{\partial \left(\overline{C}(x,\ t) + C'(x, r, t)\right)}{\partial t} \qquad (6.3.5)$$

Next, aiming to investigate the validity of Fick's law for describing the effects of velocity variation around the mean, we will observe the effects of velocity variation on the spread of the tracer as an observer moving at the mean velocity of the fluid. Under this moving coordinate system, if we prove that the solute mass flux crossing the cross-section of the pipe where we make observations is linearly proportional to the average concentration gradient, the conjecture of Eq. (6.2.11) is then validated. To do so, we first simplify Eq. (6.3.5) by recognizing $\partial \overline{C}(x, t)/\partial r$ is always zero since $\overline{C}(x, t)$ is a constant in r. Then, a coordinate transformation, using a convective coordinate, is applied to Eq. (6.3.5),

$$\xi = x - \overline{u}t \quad \text{and} \quad \text{let} \quad \tau = t. \qquad (6.3.6)$$

In Eq. (6.3.6), ξ is the moving coordinate of the solute with respect to the mean velocity, and x is a point along the flow path in the fixed coordinate. Applying the chain rule of calculus to the spatial and temporal derivatives of concentrations, we have

$$\frac{\partial C}{\partial x} = \frac{\partial C}{\partial \xi} \quad \text{and} \quad \frac{\partial C}{\partial t} = \frac{\partial \xi}{\partial t}\frac{\partial C}{\partial \xi} + \frac{\partial \tau}{\partial t}\frac{\partial C}{\partial \tau} = -\overline{u}\frac{\partial C}{\partial \xi} + \frac{\partial C}{\partial \tau}. \qquad (6.3.7)$$

In other words, the left-hand side spatial and time derivatives in the fixed coordinates are converted to the moving coordinates on the right-hand side of Eq. (6.3.7). Substituting Eq. (6.3.7) in Eq. (6.3.5), we have

$$D_m \frac{\partial^2 \left(\overline{C} + C'\right)}{\partial \xi^2} + D_m \frac{\partial^2 C'}{\partial r^2} - \left(\overline{u} + u'(r)\right) \frac{\partial \left(\overline{C} + C'\right)}{\partial \xi} \\ = -\overline{u} \frac{\partial \left(\overline{C} + C'\right)}{\partial \xi} + \frac{\partial \left(\overline{C} + C'\right)}{\partial \tau} \quad \text{or} \\ D_m \frac{\partial^2 \left(\overline{C} + C'\right)}{\partial \xi^2} + D_m \frac{\partial^2 C'}{\partial r^2} - u'(r) \frac{\partial \left(\overline{C} + C'\right)}{\partial \xi} = \frac{\partial \left(\overline{C} + C'\right)}{\partial \tau} \qquad (6.3.8)$$

In Section 6.2, the rate of spreading along the flow direction due to the velocity variation is assumed to be much greater than molecular diffusion. While this assumption is accurate, as shown later in this section, we take it for granted at this moment. We thus neglect the longitudinal diffusion term (the first term on the left-hand side of Eq. (6.3.8)). The equation thus becomes:

$$D_m \frac{\partial^2 C'}{\partial r^2} = \frac{\partial \overline{C}}{\partial \tau} + \frac{\partial C'}{\partial \tau} + u'(r)\frac{\partial C'}{\partial \xi} + u'(r)\frac{\partial \overline{C}}{\partial \xi}. \tag{6.3.9}$$

which describes the effects of D_m, C', and $u'(r)$ on $\overline{C}$. The next step is to find a general solution C' in Eq. (6.3.9) if the mean concentration, velocity perturbations, and molecular diffusion coefficient are known. However, an analytical solution to Eq. (6.3.9) is intractable. To circumvent this difficulty, G.I. Taylor made a brave and brilliant assumption that the first three terms of the right-hand side of Eq. (6.3.9) are significantly small **at large times** such that they can be neglected. The equation can thus be reduced to

$$D_m \frac{\partial^2 C'}{\partial r^2} = u'(r)\frac{\partial \overline{C}}{\partial \xi} \tag{6.3.10}$$

with boundary conditions

$$\frac{\partial C'}{\partial r} = 0 \quad \text{at} \quad r = R. \tag{6.3.11}$$

This equation states that the molecular diffusion in the radial direction, r, balances (smooths) out the effects of the velocity perturbation (the right-hand side of the equation) at large times in plain language. After integrating Eq. (6.3.10) over r twice and substituting the given boundary condition, we obtain the solution:

$$C'(\xi, r, \tau) = \frac{1}{D_m}\frac{\partial \overline{C}(\xi, \tau)}{\partial \xi}\int_0^R\int_0^R u'(r)drdr + C'(0, \tau). \tag{6.3.12}$$

The last term on the right-hand is the boundary condition. This solution describes the solute concentration deviation at any radial distance r, at any time τ, at the moving observation location ξ.

Now, you may ask why we need to solve for C'. The reason is given as follows. If we integrate each term in Eq. (6.3.9) over the cross-sectional area, we have

$$\frac{1}{\pi R^2}\int_0^R \left(D_m \frac{\partial^2 C'}{\partial r^2}\right) 2\pi r dr = \frac{1}{\pi R^2}\int_0^R \left(\frac{\partial \overline{C}}{\partial \tau} + \frac{\partial C'}{\partial \tau} + u'(r)\frac{\partial C'}{\partial \xi} + u'(r)\frac{\partial \overline{C}}{\partial \xi}\right) 2\pi r dr \tag{6.3.13}$$

After the integration, all terms containing C' or $u'(r)$ in Eq. (6.3.13) vanish because they are zero. We then have

$$\frac{\partial \overline{C}}{\partial \tau} + \frac{\partial \overline{u'(r)C'(\xi, r, \tau)}}{\partial \xi} = 0, \tag{6.3.14}$$

where the overbar indicates a cross-sectional average of the term. Eq. (6.3.14) is, in essence, describes the behavior of the averaged concentration $\overline{C}(\xi, \tau)$, due to the effects of both velocity and concentration perturbations on the moving coordinate ξ. It shows that the temporal change of $\overline{C}(\xi, \tau)$ is dictated by the spatial gradient of $\overline{u'(r)C'(\xi, r, \tau)}$ (i.e., the averaged mass flux produced by the velocity and concentration perturbations).

As we have postulated previously, if Fick's law describes this averaged mass flux due to the velocity and concentration perturbations omitted by the average velocity and concentration, this flux must be linearly proportional to the mean concentration gradient. Namely, the spatial gradient of the average mass flux must have a form similar to the diffusion equation,

$$\frac{\partial \overline{u'(r)C'(\xi, r, \tau)}}{\partial \xi} = D \frac{\partial^2 \overline{C(\xi, \tau)}}{\partial \xi^2} \tag{6.3.15}$$

where D is the dispersion coefficient.

In order to prove this, we shall calculate the term $\overline{u'(r)C'(\xi, r, \tau)}$. Since $u'(r)$ is known, $C'(\xi, r, \tau)$ is given by Eq. (6.3.12), and the overbar indicates the integration over the cross-sectional area (see Eqs. 6.3.1 and 6.3.2), this term becomes

$$\begin{aligned} \overline{u'(r)C'(\xi, r, \tau)} &= \frac{1}{\pi R^2} \int_0^R u'(r)C'(\xi, r, \tau) 2\pi r dr \\ &= \frac{2}{R^2 D_m} \frac{\partial \overline{C}(\xi, \tau)}{\partial \xi} \int_0^R u'(r) \left[\int_0^r \int_0^r u'(r) dr dr + C'(0, \tau) \right] r dr \end{aligned} \tag{6.3.16}$$

Notice that we use the dummy variable r for the upper bounds of the inner integrals since they depend on the outer integral. In addition, $\int_0^R u'(r)C'(0, \tau) r dr = 0$ because $u'(r)$ is the deviation from the mean at $R/2$ and $\int_0^R u'(r) r dr = 0$. Eq (6.3.16) represents the mass flux over the cross-sectional area M, omitted by the average velocity and the average concentration, relative to the moving coordinate system.

$$M = \left[\frac{2}{R^2 D_m} \int_0^R u'(r) \left(\int_0^r \int_0^r u'(r) dr dr \right) r dr \right] \frac{\partial \overline{C}(\xi, \tau)}{\partial \xi} \tag{6.3.17}$$

This equation further manifests that the mass flux over the cross-sectional area in the flow direction is proportional to the average concentration gradient in the flow direction at large times. **Eq. (6.3.17) is precisely the same as the Fickian mechanism in molecular diffusion.**

Recall Fick's law for molecular diffusion

$$q = -D_m \frac{\partial C}{\partial x}. \tag{6.3.18}$$

Compare Eq. (6.3.18) with the result of the shear flow analysis, Eq. (6.3.17),

$$M = -D \frac{\partial \overline{C}}{\partial \xi}. \tag{6.3.19}$$

We have

$$M = -\left[\frac{2}{R^2 D_m} \int_0^R u'(r) \left(\int_0^r \int_0^r u'(r) dr dr \right) r dr \right] \frac{\partial \overline{C}}{\partial \xi} \tag{6.3.20}$$

or

$$D = \frac{2}{R^2 D_m} \int_0^R u'(r) \left(\int_0^r \int_0^r u'(r) dr dr \right) r dr \tag{6.3.21}$$

where D is called the longitudinal dispersion coefficient [L^2/T], a constant in time and space. It represents the solute's spreading rate along the flow direction per unit cross-sectional area.

Thus, the one-dimensional dispersion equation for the average concentration in the moving coordinate system is:

$$\frac{\partial \overline{C}}{\partial \tau} = D \frac{\partial^2 \overline{C}}{\partial \xi^2}. \tag{6.3.22}$$

Converting Eq. (6.3.22) to the fixed coordinate system, we have

$$\frac{\partial \overline{C}}{\partial t} + \overline{u} \frac{\partial \overline{C}}{\partial x} = D \frac{\partial^2 \overline{C}}{\partial x^2}. \tag{6.3.23}$$

This equation is the same as Eq. (6.2.11) and is called an advection–dispersion equation (ADE) or convection–dispersion equation (CDE). It has the same form as an advection–diffusion equation, except the diffusion coefficient is replaced by the dispersion coefficient. In other words, Taylor's analysis proved that the effects of the velocity and concentration variations, due to fluid viscosity and roughness of the pipe's wall, over the pipe's cross-sectional area can be represented by Fick's law **after the tracer has migrated for a sufficient time**. In particular, the mass flux ascribed to the fluid–dynamics–scale variation is linearly proportional to the mean concentration gradient, like in molecular diffusion analysis. We, therefore, conclude that the one-dimensional advection–dispersion equation is appropriate to describe the cross-sectional-area average concentration, but not the concentration at every point in the pipe. **Notice that this conclusion is valid only if the tracer has traveled sufficient distances**.

We reiterate the concept that if the velocity perturbations (or detailed velocity variation) are known, the cross-sectional area average is the spatial or volume average. Otherwise, the average is the ensemble average, implying the ADE is an ensemble mean equation. The equivalence between the spatial average and ensemble average would depend on the fulfillment of the ergodicity assumption (see Chapter 1).

A summary of Taylor's analysis is narrated below:

(1) After the solute has traveled **a sufficiently long time, and when the effects of fluid–dynamics–scale velocity variation are smeared (or balanced) by the molecular scale velocity variation,** the cross-sectional averaged concentration follows the normal distribution and reaches the Fickian regime. Otherwise, the average concentration distribution does not follow Fick's law, or its behavior is non-Fickian: it could be any distribution dictated by the fluid–dynamics–scale velocity profile. If the velocity profile is unknown, the ADE predicts the most likely distribution in the ensemble sense, although the predicted result may significantly deviate from the actual one.

(2) In the Fickian regime, the average concentration distribution from a slug input is a normally distributed cloud moving at the mean velocity, $\overline{u}$, and continuing to spread at the rate according to

$$\frac{d\sigma_x^2}{dt} = 2D \tag{6.3.24}$$

In this equation, σ_x^2 is the spatial variance of the solute cloud (a statistical measure of the size of the cloud.

(3) The longitudinal dispersion coefficient, D, is invariant in time and space.

(4) Lastly, the analysis assumes the pipe is straight, the roughness of the pipe does not change, and the flow maintains a steady and uniform pace.

Lastly, Taylor's analysis thus unravels the enigma of the differences between the cross-section averaged concentration snapshots and the solution of an ADE in Fig. 6.2b.

Time to reach the Fickian's regime. The next question we must address is: how long is long enough to reach the Fickian regime? Chatwin (1970) has shown that the averaged concentration reached a normal distribution when the injected tracer traveled over time $t > 0.4R^2/D_m$, which indicates that the time to reach the Fickian regime increases with the pipe's radius and decreases with the molecular diffusion coefficient's increase.

Let us examine some examples based on the above analysis. Taylor (1953) investigated the dispersion of a solute in a laminar flow in a tube. The velocity distribution $u(r) = u_0(1 - r^2/a^2)$; a is the tube's radius of the tube; u_0 is the maximum velocity. Using Eq. (6.3.21), he found the dispersion coefficient as

$$D = a^2 u_0{}^2/192D_m.$$

Suppose the tracer is a salt solution. Molecular diffusion of salt solution in water is $D_m = 10^{-5}$ cm^2/sec, $u_0 = 1$ cm/sec, and $a = 2$ mm. Then $D = 20$ cm^2/sec: $D >>> D_m$, as we speculate in the early sections. The initialization time for the Fickian regime is

$$0.4\frac{4a^2}{D_m} = 1600 \text{ sec}.$$

During this time, a slug of the tracer would travel 1600 cm. Before this distance and time, the one-dimensional ADE equation with a constant dispersion coefficient cannot describe the tracer's spread but only in the ensemble sense with significant uncertainty.

6.4 Solute Spreading Without Molecular Diffusion

Overall, Taylor's analysis manifested that at the late time of solute migration, the molecular diffusion balanced out the difference in concentration induced by the fluid–dynamics–scale velocity variation in the pipe. To investigate the spread of a trace before reaching the Fickian regime, consider two parallel plates of infinite extent, separated by a distance h. Water fills the space between the two plates. Suppose a tracer is instantaneously distributed as a line between the two plates. Afterward, the top plate is moving with velocity u_m to the right, the bottom remains still, and the velocity profile is linearly distributed from top to bottom,

$$u(y) = u_m y/h. \tag{6.4.1}$$

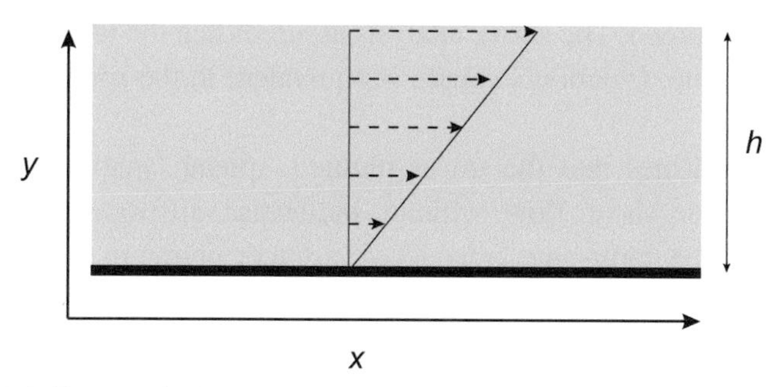

Figure 6.4 A linear velocity profile.

The vertical coordinate is denoted by y, and the distance (thickness of water) between the two plates is h (Fig. 6.4). Now, we like to calculate how the ink line stretches as a function of time.

To do so, we determine the average velocity over the thickness is

$$\overline{u} = \frac{1}{h}\int u(y)dy = \frac{1}{h}\int \frac{u_m y}{h}dy = \frac{u_m}{h^2}\frac{y^2}{2}\bigg|_0^h = \frac{u_m}{2} \tag{6.4.2}$$

The velocity deviation along the y axis from the average is

$$u'(y) = u(y) - \frac{u_m}{2} = \frac{u_m y}{h} - \frac{u_m}{2} = u_m\left[\frac{y}{h} - \frac{1}{2}\right] \tag{6.4.3}$$

The deviation of the ink position at the top $(y = h)$ from the average position at time t:

$$x' = u't = \frac{u_m}{2}t \tag{6.4.4}$$

We square Eq. (6.4.4) and have the spatial deviation from the average position square:

$$\delta^2 = (x(y) - \overline{x})^2 = \left[\frac{u_m}{2}\right]^2 t^2 \tag{6.4.5}$$

It shows that the square of the deviation grows with t square. If we assume that the rate of growth is two times the diffusion coefficient as articulated in Chapter 4 (Fick's law), we have

$$2D = \frac{d\delta^2}{dt} = \frac{(u_m)^2}{2}t = \frac{u_m}{2}u_m t. \tag{6.4.6}$$

In other words, the dispersion coefficient D increases with time and the ink travel distance $(x_m = u_m t)$ – scale-dependent dispersion, although the concentration distribution is unknown. The above analysis assumes that the linear velocity profile is known. If not, the deviation's square is equivalent to the spatial variance in the ensemble sense.

Eq. (6.4.6) indicates that the tracer plume's spread (spatial variance) grows quadratically under shear flow without molecular diffusion. In essence, this behavior depicts the early-time behavior of a tracer plume in the shear flow in a pipe before molecular diffusion can modify the fluid–dynamics–scale velocity variation. In plain language, the ink's spread in early times grows nonlinearly.

Such early- and late-time behaviors are consistent with the previous analysis of Brownian particles' spread in static fluids. In Brownian motion, the molecular-scale velocity variation is induced by the random collision of molecules. The persistence time (the time the particle remembers the previous velocity) for particles of colloidal sizes is about thousandths of a second. On the other hand, the viscosity and friction in the shear flow dispersion initiate the fluid-dynamics-scale velocity variation, which has a persistent time of several thousand seconds, depending on the tube's diameter and molecular diffusion coefficient. This long persistent time means that a solute must travel long distances under a constant flow field to reach the Fickian regime in which the advection–dispersion equation is valid.

Before the persistent time, both the snapshots of point and cross-sectional averaged concentrations of a solute cloud echo the distribution of fluid-dynamics-scale velocity. Therefore, delineating the fluid-dynamics-scale velocity distribution in each scenario is the key to predicting the solute transport at early times. Of course, it can always be "approximated" by the ADE with either constant or variable dispersion coefficients but with significant uncertainty.

6.5 Conclusion

Taylor's analysis demonstrated that the concentration averaged over the pipe's cross-sectional area will reach the Fickian's regime, having a bell-shaped normal spatial distribution, after the tracer plume has traveled long distances. Most importantly, the point concentrations may not conform to the distribution since the analysis is based on a volume or ensemble average approach.

Taylor's analysis implies that the average concentration distribution in the ensemble sense at early times does not follow the normal distribution. A comprehensive characterization of the fluid-dynamic-scale velocity field and a fully three-dimensional advection–diffusion model could lead to predictions close to reality.

Thinking Points:

- Taylor's analysis assumes that the pipe is straight with uniform roughness, and the flow is steady and laminar. How do these assumptions affect the development of the Fickian regime in solute migration in a pipe?
- Consider steady-state water flow between two frictionless parallel plates. The upper portion of the water flows at a velocity V_1 much faster than V_2 at the lower portion. A uniform slug of a trace is released between the plates at time zero. Plot the snapshot of the depth-averaged concentrations at some early and large times, considering the effects of molecular diffusion.

7

Solute Transport in Soil Columns

7.1 Introduction

Previous chapters have introduced the control volume, CV, which is much larger than molecules but smaller than the entire fluid. This CV defines the average molecule velocity controlling the fluid's dynamics. The molecular diffusion concept and Fick's law, in turn, resolve the molecular scale velocity variations (i.e., random motion of molecules) omitted in the average velocity. Chapter 6 explains upscaling the diffusion concept and Fick's law to accommodate the effects of uncharacterized fluid-dynamic-scale velocity variations around the average velocity over the cross-section area of the pipe under the shear flow. Because these velocity variations are much more significant than the molecule's random velocity, the mechanic dispersion concept and the advection–dispersion equation were contrived to analyze solute transport in a pipe. This chapter introduces the advection–dispersion concept and equation for solute transport in soil column porous media. They extend the molecular diffusion and mechanic dispersion theories to accommodate the average velocity of the CVs for porous media and velocity variations due to pore-scale heterogeneity.

7.2 Advection and Dispersion Concepts in Solute Transport in Porous Media

As Yeh et al. (2015) discussed, Darcy's law is an ensemble mean law, as is Fick's law. Darcy's law describes the velocity of water averaged over a CV of a porous medium, which comprises many pores. Because of the unknown intricate pore distribution, applications of the law to a core sample (a porous medium) depict the statistically averaged (or ensemble average) head and velocity fields, which are different from the head and velocity in each pore. We will call this ensemble-averaged velocity and head the Darcy-scale velocity and head. This situation is comparable to the average velocity in a pipe, which neglects the fluid-dynamic

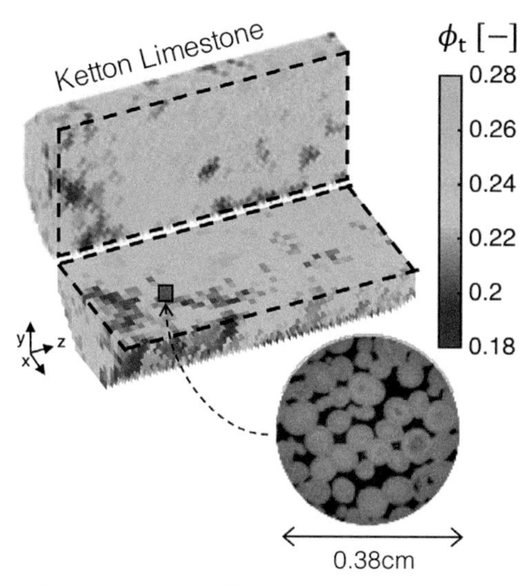

Figure 7.1 3D porosity maps of the Ketton Limestone sample. Both vertical and horizontal cross-sections along the length of the samples are shown. High-resolution 2D gray-scale tomograms acquired on smaller 0.38 cm-diameter sister plugs are also shown to highlight the similarities in the granular microstructure of the sample (Φ_t is the porosity).

scale velocity perturbations due to the viscosity and roughness of the pipe wall. So does the Darcy-scale velocity, which ignores the pore-scale velocity perturbations caused by intricate pore heterogeneity.

Before discussing theories for solute transport in porous media, let us examine existing works on solute transport in a rock core observed using high-resolution scanning. Kurotori et al. (2019) used positron emission tomography (PET), a nuclear imaging technique, to investigate miscible solute displacements at the core scale in a porous limestone. They used a cylindrical Ketton Limestone rock core (diameter 5 cm, length 10 cm) for a set of pulse-tracer experiments. Ketton Limestone comprises smooth spherical grains (average particle diameter ~0.6 mm). The porosity at the sub-core scale shows a significant degree of variability ($n = 0.23 \pm 0.04$), Fig. 7.1.

In the pulse-tracer experiments, two isotopes were used. The PET data acquired during a pulse-tracer test are illustrated in Fig. 7.2. These are presented in the form of three-dimensional tracer concentration maps at three different times, corresponding to the injection of 0.25, 0.5, and 0.75 pore volumes (i.e., the porosity times the total volume of the core).

The images show that the tracer plume moves nonuniformly through the core due to pore-scale heterogeneities (pore-size variation, intricate pore connectivity, and pore distribution). The plume center moves with time, with margins spreading out, and the entire plume is diluted at the observation scale larger than pores.

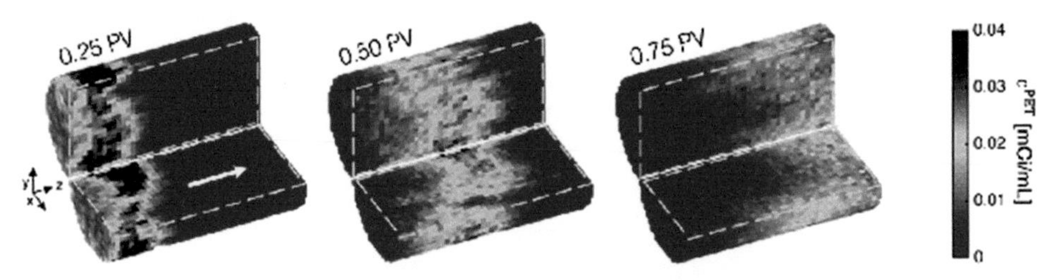

Figure 7.2 3D concentration maps of the tracer plume in Ketton Limestone at 025 PV (pore volume), 0.60 PV, and 0.75 PV. The experimental data are obtained from PET scans at a flow rate 4 mL/min.

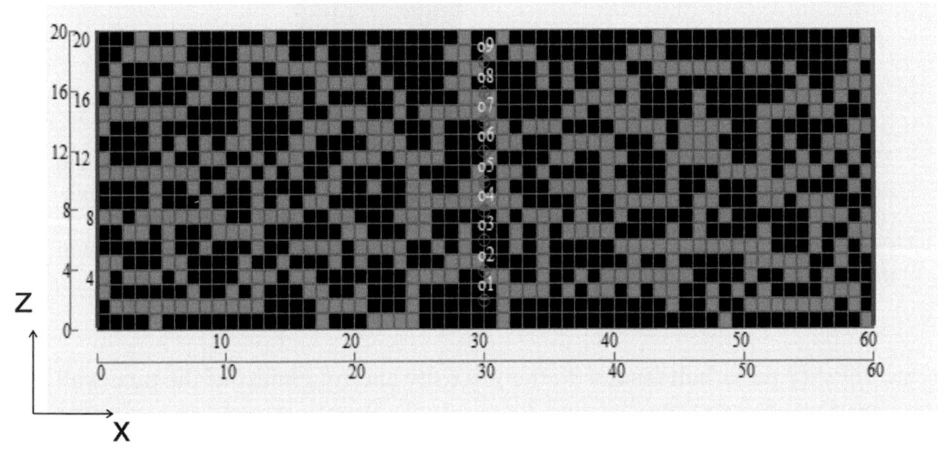

Figure 7.3 The synthetic pore network. The inner axis labels are the element number, and the outer labels are the distance.

To better understand the physics behind the above images, we use a numerical model (Chapter 5) to simulate flow and solute transport in a horizontal plane cross-section parallel to the longest dimension of a synthetic core sample (Fig. 7.3). In the figure, the red block represents the void, and the black block denotes the solid. A steady flow is imposed from the left to the right of the cross-section and is simulated using the Hagen–Poiseuille law (Yeh et al., 2015), assuming the flow is laminar. A slug of tracer is then released from the left to migrate with the flow. Nine sampling points (from No. 01 to No. 09) are installed at x = 30 units at various z elevations. The examples are for illustration purposes. As such, the units used in the example are arbitrary without loss of generality.

Simulated head field (color contours) and streamlines (black lines) in Fig. 7.4 show tortuous streamlines, reflecting complex pore-scale velocity. The head contours are irregular, and flow concentrates on a few flow paths. Some stagnant zones also exist (black circular areas or dead-end pores). In contrast, Fig. 7.5 is the

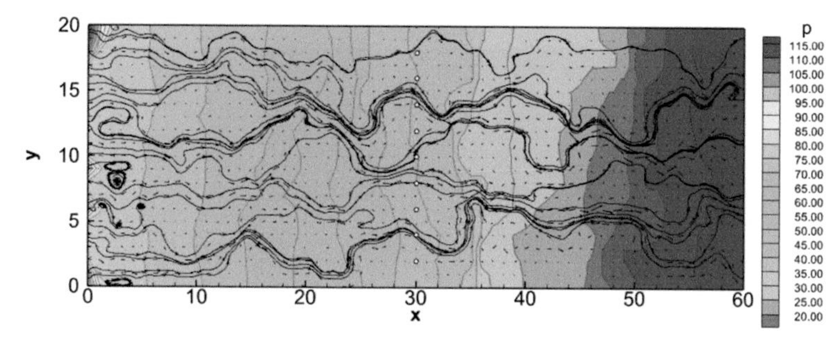

Figure 7.4 Simulated head field and streamlines in a complex pore network.

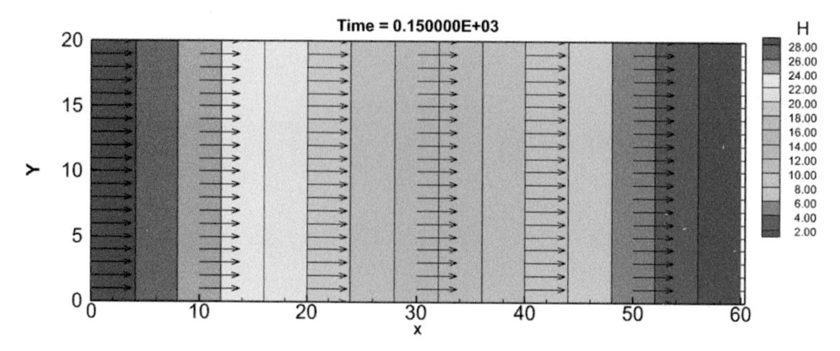

Figure 7.5 Simulated head field (color contours) and streamlines (lines with arrows) using Darcy's law with a uniform hydraulic conductivity field.

smooth head field (color contour lines) and uniform streamlines of the Darcy-scale velocity (lines with arrows), simulated using Darcy's law with a uniform hydraulic conductivity under the same hydraulic gradient boundary conditions.

The contrast between Figs. 7.4 and 7.5 manifests that the head field is less sensitive to the pore-scale heterogeneity than the velocity field and streamlines. The flow field derived from the Darcy-scale velocity of the core is significantly different from that at a scale smaller than the Darcy scale (i.e., pore scale). Such differences imply that using the Darcy-scale velocity alone is insufficient to capture the evolution and migration of solutes in the core sample.

The comparison of simulated tracer movement (advection only) snapshots at time = 150 using the pore-scale (Fig. 7.6) and the Darcy-scale (Fig. 7.7) velocity fields supports the above conjecture. The simulations used the methods of characteristics (MOC, Chapter 5) most appropriate for modeling advection-dominant solute transport. MOC allows omission of molecular-scale diffusion and pore-scale dispersion, except for numerical dispersion, which is kept minimal by obeying the Courant and Peclet number criteria (Chapter 5).

The snapshot of the plume resulting from the pore-scale velocity (Fig. 7.6) shows the split of the tracer slug into two major parts, following the distinct

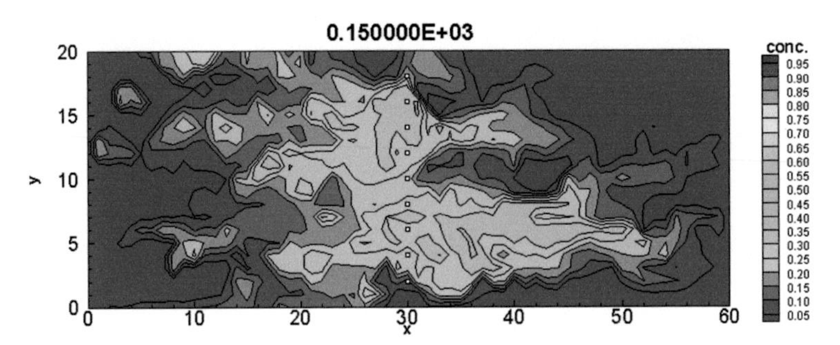

Figure 7.6 A snapshot of the concentration distribution in a core sample at t = 150. The small white circles are the breakthrough sampling points at x = 30 but at different z.

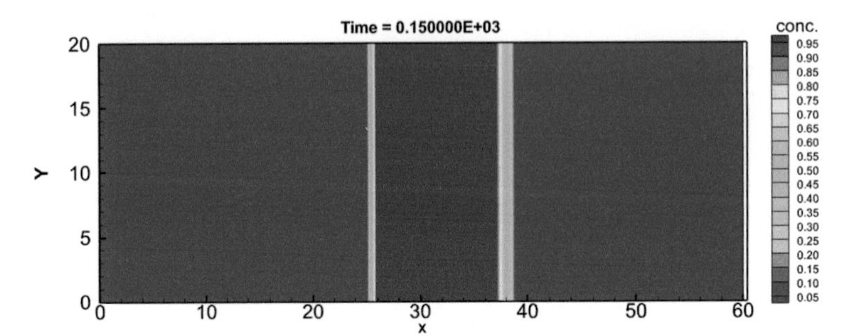

Figure 7.7 A snapshot of the concentration distribution based on Darcian velocity at t = 150.

streamlines (Fig. 7.4) and entrapment of some residual tracers in the stagnant (immobile) zones. The solute plume spreads widely, exhibiting significantly earlier and later arrival of some solutes than the average solute position depicted in Fig. 7.7. The variable and tortuous pore-scale flow paths are the primary cause of the spread, besides the molecular diffusion and numerical dispersion. On the other hand, the snapshot of the tracer plume simulated by the Darcy-scale velocity (Fig. 7.7) retains a shape similar to the input slug. The plume shows a piston-type displacement, with slight smearing before and behind the slug due to numerical dispersion, resulting from the finite grid size used in the numerical model (Chapter 5).

In addition, we show the breakthrough curves at $x = 30$ of the core in Fig. 7.8. In the figure, thin lines of different colors are the BTCs in the pore-scale network at different z locations at $x = 30$. Each line at each z behaves differently, indicating that the solute has traveled through different flow paths, encountering various pore heterogeneity to reach each location. The thick blue line is the average of all the thin lines over the vertical locations at $x = 30$, analogous to the BTC collected over

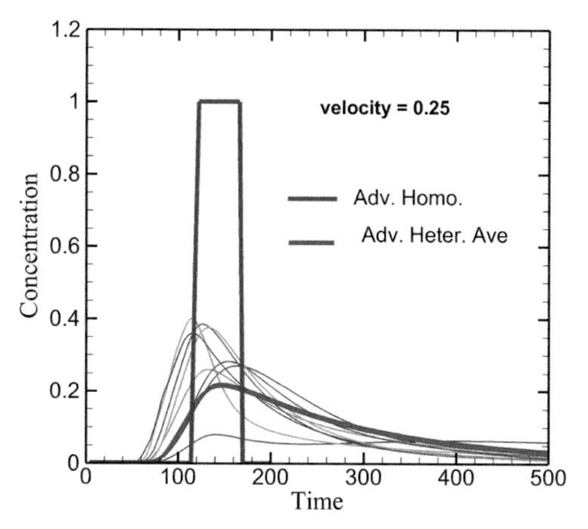

Figure 7.8 The BTC of solute moving through the core with Darcy-scale velocity is indicated by the thick red line. The thick blue line is the average simulated solute BTCs (thin lines of different colors) based on pore-scale velocity in the core at different vertical locations at $x = 30$.

the core cross-section. The thick red line is the BTCs simulated with Darcy-scale velocity alone at $x = 30$. Since the Darcy-scale velocity is uniform, each BTC at different y locations at $x = 30$ is identical to the averaged BTC.

These plots show that the averaged BTC in the pore network is different from the BTC solely based on Darcy-scale velocity. Its peak is attenuated substantially and spreads out and tails extensively (i.e., late arrivals of large amounts of solutes). These behaviors are unaccounted for by the Darcy-scale velocity itself. Considering the pore-scale velocity variations and some processes in addition to the Darcy-scale velocity becomes necessary. Viewing that the spreading and tailing are analogous to molecular diffusion and mechanic dispersion in the shear flow dispersion in the pipe, scientists and engineers thus have exploited the diffusion and dispersion concepts to capture these unaccounted behaviors.

A marathon race is a good analog of solute migration through a porous medium. During a race, we frequently report runners using average speed (i.e., Darcy-scale velocity). Of course, some runners run faster (i.e., the pore-scale velocity at some pores is larger than the others), and some are slower than the majority who move at the average speed. The faster runners arrive at the destination earlier, and slow ones arrive later than the majority. As a result, the runners spread out during the race and arrive at the destination at different times. In solute transport analysis, the runners' spread out is called the mechanic dispersion.

The above discussions point to the following facts. First, we attempt to overlook complex pore-scale velocity for practical purposes and adopt the Darcy-scale velocity. As such, we treat the solute movement due to Darcy-scale velocity as advection and pore-scale velocity variations around the Darcy-scale velocity as dispersion. Second, the dispersion concept is unnecessary if the pore-scale velocity can be determined. Finally, the BTC at the end of a core sample accumulates the effects of pore-scale heterogeneity along the streamline the solute has traveled. This BTC does not reveal how the solute distributes inside the porous medium. Analogous to the marathon race, the number of runners arriving at the destination at different times does not reveal the runners' distribution and what they have experienced throughout the race.

The above example demonstrates that fluids move at a multiscale velocity (Darcy, pore, and molecular scales exluding numerical dispersion in the simulation). The definitions of advection, diffusion, and dispersion phenomena depend on our model scale (e.g., Darcy, pore, fluid-dynamic, or molecular scale), observation scale (the volume sampled by our measurement devices), and the scale of our interests (one molecule or a volume of solutes). Hereafter, the advection refers to the Darcy-scale velocity derived from Darcy law assuming the soil column is homogenous, and dispersion is merely a parsimonious term describing the effects of local- or pore-scale velocity variations unaccounted by the Darcy-scale velocity.

7.3 Mathematical Formulation of Solute Transport in Porous Media

Next, we will formulate the above concept into a mathematical equation to analyze or predict solute plume migration in porous media in a soil column or core sample.

Consider the mass balance of a control volume of solute moving with a fluid in a porous medium.

The mass balance equation for the solute over a control volume of an arbitrary shape (Fig. 7.9) takes the form

$$\frac{\partial}{\partial t}\int_V \rho C\theta d\Omega = -\int_A (\rho C\mathbf{q})\cdot \mathbf{N}dA - \int_A \mathbf{J}\cdot \mathbf{N}dA - \int_V \rho\gamma\theta d\Omega \tag{7.3.1}$$

where V is the volume of the porous medium (a CV). The solute's concentration (mass of solute/mass of solvent) is denoted by C. The solvent's mass per volume is ρ, and θ is the porosity (n) for a fully saturated medium and is the moisture content for partially saturated porous media. The term on the left side of Eq. (7.3.1) is the rate of change of the total mass of the solute in the CV (a volume integral), resulting from the sum of mass fluxes from three processes on the left-hand side of the equation. The first process is the net mass flux vector ($\rho C\mathbf{q}$) [M/L^{2}T] due to advection ($\mathbf{q}$ is the specific discharge vector based on Darcy's law) entering and exiting the surface area (A) that encompasses the CV (a surface integral). Note that $\mathbf{q} = n\mathbf{u}$ for saturated media or $\mathbf{q} = \theta\mathbf{u}$ for unsaturated media, and $\mathbf{u}$ is the

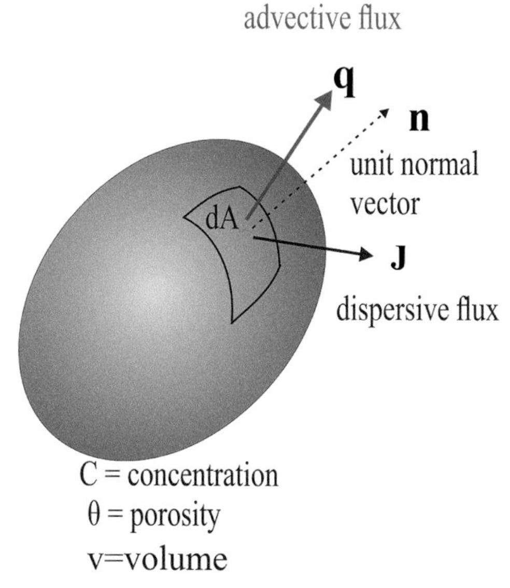

Figure 7.9 Advective and dispersive fluxes crossing the surface of a unit volume of a porous medium.

Darcy-scale velocity vector mentioned previously. The second term (a surface integral) accounts for the net mass flux due to dispersion (**J**: dispersive flux vector) crossing the surface of this CV. This flux is excluded from the Darcy-scale advective flux, as explained previously. Since these two fluxes enter and leave through the irregular-shaped surface, their dot product ($\cdot$) with the unit normal vector (**N**) projects them to the direction perpendicular to the surface. The last term in Eq. (7.3.1) stands for the mass change due to the chemical reaction (a sink or a source) within the CV, where γ is the chemical reaction rate [1/T].

Notice that Eq. (7.3.1) contains both the volume and surface integrals. We need to convert the surface integrals to the volume integrals so that all terms in Eq. (7.3.1) are defined over the unit volume of the porous media, and the mass balance is calculated appropriately. This conversion requires using Gauss's theorem, also called the divergence theorem in vector calculus. The theorem states that the surface integral of a vector field over a closed surface (i.e., the flux through the surface) is equal to the volume integral of the divergence of the same flux over the region inside the surface. In plain language, the sum of influx and outflux on the surface of a CV equals the net flux of the entire volume of the object. Mathematically, the theorem is

$$\int_A \mathbf{U}\cdot\mathbf{N}dA = \int_V \nabla\cdot\mathbf{U}d\Omega \tag{7.3.2}$$

where **U** is a given flux vector. The application of the theorem to the first two terms on the right-hand side of Eq. (7.3.1), which involve the surface integrals, leads to an equation with consistent volume integrals:

$$\int_V \left[\frac{\partial}{\partial t}(\rho C\theta) + \nabla\cdot\rho C\mathbf{q} + \nabla\cdot\mathbf{J} + \rho\gamma\theta\right] d\Omega = 0. \tag{7.3.3}$$

If we let the CV or V approach infinitesimally small, we have

$$\frac{\partial}{\partial t}(\rho C\theta) + \nabla\cdot\rho C\mathbf{q} + \nabla\cdot\mathbf{J} + \rho\gamma\theta = 0 \tag{7.3.4}$$

for the continuous porous media. We then expand Eq (7.3.4) to form the following equation:

$$C\frac{\partial\rho\theta}{\partial t} + (\rho\theta)\frac{\partial C}{\partial t} + C\nabla\cdot\rho\mathbf{q} + \rho\mathbf{q}\cdot\nabla C + \nabla\cdot\mathbf{J} + \rho\gamma\theta = 0. \tag{7.3.5}$$

Notice that the first and the third terms on the left-hand side of Eq. (7.3.5) are the product of concentration C and the continuity equation for the water (i.e., groundwater flow equation):

$$\frac{\partial\theta\rho}{\partial t} + \nabla\cdot\rho\mathbf{q} = 0. \tag{7.3.6}$$

As a result, the sum of the first and third terms in Eq. (7.3.5) equals zero, Eq. (7.3.5) then becomes

$$(\rho\theta)\frac{\partial C}{\partial t} + \rho\mathbf{q}\cdot\nabla C + \nabla\cdot\mathbf{J} + \rho\gamma\theta = 0. \tag{7.3.7}$$

Notice that the above equation is general, suitable for steady or transient, compressible fluids, variably saturated, deformable, and heterogeneous porous media since the water balance Eq. (7.3.6) includes all these conditions. In Eq. (7.3.7), C is defined as the mass of solute per mass of solvent (ppm), and ρ is the mass of solvent per volume of water.

If we redefine the concentration as the mass of solute per volume of solvent/ water:

$$\widehat{C} = \rho C \tag{7.3.8}$$

which has a dimension of M/L^3 (e.g., mg/L), and we apply it to Eq. (7.3.4). We have

$$\frac{\partial\widehat{C}\theta}{\partial t} + \nabla\cdot\widehat{C}\mathbf{q} + \nabla\cdot\mathbf{J} + \rho\gamma\theta = 0. \tag{7.3.9}$$

Eq. (7.3.9) is a new form of the mass balance equation for the solute. Notice that Eq. (7.3.9) is also valid for compressible or incompressible fluids in heterogeneous or homogeneous, variably saturated, deformable, or rigid porous media. Precisely, both θ and $\mathbf{q}$ could vary spatially as well as temporally. If the fluid is compressible, i.e., fluid density varies in time and space, the concentration $\widehat{C}$ will depend on the density.

For incompressible fluids, Eq. (7.3.5) can be simplified. In this case, ρ is constant; the continuity equation for water is

$$\frac{\partial \theta}{\partial t} + \nabla \cdot \mathbf{q} = 0. \tag{7.3.10}$$

By substituting Eq. (7.3.10) into Eq. (7.3.9), the chemical mass balance Eq. (7.3.9) then takes a new form

$$\theta \frac{\partial \widehat{C}}{\partial t} + \mathbf{q} \cdot \nabla \widehat{C} = -\nabla \cdot \mathbf{J} - \rho \gamma \theta. \tag{7.3.11}$$

We remind the reader that this equation is correct only for an incompressible fluid and transient flow.

For hydrologic analysis of solute transport in porous media, water is generally considered incompressible, Eq. (7.3.11) can be written in the Cartesian coordinates:

$$\frac{\partial \widehat{C}}{\partial t} + u_x \frac{\partial \widehat{C}}{\partial x} + u_y \frac{\partial \widehat{C}}{\partial y} + u_z \frac{\partial \widehat{C}}{\partial z} = -\frac{1}{\theta} \left[\frac{\partial J_x}{\partial x} + \frac{\partial J_y}{\partial y} + \frac{\partial J_z}{\partial z} \right] - \rho \gamma \tag{7.3.12}$$

where $u_x = q_x/\theta$, $u_y = q_y/\theta$, and $u_z = q_z/\theta$.

Similarly, Eq. (7.3.7) can be written as

$$\frac{\partial C}{\partial t} + u_x \frac{\partial C}{\partial x} + u_y \frac{\partial C}{\partial y} + u_z \frac{\partial C}{\partial z} = -\frac{1}{\theta \rho} \left[\frac{\partial J_x}{\partial x} + \frac{\partial J_y}{\partial y} + \frac{\partial J_z}{\partial z} \right] - \gamma. \tag{7.3.13}$$

Again, even though the velocity components are outside the spatial derivatives, they could vary in space or time – a direct result of the inclusion of the water continuity equations, Eqs. (7.3.6) and (7.3.10), in the solute transport equations. Next, we define the dispersive flux vector $\mathbf{J}$.

7.4 Dispersion in Porous Media at Darcy Scale

The dispersive flux ($\mathbf{J}$) arises from the velocity variations at many scales including (1) velocity variations within a pore and (2) velocity variation among pores.

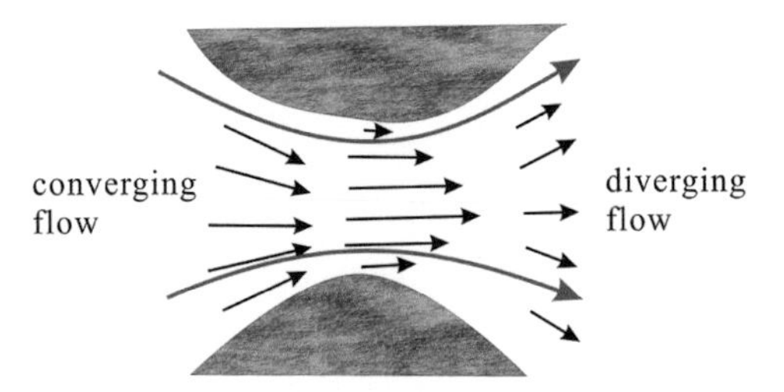

Figure 7.10 An illustration of velocity variations in a pore.

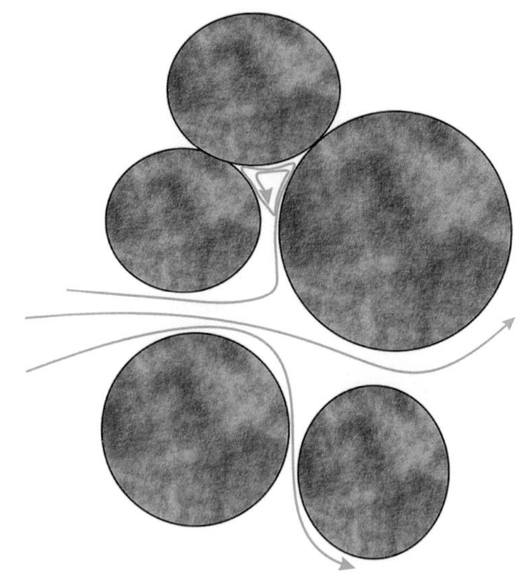

Figure 7.11 An illustration of velocity variation among pores (tortuosity, dead-end pore, branching, and interfingering).

Within a pore, the velocity variation could be due to the roughness of the surface, size and shape of the pore, the viscosity of the fluid, and random motion of molecules (Fig. 7.10). At a scale larger than one pore, the velocity variation could result from the velocity changes between pores of different sizes, tortuous pore channels, and the dead-end pores (see Fig. 7.11).

While dead-end pores entrap and slow the solute movement, tortuosity, branching, and interfingering of pore channels in a porous medium induce crossflow, enhancing the spread or mixing of solutes. Furthermore, our solute sampling devices generally sample a porous volume much larger than many pores, and the averaged value obtained over the device's sampling volume (device averaging) also promotes mixing. Consequently, the smearing around Darcy-scale velocity at our observation scale resembles that due to random motion of molecules or molecular diffusion. These reasons seemly justify the extension of the diffusion equation or Fick's law to represent solute spreading in the porous medium due to pore-scale velocity variation. Furthermore, after solutes travel a distance greater than many pores (or grains), they likely migrate randomly or independently of previous velocities. The spreading of solutes reaches the Fickian regime (see Chapters 4 and 6). In the Fickian regime, the spreading rate of a solute plume becomes constant, and a slug of tracer plume shape evolves to the normal distribution. Fick's law, therefore, is deemed adequate to represent the dispersive fluxes due to pore-scale velocity variations. Accordingly, dispersion is often related to grain size in a uniformly packed soil column (Bear, 1972. Saffman, 1959).

7.5 Mathematical Models for Dispersive Flux

According to the reasons discussed above, the dispersive flux in the flow direction (e.g., x-direction) thus can be written as

$$J_x = -\rho\theta D_L \frac{\partial C}{\partial x} \tag{7.5.1}$$

where $\partial C/\partial x$ is the concentration gradient [1/L] in the x-direction. D_L is the longitudinal dispersion coefficient [L^2/T], which is the solute plume's spreading rate in the direction of the Darcy-scale flow. The sum of the molecular diffusion coefficient D_m and pore-scale mechanical dispersion coefficient represents the mixing due to pore-scale velocity variation.

$$D_L = D_m + D_{\text{mech}}. \tag{7.5.2}$$

The molecular diffusion D_m is generally neglected because of its small magnitude compared to the mechanical dispersion. Nevertheless, this dispersion coefficient includes the two-scale velocity variations.

Likewise, the transverse dispersive fluxes in the directions perpendicular to the average flow direction (e.g., in the y and z directions) are expressed as

$$J_y = -\rho\theta D_T \frac{\partial C}{\partial y}, \quad J_z = -\rho\theta D_T \frac{\partial C}{\partial z} \tag{7.5.3}$$

where D_T is the transverse dispersion coefficient. Notice that $\widehat{D} = \theta D$ is often used in soil physics (see Bredehoff & Pinder, 1973).

7.6 Advection–Dispersion Equation for Solute Transport in Porous Media

Substitutions of Eqs. (7.5.1) and (7.5.3) into Eqs. (7.3.7) and (7.3.11) yield the following advection–dispersion equation (ADE) for solute transport in porous media. If the fluid is compressible and the concentration is defined as mass per mass, we have

$$(\rho\theta)\frac{\partial C}{\partial t} = \nabla\cdot(\rho\theta\mathbf{D}\nabla C) - \rho\mathbf{q}\nabla C - \rho\gamma\theta. \tag{7.6.1}$$

If the fluid is incompressible and the concentration is defined as mass per volume, we have

$$\theta\frac{\partial \widehat{C}}{\partial t} = \nabla\cdot\left(\theta\mathbf{D}\nabla\widehat{C}\right) - \mathbf{q}\cdot\nabla\widehat{C} - \rho\gamma\theta. \tag{7.6.2}$$

In the analysis of solute transport in aquifers or the vadose zone, water is the fluid that carries the solute. Water is generally treated as an incompressible fluid, and the density is equal to 1 g/cm^3. For this reason and to avoid the confusion between the above two equations, we unify them as one form of advection and dispersion equation:

$$\theta\frac{\partial C}{\partial t} = \nabla\cdot(\theta\mathbf{D}\nabla C) - \mathbf{q}\cdot\nabla C - \gamma\theta. \tag{7.6.3}$$

It can be expressed in the Cartesian coordinate systems as

$$\theta\frac{\partial C}{\partial t} = \frac{\partial}{\partial x}\left(\theta D_x\frac{\partial C}{\partial x}\right) + \frac{\partial}{\partial y}\left(\theta D_y\frac{\partial C}{\partial y}\right) + \frac{\partial}{\partial z}\left(\theta D_z\frac{\partial C}{\partial z}\right) - q_x\frac{\partial C}{\partial x} - q_y\frac{\partial C}{\partial y} - q_z\frac{\partial C}{\partial z} - \gamma\theta \tag{7.6.4}$$

In the above two equations, the concentration unit could be either mass of solute per volume of solvent or mass of solute per mass of solvent since the density of water is assumed to be constant and equal to 1.

7.7 Properties of Dispersion Coefficients

The dispersion coefficient is considered a tensor (a matrix in algebra) and could be isotropic or anisotropic, as is the molecular diffusion coefficient. The dispersion tensor is defined according to the direction of flow. The dispersion coefficient in the direction parallel to the flow direction is the longitudinal dispersion coefficient (D_L) and in the direction perpendicular to the flow direction is the transverse dispersion coefficient (D_T). Generally, $D_L \gg D_T$, even if porous media have the isotropic hydraulic conductivity. Furthermore, the dispersion coefficient varies with velocity. As articulated previously, the dispersion coefficient represents the effect of pore-scale velocity variations. These variations increase with the Darcy-scale velocity and depend on the hydraulic conductivity, porosity, and hydraulic gradient. In other words, the dispersion coefficients, D_L and D_T, are not sole properties of porous

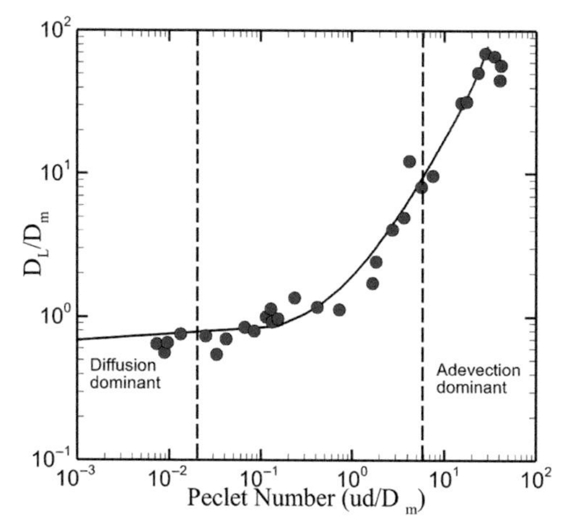

Figure 7.12 The relation between the normalized longitudinal dispersion coefficient and the Peclet number. Red dots are data, and the solid line is the fitted trend. (reproduced from Marinov, 2014).

media; they vary from one experiment to another. Therefore, the dispersion coefficient value determined from a medium under one flow scenario does not apply to different flow scenarios in the same medium. Specifically, the dispersion coefficient is a scenario-dependent property of a medium.

Over the past decades, many laboratory experiments on porous media have documented the flow dependence of the dispersion coefficient. They show power-law relations between the normalized dispersion coefficient (D_L/D_m) and the Peclet number (P_e) (Bear, 1972) (Fig. 7.12). The Peclet number is a dimensionless number that measures the relative strength of the advection to the molecular diffusion process. Specifically, it is

$$P_e = \frac{ud}{D_m} \tag{7.7.1}$$

where u is the Darcy-scale velocity (q/θ); d is a characteristic length (mean grain diameter), and D_m is the molecular diffusion coefficient.

According to Fig. 7.12, a power-law relationship between the normalized dispersion coefficient (D_L/D_m), parallel to the flow lines (or longitudinal direction), and the Peclet number is

$$\frac{D_L}{D_m} \propto \left[\frac{ud}{D_m}\right]^n. \tag{7.7.2}$$

This relationship shown in the figure can be grouped into two regimes. Namely, molecular diffusion dominated regime $(P_e < 10)$ and advection dominated regime $(P_e > 10)$. In the first regime, the random motion of molecules controls the spread of the solutes since the Darcy-scale velocity (advection) is negligible. On the other hand, in the advection-dominated regime, the value of the power n approaches unity. That is,

$$\frac{D_L}{D_m} = c\left[\frac{ud}{D_m}\right] \quad \text{or} \quad D_L = cdu = \alpha_L u. \tag{7.7.3}$$

As a result, the longitudinal dispersion coefficient is linearly related to the Darcy-scale velocity via the longitudinal dispersivity α_L [L]. The dispersivity is independent of the Darcy-scale velocity but is related to the mean grain diameter d and the coefficient, c, i.e., $\alpha_L = cd$. The dispersivity, thus, is generally considered a transport property of a porous medium.

Laboratory experiments have also shown that the longitudinal dispersivity is about 20 to 30 times the transverse dispersivity (α_T). For example, $\alpha_L = 20\alpha_T$ for unconsolidated sand and $\alpha_L = 1$ cm or less, varying with grain size. In other words, the transverse dispersivity is generally very small.

Finally, for a working coordinate not aligned with (or perpendicular to) the Darcy velocity vector, the dispersivity tensor can be expressed as

$$D_{kl} = \alpha_{ijkl}\frac{u_i u_j}{|\mathbf{u}|} \tag{7.7.4}$$

where $\alpha_{ijk\ell}$ is a fourth-rank symmetrical tensor; u_i are the components of the Darcy-scale velocity $\mathbf{u} = \mathbf{q}/\theta$ (see Bear, 1972). Despite the complexity of Eq. (7.7.4), a practical formulation of the dispersion coefficient is that $D_{xx} = \alpha_L|\mathbf{u}|$, $D_{yy} = \alpha_T|\mathbf{u}|$, and $D_{zz} = \alpha_T|\mathbf{u}|$ if the Darcy-scale flow is along the x-axis of the working coordinate (Bear, 1979).

7.8 The Equation for Solution Transport in Laboratory-Scale Porous Media

The governing equation for solute transport in a soil column in the one-dimensional flow then takes the following forms:

$$\theta\frac{\partial C}{\partial t} = \frac{\partial}{\partial x}\left[\theta D\frac{\partial C}{\partial x}\right] - q\frac{\partial C}{\partial x} - \gamma\theta \text{ where } D = \alpha u. \tag{7.8.1}$$

Flow and solute transport in a soil column is always three-dimensional. Applying one-dimensional models to the column, we implicitly assume that the concentration is averaged over the cross-section area.

Note that if the density, porosity or moisture content, and specific discharge are spatially invariant in Eq. (7.8.1), the equation is identical to the classical advection–diffusion equation discussed in Chapter 4. Many analytical solutions for a variety of conditions are available (Chapter 4). More versatile numerical solutions to this equation are presented in Chapter 5.

7.9 Non-Fickian Solute Transport

The ADE (Eq. 7.8.1) generally describes satisfactorily solute transport phenomena in uniformly packed sand columns or sandboxes. However, there are cases in which the ADE does not capture the plume distributions. For example, Herr et al. (1989) conducted laboratory tracer experiments in a soil column under steady-state saturated flow. A salt solution of 1g/L was released as a step function at the column inlet. The BTCs were collected at the column outlet. The column had a length of 1m and a diameter of 0.1 m and was filled with a mixture of uniform sand and porous ceramic cubes. The permeability and the dispersivity of the sands and the permeability of the ceramic materials (Table 7.1) were measured separately. The mixture of both media were randomly distributed along the column.

The experiments were conducted with two different types of sand as the background and different types of heterogeneity. Table 7.1 lists K_h/K_s, θ_h *and* θ_s values, where θ_s is the volume of the sand per total volume of the column while θ_h is the total pore volume of the embedded heterogeneities per total column volume. K_s is the hydraulic conductivity of the sand, K_h is that of the heterogeneities (in meters per second). The porosity of the ceramic cubes was constant (0.4). PVC-cubes ($K_h/K_s = 0$) represent impermeable heterogeneities.

Ceramic plates (80 × 50 × 10 mm) replaced the PVC cubes in some experiments. The cube sizes were equal for each experiment but varied from 15 mm to 18 mm in run 1 and runs 3–9 (Table 7.1) and from 21 mm to 23 mm for runs 10–14. Each experimental column contained approximately 200 cubes. For experiment runs 10–14, the sand used as background filling differed from those for runs 1–9 (see Table 7.1).

Figure 7.13 shows the measured BTCs (black dots) for the cases where $K_h > K_s$. The solid lines denote the BTCs calculated by the ADE after calibrating it to the data. Figure 7.14 shows the BTCs for the cases where $K_h < K_s$. In this figure, the solid lines are regressed with the dual-domain model (see Section 7.9.1). In both figures, measured concentrations are normalized to the initial concentration C_0. The time axis refers to the average travel time calculated by dividing the column length L with the average Darcy-scale velocity.

The BTCs in the sand column with highly permeable heterogeneities ($K_h > K_s$) show a significant increase in the dispersion zone at the tracer front and tail as the ratio of K_h/K_s increases. Clear evidence that a greater degree of heterogeneity increases dispersion. Notice that the 0.5 relative concentrations arrive at the same normalized time regardless of the heterogeneity. The BTCs generally conform to the ADE.

Table 7.1. *The cases and material properties in the soil column experiments by Herr et al. (1989).*

Characteristic of the background sand	*Mean grain diameter* $d_m = 5.7\times10^{-4}$ (m) Hydraulic conductivity $Ks = 1.5\times 10^{-3}$ (m/s) Dispersivity $\alpha_L = 0.0006$ (m)									$d_m = 2.8\times 10^{-4}$ (m) Ks = 2.2×10^{-4} (m/s) $\alpha_L = 0.0003$ m				
	Ks>>Kh									Ks<<Kh				
Run	1	2	3	4	5	6	7	8	9	10	11	12	13	14
Geometry of ceramic heterogeneity	cubes*	plates	cubes	cubes	cubes	cubes	cubes	cubes	cubes	cubes	cubes	plates	cubes	cubes
K_h/K_s	0.0	0.013	0.013	0.053	0.053	0.2	0.33	0.33	0.66	1.33	2.5	2.5	3.0	4.0
θ_h	0.0	0.049	0.074	0.094	0.073	0.059	0.112	0.073	0.072	0.059	0.059	0.051	0.059	0.059
θ_s	0.332	0.331	0.315	0.304	0.324	0.320	0.273	0.311	0.327	0.334	0.334	0.342	0.334	0.334

* PVC-cubes (permeability = zero)

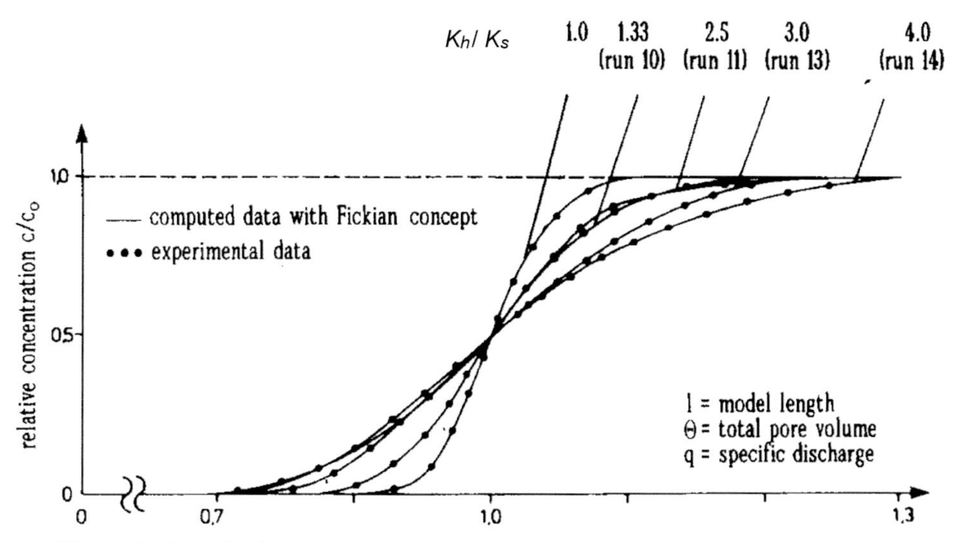

Figure 7.13 BTCs from experiments with $K_h > K_s$ (Table 7.1) (from Herr et al. 1989).

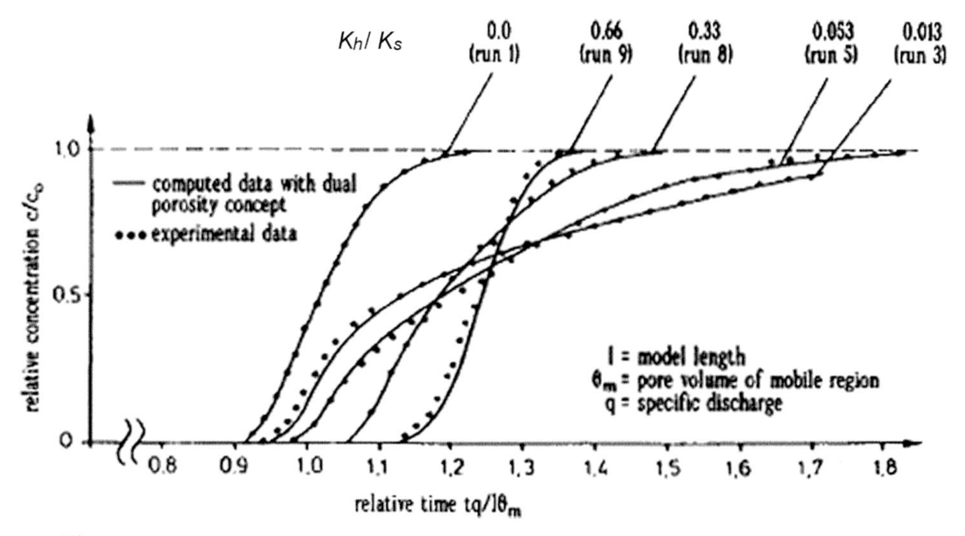

Figure 7.14 BTCs from experiments with $K_h < K_s$ (Table 7.1) (from Herr et al. 1989).

The BTCs for cases ($K_h < K_s$) illustrate that the arrival times of the 0.5 concentration become earlier, and long tailings become more profound than the cases ($K_h > K_s$) as the ratio K_h/K_s decreases (the heterogeneity becomes less permeable). The ADE could not fit the observed BTCs well, and a dual-domain model (section 7.9.1) yielded better results, indicating that the solute migration did not reach the Fickian regime–non-Fickian transport.

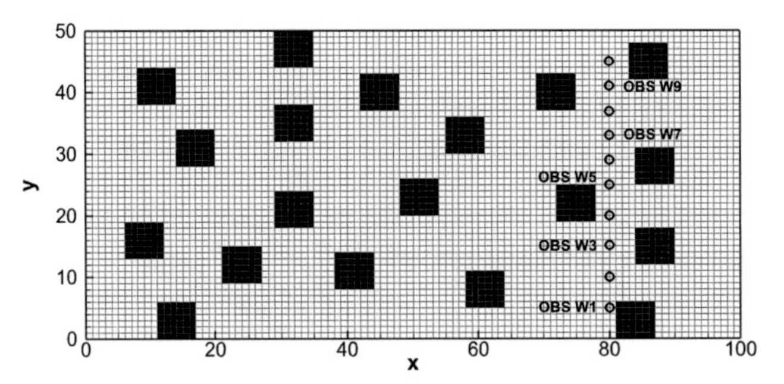

Figure 7.15 Setup for the numerical experiments to illustrate the effects of heterogeneity on solute transport.

The experiments by Herr et al. (1989) reveal the effects of different heterogeneity on the BTCs at the outlet of the soil column. To investigate the solute behavior inside the soil column under these cases, we conducted numerical flow and tracer experiments in media with a similar setup as Herr et al. (1989) using VSAFT2 (Variably Saturated Flow and Transport model in 2-D, available at www.tian.hwr.arizona.edu) (Yeh et al., 1993). The numerical experiments and results are discussed below.

The 2-D, synthetic, horizontal, sandy aquifer with arbitrarily distributed square inclusions for this numerical experiment is shown in Fig. 7.15. The aquifer is discretized into 50×100 (1 cm $\times$ 1 cm) blocks. Each block has a different K value according to the distribution of the sand and inclusions (i.e., a highly parameterized heterogeneous model, HPHM). Flow and solute transport within each block are assumed to follow Darcy's law and the ADE. The sand was assumed to have a K of 1 cm/sec; the K of the inclusions was 5 cm/sec for Case 1 (or the High K case) and 0.2 cm/sec for Case 2 (the Low K case). The steady flow from the left side of the aquifer to the right side was simulated with impermeable boundaries on the other two sides.

A slug of tracer was released uniformly from the left-side boundary. A numerical simulation of the slug of tracer migration through the aquifer was conducted, assuming that the pore-scale dispersivity is zero or the pore-scale dispersion is represented by numerical dispersion. Observation wells shown in the figure collected BTCs.

Figure 7.16 shows the simulated steady Darcy-scale velocity fields and snapshots of the tracer distributions for the two cases at 20 seconds after the release of the tracer. For Case 1, the flow converges toward the high K inclusions and diverges afterward, and vice versa for Case 2. Snapshots of the 2-D tracer distributions for the two cases reveal that the overall movement of the tracer plume in Case 1 is faster than in Case 2; the peak concentrations in Case 1 are relatively higher than in Case 2. Moreover, the plume spreading is greater in Case 2 than in

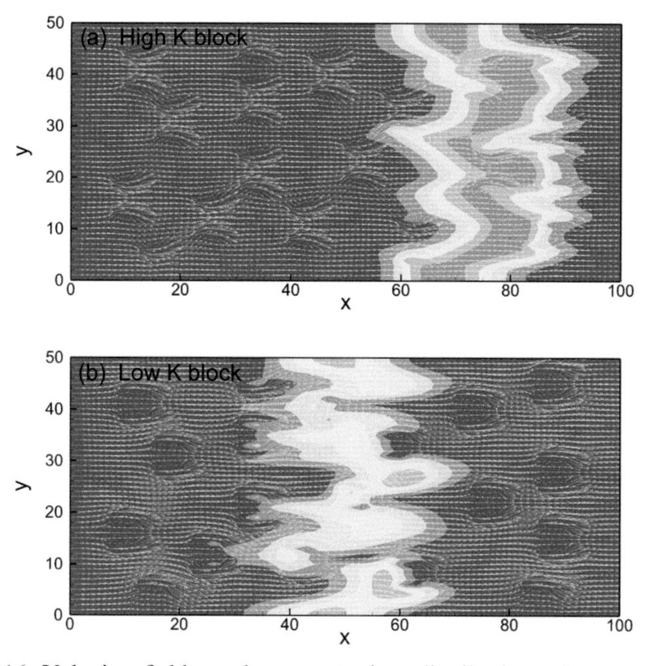

Figure 7.16 Velocity fields and concentration distributions for case 1 (High K blocks) and case 2 (Low K blocks) at a fixed simulation time.

Case 1 because tracers at low concentrations are trapped in the low K inclusions and move behind the fast-moving main plume.

Plots of the snapshots along x at different y locations are shown in Figs. 7.17 (Case 1) and 7.18 (Case 2). An averaged snapshot over the cross-section (the average over the y-direction) is also shown. This averaging conforms to the cross-sectional average concentration inherent in one-dimensional ADE. For Case 1, the concentration distributions along these selected horizontal lines are approximately normal, although some exhibit multiple peaks and skew distributions. However, the distribution of the averaged concentration is close to a Gaussian distribution, suggesting the validity of Fick's law – a testimony of the ensemble mean nature of the ADE.

On the other hand, Fig. 7.18 is the snapshots in Case 2, where heterogeneity has a lower hydraulic conductivity than the background sand. The figure shows that multiple peaks and tailing of the tracer plume (i.e., significant low concentrations lagged behind the majority of the tracer.) are apparent for most of the concentration distributions, even though the Gaussian distribution closely approximates some distributions. The cross-section averaged concentration distribution nevertheless displays the fast movement of the majority of the tracer and the slow movement of some tracers. In other words, ADE fails to quantify the migration of the tracer.

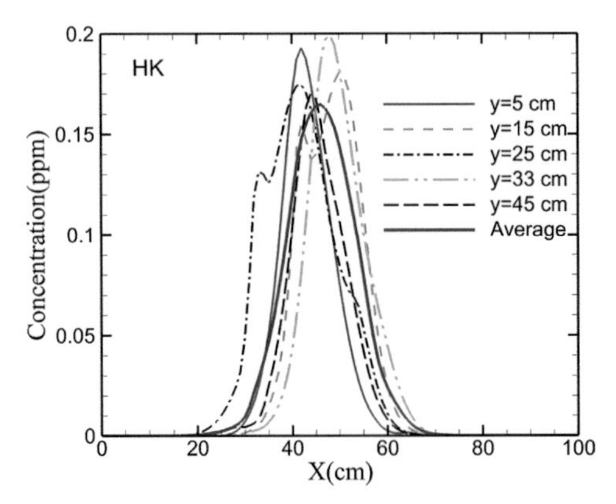

Figure 7.17 Snapshots of the solute distribution along the horizontal lines at different y locations for Case 1. The thick solid red line denotes the average of all the snapshots along all the horizontal lines. H.K. means the heterogeneity has a higher K values than the background sand.

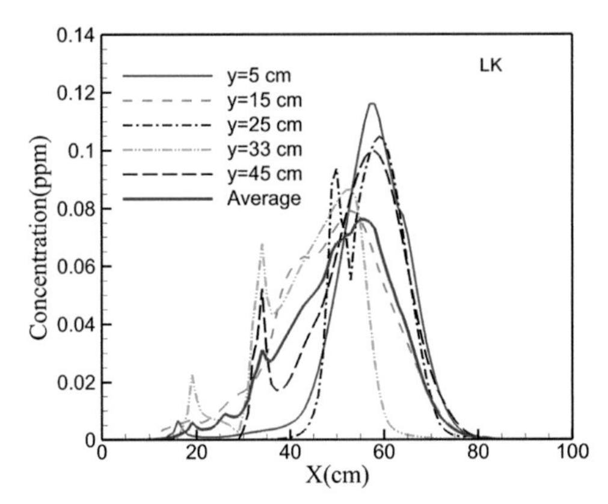

Figure 7.18 Concentration BTCs along selected horizontal lines for Case 2 as well as the average BTC. L.K. denotes $K_h<K_s$.

Breakthroughs for selected observation wells and the average over all the observation wells for cases 1 and 2 are presented in Figs. 7.19 and 7.20, respectively. Comparisons of the two figures suggest that when the embedded heterogeneities have high hydraulic conductivity values (Case 1), the breakthrough curve at each observation well behavior according to Fick's law so does the average one. However, although smooth, the breakthroughs and their average are highly skewed with long tails in the low conductivity block case (Case 2), indicative of non-Fickian behaviors.

The non-Fickian behaviors arise in Case 2 because some tracers are trapped in the isolated low hydraulic conductivity blocks and slowly migrate out. In other words, the differential tracer distribution due to velocity variation has not been smeared out

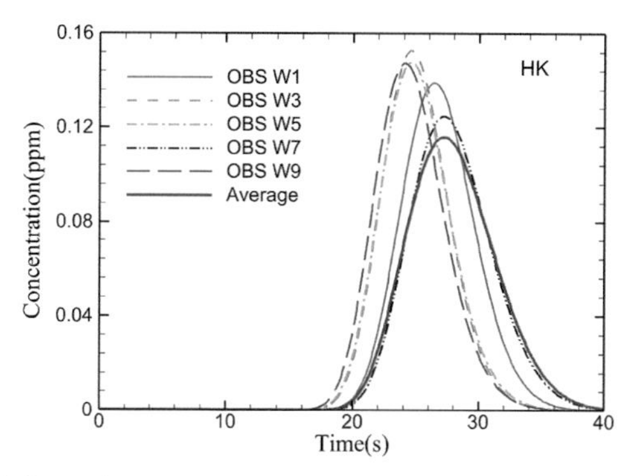

Figure 7.19 Breakthrough curves at selected observation wells and the average one over all the observation wells for Case 1.

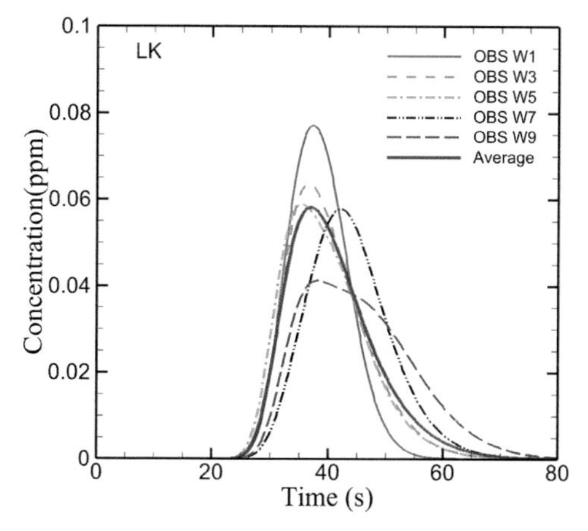

Figure 7.20 Breakthrough curves at selected observation wells and the average one over all the observation wells for Case 2.

by the crossflow and molecular diffusion (see section 6.3, in Chapter 6). As a result, we may say that tracer migration due to hydraulic conductivity contrast does not behave like random walks of molecules in pure molecular diffusion (anomalous dispersion as some calls it). Such deviations from molecular diffusion may be due to the lack of ergodicity (see discussion points below).

Discussion Points:

- *Speculate the solute snapshots and breakthroughs if the simulation domain increases and so does the number of the high or low K blocks. Would the solute plume reach the Fickian regime? (hint: ergodicity or central limited theorem).*
- *If only one or two high or the low K block exist, what would the snapshot and BTC look like?*

- *If the simulation domain remains the same but the discretation grid size increases (numerical dispersion becomes large), how would the result change?*

7.9.1 Dead-End Pore (or Dual-Domain) Model

To replicate the early arrival and tailing characteristics of some BTC in laboratory experiments, without knowledge of the heterogeneity, Coast and Smith (1964) introduced a dead-end pore model. Later, this model was renamed as mobile/ immobile zone (e.g., DeSmedt and Wierenga, 1979) or dual-porosity or dual-domain model. It aims to overcome the difficulty of ADE to model the observed non-Fickian behavior of tracer BTCs. The dead-end pore model contemplates that the classical ADE should include a source/sink term because of the unknown low hydraulic conductivity zones. This term can replicate the phenomenon that some solutes may be trapped in dead-end pores (i.e., low permeable blocks) and slowly communicate with the main flow, creating long tails in solute BTCs.

Implicitly, this dead-end pore model takes the ensemble-mean approach, as does the ADE, since the complex pore distribution is intractable. It avoids mapping the dead-end pore distribution in the medium but conceptualizes that the medium consists of two overlapping regions: mobile and immobile in the ensemble sense (Fig. 7.21). The dead-end pore model lumps all the well-connected pores into a mobile zone, ignoring their spatial distribution. Similarly, it omits the spatial distribution of the poorly connected pores and assembles them into an immobile zone,

The dead-end pore model assumes that the solute transport process in the mobile zone (or region) is described by an ADE with a source or sink term, representing the interaction between the two regions.

$$\frac{\partial C}{\partial t}+u\frac{\partial C}{\partial x}=D_L\frac{\partial^2 C}{\partial x^2}-(1-f)\frac{\partial C_d}{\partial t}. \tag{7.9.1}$$

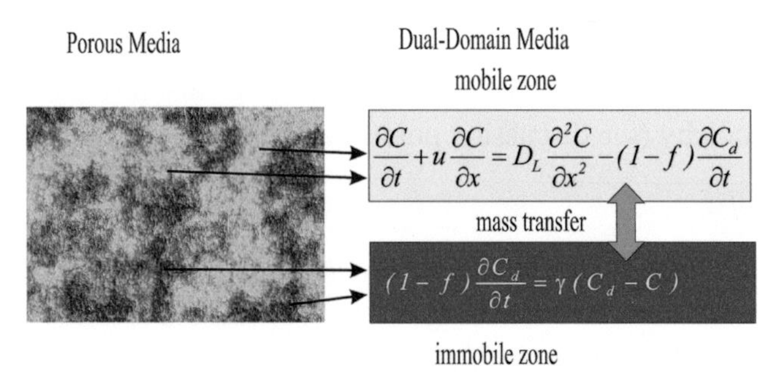

Figure 7.21 Porous media (well-connected pores, yellow; dead-end pores, purple). Right is the dual-domain media conceptual model.

The terms $C(x, t)$ and $C_d(x, t)$ denotes the concentration in the fluid in the mobile region and the immobile region, respectively. The last term of Eq. (7.9.1) is the sink and source term, in which f is the volume fraction of the total pore volume occupied by the moving fluid (mobile region). Notice that when $f = 1$, the equation is the classical ADE. Specifically, this sink or source term accounts for the contribution from the mobile region to the immobile one or vice versa. A mass transfer model quantifies this contribution. Thus,

$$(1 - f)\frac{\partial C_d}{\partial t} = \gamma(C_d - C) \tag{7.9.2}$$

where γ is the mass transfer coefficient between the two regions. Eq. (7.9.2) states that the mass moving into or out from the immobile region is linearly proportional to the difference in the concentration of the two overlapping regions. This mechanism is the same as the linear reservoir model used to describe the interaction between a stream and an aquifer in the well-mixed model or a non-equilibrium linear model used to describe chemical reactions between solids and fluids (Chapter 3). As shown in Chapter 3, this mass transfer model is a diffusion process. If the transfer coefficient is small, slow diffusion of tracers from the dead-end pores explains some tracers' late arrival and long trails of a BTC. Likewise, the volume of the immobile zone reduces the volume of the mobile zone where the main flow travels. Consequently, the tracer in the mobile zone arrives earlier than in the porous medium without the immobile zone (Figs. 7.14 and 7.20).

Equations (7.9.1) and (7.9.2) are ensemble-averaged equations since the fraction and locations of immobile zones are unknown in addition to the mass transfer coefficient value. As a result, the contribution from the immobile zone could be from anywhere in the porous medium. These equations do not yield any spatial distribution details but ensemble-averaged ones.

As a matter of fact, the dispersion concept ingrained in ADE has incorporated the effects of dead-end pores. However, the effects of other types of heterogeneity overwhelm the dead-end pore effects once the solute plume has encountered sufficient different heterogeneity reaching the Fickian regime. In other words, ADE is the most likely transport mechanism without detailed information about the aquifer heterogeneity. Thus, the general ensemble statistics of pore-scale heterogeneity in the column can characterize the dispersivity of ADE. The dual-domain model is merely a particular case of the ADE model. For this reason, the dual-domain model, Eqs. (7.9.1) and (7.9.2) has to be fitted (or conditioned) to observed BTCs to ascertain the mobile region's dispersivity, mass transfer coefficient, and volume fraction.

Because the dead-end pore (or dual-domain) model has additional parameters f and γ for adjustment, it could fit the non-Fickian BTCs from column experiments well. Zhang and Gable (2008) and Zhang and Zhang (2015) reached the same assessment. On the other hand, because f and γ are fitting parameters, if the tracer slug is released from different locations of the soil column, the fraction of immobile zones encountered along the tracer's pathway could be different, leading to different BTCs. Likewise, Herr et al. (1989) showed that the value of γ varies with flow rate. That is to say, calibrating the model for each scenario is necessary – a phenomenological model (not the most-likely or general one).

The dual-domain model is analogous to the parsimonious delayed yield model (Boulton, 1954) or the gravity delayed yield model (Neuman, 1972) that reproduces the S-shaped drawdown-time curves observed in a water-table aquifer during a pumping test. As Mao et al. (2011) and Yeh et al. (2015) explained, the change in the water table and the unsaturated flow in the vadose zone cause the S-shaped drawdown-time curve. A classical variably saturated (unsaturated/saturated) flow model without new theories spontaneously reproduces the phenomenon. In other words, once heterogeneity in a soil column is mapped sufficiently, ADE considering only molecular diffusion, automatically addresses either the Fickian or non-Fickian solute transport process.

Finally, we emphasize that both Fickian dispersion and dead-end pore models rest upon the ensemble-mean behavior concept; the former assumes the Gaussian probability distribution and the latter the non-Gaussian one. Validating both probabilistic distributions requires ergodicity. That is, the cross-sectional concentration must sample sufficient point concentrations to represent its intended distribution. If not, neither the ADE nor the dead-end pore model is a statistically valid model, as demonstrated in Figs. 7.17 to 7.20. For this reason, applications of these models to a field-scale problem where the nonstationary heterogeneity exists may find that the effects of spatial variability of hydraulic properties at the scales larger than the columns are more profound than the dispersion and dead-end pores from the soil column experiments (Chapter 11).

7.10 Reactive Solute Transport

This section formulates the governing equation for the transport of reactive solutes in porous media based on ADE, which is

$$\theta \frac{\partial C}{\partial t} = \frac{\partial}{\partial x}\left[\theta D \frac{\partial C}{\partial x}\right] - q\frac{\partial C}{\partial x} - \gamma\theta \tag{7.10.1}$$

Again, the last term $\gamma\theta$ is the sink or source, similar to the last term in the dual-domain model Eq. (7.9.1), but this term represents the chemical source or sink.

Chapter 3, section 3.5 shows that the reactive rate can be related to the solute concentration change rate in the solid phase.

$$\gamma = \frac{\rho_b}{n\rho}\frac{dC_S}{dt} \tag{7.10.2}$$

In which ρ_b and n are the bulk density and porosity (or moisture content under unsaturated conditions) of the porous medium, respectively. The chemical concentration in the solid phase is C_S. This rate of change can be tailored to either equilibrium or non-equilibrium reactions.

7.10.1 Equilibrium Reaction

For equilibrium reactions, the exchange of the chemical in the solid matrix and the fluid is instantaneous and is linear

$$C_s = aC \tag{7.10.3}$$

The coefficient a is the dimensionless partition coefficient (i.e., the fraction of the concentration in the liquid). Using this relationship, we rewrite Eq. (7.10.2) as

$$\gamma = \frac{\rho_b}{n}\frac{a}{\rho}\frac{dC}{dt} = \frac{\rho_b}{n}K_d\frac{dC}{dt} \tag{7.10.4}$$

where K_d is the distribution coefficient $\left[L^3/M\right]$. Substituting this relationship, we can rewrite Eq. (7.10.2) as

$$\theta\frac{\partial C}{\partial t} = \frac{\partial}{\partial x}\left[\theta D\frac{\partial C}{\partial x}\right] - q\frac{\partial C}{\partial x} - \theta\frac{\rho_b}{n}K_d\frac{dC}{dt} \tag{7.10.5}$$

Rearranging the equation lead to

$$\theta\left(1 + \frac{\rho_b}{n}K_d\right)\frac{\partial C}{\partial t} = \frac{\partial}{\partial x}\left[\theta D\frac{\partial C}{\partial x}\right] - q\frac{\partial C}{\partial x}, \tag{7.10.6}$$

which can be further simplified as

$$\theta R\frac{\partial C}{\partial t} = \frac{\partial}{\partial x}\left[\theta D\frac{\partial C}{\partial x}\right] - q\frac{\partial C}{\partial x} \quad \text{and} \quad R = 1 + \frac{\rho_b}{n}K_d \tag{7.10.7}$$

The R is the retardation factor (Chapter 3). Diving R through Eq. (7.10.7), we see that if $R>1$, the dispersion coefficient D and q will be reduced. The reactive tracer consequently will arrive later and spread less than the non-reactive tracer.

7.10.2 Non-Equilibrium Reaction

If the chemical reaction undergoes the linear non-equilibrium reaction, Eq. (7.10.1) becomes

$$\theta\frac{\partial C}{\partial t} = \frac{\partial}{\partial x}\left[\theta D\frac{\partial C}{\partial x}\right] - q\frac{\partial C}{\partial x} - \theta\frac{\rho_b}{n\rho}\frac{dC_S}{dt} \tag{7.10.8}$$

Solving this equation requires an additional equation, which prescribes the behavior of the last term in the equation. This behavior is conceptualized as a transient mass transfer process, representing the interaction between the chemical in the fluid and the solid phases.

$$\frac{dC_S}{dt} = \beta(C_s - aC) \tag{7.10.9}$$

The mass transfer coefficient is β [1/T], denoting the transfer rate between the fluid and solid, which is linearly proportional to the difference between the current concentration in the fluid phase and the equilibrium concentration in the solid phase.

We should notice that mathematically and conceptually, the non-equilibrium reaction model is identical to the dual-domain model in the previous section. However, the chemical non-equilibrium model, instead of using mobile and stagnant liquid regions, uses the porosity or moisture content as the fraction where the chemical reaction occurs. Under fully saturated conditions, the chemical reaction occurs within pores, and under unsaturated conditions, it is controlled by the moisture content distribution. Generally, the porosity is assumed constant, but moisture content varies under transient flow conditions. Further, the detailed distribution of the chemical concentration in the solid phase could vary and is unknown. Therefore, this model is again an ensemble mean equation. Specifically, even if the average ensemble porosity and mass transfer coefficient are known, this non-equilibrium model's validity still depends on if the solute has met the ergodicity assumption ingrained in these equations. Deviations of the observed BTCs from the theoretical model is expected.

Discussion points:

- If the soil column is sufficiently long, allowing the tracer to experience different pore heterogeneity, would the dead-end pore effects disappear and the BTC reach the Fickian behavior?
- While the above discussion of dispersion is limited to fully saturated porous media, speculate the effects of unsaturated flow on dispersion? (Hint: see Padilla et al., 1999)

- Suppose the 1-D ADE or the dual-domain model is expanded to multi-dimensional field problems with homogeneous assumptions. Would this approach satisfy the scale of our interest? (Chapters 10 and 11).

7.11 Homework

Determine dispersivity values for breakthrough data (Fig. 7.13) of Herr et al. (1989) using the parameter estimation techniques in Chapter 8. Then, determine the relationship between the dispersivity and the ratio of K_h/K_s. Draw some conclusions from this relationship.

8

Parameter Estimation

8.1 Introduction

This chapter introduces practical graphical approaches to estimate advection velocity and dispersivity of solute migration through homogeneous soil columns, using one-dimensional ADE discussed in previous chapters. Methods of spatial and temporal moments are also introduced for solute concentration breakthroughs in one-dimensional transport and snapshots of the multidimensional solute migrations, respectively. Instead of automatic nonlinear regression analysis, these methods use physical insights and analytical solutions to illustrate the logic behind the estimating these parameters. The automatic regression analysis (such as Microsoft Excel introduced in Chapter 1) may find the parameters that fit the solution to the data well. However, the parameter values may not be physically possible if the estimation problem is not correctly defined (see examples in Chapter 11).

8.2 Estimation of Parameters Using Analytical Solution

Consider soil column tracer experiments under steady-state flow conditions. According to Chapter 7, the governing ADE is

$$\frac{\partial C}{\partial t} + u\frac{\partial C}{\partial x} = D\frac{\partial^2 C}{\partial x^2} = \alpha_L u\frac{\partial^2 C}{\partial x^2}. \tag{8.2.1}$$

Eq. (8.2.1) assumes that the velocity (u) and the longitudinal dispersivity (α_L) are constant in space and time for simplicity. Most soil column experiments adopt these assumptions. Further, we consider the solute is conservative (i.e., no chemical reaction) and the domain is unbounded (i.e., no boundary effects). An impulse input and a step input of solute are discussed, where analytical solutions are available.

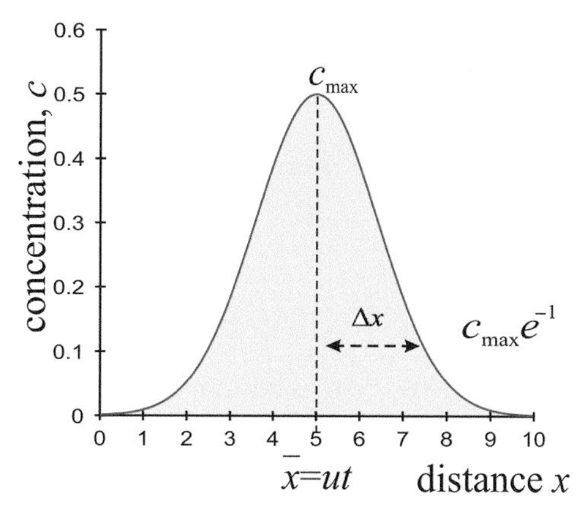

Figure 8.1 A snapshot of a concentration plume spatial distribution at time = 5, u = 1, $\alpha_L = 1$, normalized mass 1.

8.2.1 Impulse input

If a conservative tracer of a mass, M, is released at $t = 0$ and $x = 0$ in an unbounded soil column ($-\infty < x < +\infty$), the analytical solution to Eq. (8.2.1) is:

$$C(x,t) = \frac{M}{\rho n A\sqrt{4\pi D t}}\exp\left[-\frac{(x-ut)^2}{4Dt}\right] = \frac{M}{\rho n A\sqrt{4\pi(\alpha_L u)t}}\exp\left[-\frac{(x-ut)^2}{4(\alpha_L u)t}\right]$$
$$= \frac{M}{\rho n A\sqrt{4\pi\alpha_L\bar{x}}}\exp\left[-\frac{(x-\bar{x})^2}{4\alpha_L\bar{x}}\right] = C_{max}(\bar{x})\exp\left[-\frac{(x-\bar{x})^2}{4\alpha_L\bar{x}}\right]. \quad (8.2.2)$$

In the equation, A is the column's cross-sectional area, n is the porosity and $\bar{x} = ut$ is the location where the maximum concentration, C_{max}, resides. The plot of $C(x, t)$ as a function of x at a given time t (i.e., a snapshot) is displayed as the red line in Fig. 8.1. The snapshot is a bell-shaped, symmetrical, normal distribution.

8.2.1.1 Snapshots

Suppose a snapshot of a tracer plume at the time, t^*, after the tracer was released, is available (the red line in Fig. 8.1), and the location where the tracer was released is known. Under this situation, the location of the maximum tracer concentration can be used to estimate the groundwater flow velocity:

$$u = \frac{\bar{x}}{t^*} \quad (8.2.3)$$

At time, t^*, when the snapshot is taken, the spatial concentration distribution can be expressed in terms of the maximum concentration:

$$C(x,t^*) = C_{\max}(\overline{x})\exp\left[-\frac{(x-\overline{x})^2}{4\alpha_L\overline{x}}\right] \quad (8.2.4)$$

If n, A, ρ and M are known, one can estimate α_L directly from $C_{\max}$ and $\overline{x}$ by solving the following equation.

$$C_{\max}(\overline{x}) = \frac{M}{\rho nA\sqrt{4\pi\alpha_L\overline{x}}}. \quad (8.2.5)$$

Otherwise, one can select two concentration levels from the snapshot (i.e., $C(x,t^*) = C_{\max}e^{-1}$), and let the distance between the two concentration levels be Δx as illustrated in Fig. 8.1. According to Eq. (8.2.4), at this concentration level, the argument in the exponent term of the equation must equal 1:

$$\frac{(x-\overline{x})^2}{4\alpha_L\overline{x}} = 1. \quad (8.2.6)$$

Therefore, the distance between the locations of the peak concentration $C_{\max}e^{-1}$ is

$$\Delta x = x - \overline{x} = 4\sqrt{\alpha_L\overline{x}}. \quad (8.2.7)$$

Consequently, the dispersivity, [L], is

$$\boxed{\alpha_L = \frac{\Delta x^2}{16\overline{x}}} \quad (8.2.8)$$

*8.2.1.2 Breakthrough Curves (*C – t *data)*

If a concentration breakthrough at $x = L$ from a release of a pulse of a tracer is collected under a steady-state flow condition, Fig. 8.2 illustrates the corresponding breakthrough curve as the red line.

We select two concentrations from the BTC, where the concentration is equal $C_{\max}e^{-1}$ and determines the time corresponding to the two points t_1 and t_2. At time $= t_1$, the analytical solution for the concentration is

$$C_{\max}e^{-1} = C_{\max}e^{-(\mathrm{L}-ut_1)^2/4\alpha_L ut_1}. \quad (8.2.9)$$

That is,

$$1 = (L-ut_1)^2/4\alpha_L ut = (u\overline{t}-ut_1)^2/4\alpha_L ut_1 = u^2(\overline{t}-t_1)^2/4\alpha_L ut_1 \quad (8.2.10)$$

in which $\overline{t}$ stands for the averaged arrival time of the breakthrough curve:

$$\overline{t} = \int_0^\infty tC(x,t)dt. \quad (8.2.11)$$

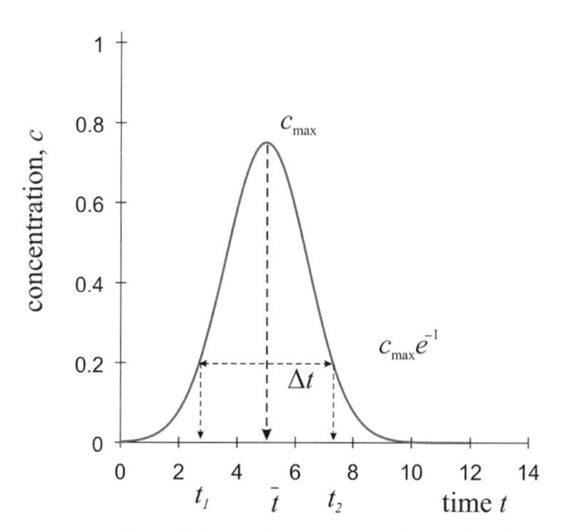

Figure 8.2 A concentration breakthrough curve at the end of a soil column with a length of L.

The average arrival time may not be the same as the arrival time of the maximum concentration at L. (since this is a breakthrough, not a snapshot). Let $\Delta t = 2|\bar{t} - t_1|$ in Eq. (8.2.10). We have

$$1 = \frac{u\Delta t^2}{16\alpha_L t_1} = \frac{\Delta t^2}{16\alpha_L t_1 \bar{t}} u\bar{t}.$$

This leads to the following formula for dispersivity

$$\alpha_L = \frac{1}{16}\frac{\Delta t^2}{t_1 \bar{t}} L. \tag{8.2.12}$$

This approach is merely an approximation unless L is sufficiently long, so the BTC approaches a normal distribution. In other words, this approach assumes that the maximum concentration's arrival time coincides with the tracers' mean arrival time. Note that this limitation has nothing to do with the non-Fickian transport. As remarked in Chapter 4, at a fixed location, the solute arrival is a combined effect of the groundwater velocity and the spread of the solute. As a result, BTC is always non-symmetrical. Fitting the analytical solution to the entire breakthrough curve is more desirable.

8.2.2 Step Input

Consider the case where input concentration is a step input under a steady flow in the column. Specifically, a tracer solute of a given concentration C_0 is continuously

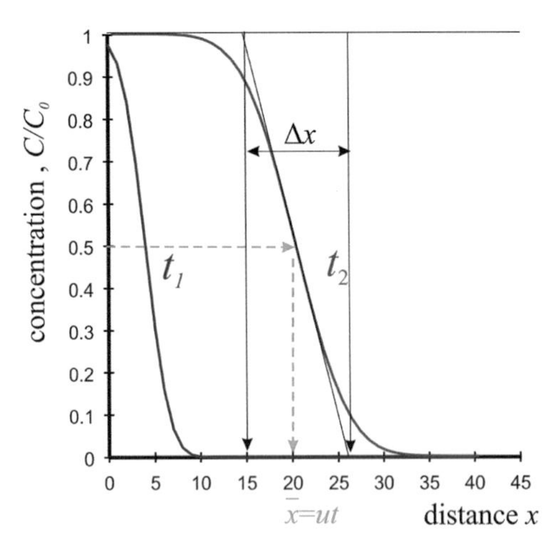

Figure 8.3 The snapshot of a step input to a soil column.

released at the front end of the soil column, where the water flow is steady. The initial conditions are $t = 0, C = C_0, x \leq 0,$ and $C = 0, x > 0$; boundary conditions are $C = 0, x \to \infty$, and $C = C_0, x \to -\infty$. The analytical solution for this case is

$$C(x, t) = \frac{C_0}{2} erfc\left(\frac{x - ut}{\sqrt{4Dt}}\right) \tag{8.2.13}$$

where $erfc(z) = 1 - erf(z)$ is the complementary error function.

8.2.2.1 Snapshots

The snapshots of a solute plume due to this type of input at two different times are shown in Fig. 8.3, where the normalized concentration C/C_0 is illustrated as a function of distance x. The location $\bar{x}$, where C/C_0 equal to 0.5, is equivalent to the maximum concentration in the snapshot of an impulse input. That is, $\bar{x} = ut$ where t is the time the snapshot is taken. Therefore, the velocity of the groundwater low can be determined if the origin of $\bar{x}$ is the input location. Otherwise, at least two snapshots at t_1 and t_2 must be used. The velocity is then determined by

$$u = (\bar{x}_1 - \bar{x}_2)/(t_1 - t_2)$$

Estimation of the Dispersivity We choose the snapshot at t_2 as an example. We draw a straight line tangent to the curve at the location $\bar{x}$, where the concentration is 50% of the input concentration. Afterward, we determine the distance between the two intercepts created by the straight line (i.e., Δx), as illustrated in Fig. 8.3.

The reciprocal of the distance is the slope, which can be used to estimate the dispersivity. Specifically, according to Eq. (8.2.13), the slope of the straight line at $C/C_0 = 0.5$ is

$$\left[\frac{-\partial}{\partial x}\left(\frac{C}{C_0}\right)\bigg|_{x=ut}\right] = \frac{1}{\Delta x} \tag{8.2.14}$$

which yields

$$\Delta x = \sqrt{4\pi\alpha_L x} = \sqrt{4\pi\alpha_L ut}. \tag{8.2.15}$$

Then, dispersivity is

$$\boxed{\alpha_L = \frac{\Delta x^2}{4\pi\overline{x}}} \tag{8.2.16}$$

Below details the procedure for finding the slope of the straight line.

$$\frac{C}{C_0} = \frac{1}{\sqrt{4\pi Dt}} \int_{\xi=-\infty}^{0} e^{\frac{-(x-\xi-ut)^2}{4Dt}} d\xi \tag{8.2.17}$$

$$\begin{aligned} \frac{\partial C/C_0}{\partial x} &= \frac{1}{\sqrt{4\pi Dt}} \int_{-\infty}^{0} \left(\frac{-2(x-\xi-ut)}{4Dt} e^{\frac{-(x-\xi-ut)^2}{4Dt}}\right) d\xi \\ \frac{\partial C/C_0}{\partial x}\bigg|_{x=ut} &= \frac{1}{\sqrt{4\pi Dt}} \int_{-\infty}^{0} \left(\frac{2\xi}{4Dt} e^{\frac{-\xi^2}{4Dt}}\right) d\xi \end{aligned} \tag{8.2.18}$$

$$\frac{1}{\sqrt{4\pi Dt}}\left(-\int_{-\infty}^{0} e^{\frac{-\xi^2}{4Dt}} \frac{d\xi^2}{4Dt}\right) = \frac{1}{\sqrt{4\pi Dt}}(-1)e^{\frac{-\xi^2}{4Dt}}\bigg|_{-\infty}^{0} = \frac{-1}{\sqrt{4\pi Dt}} \tag{8.2.19}$$

8.2.2.2 Breakthrough Curves

Consider a breakthrough curve collected at a location, x, as shown in Fig. 8.4. To estimate the velocity, we first find the 50% level of C/C_0 position of the curve to determine $t_{0.5}$ (the averaged arrival time of the breakthrough). If the origin of x is zero, the velocity is determined $u = x/t_{0.5}$. Afterward, draw a line tangent to the $t_{0.5}$ position, intercepting 100% and 0 % of C/C_0, and then determine Δt: the time lapse between the two intercepts.

Mathematically, this Δt is

$$\Delta t = \left(\frac{\partial C/C_0}{\partial t}\bigg|_{x=ut}\right)^{-1} \tag{8.2.20}$$

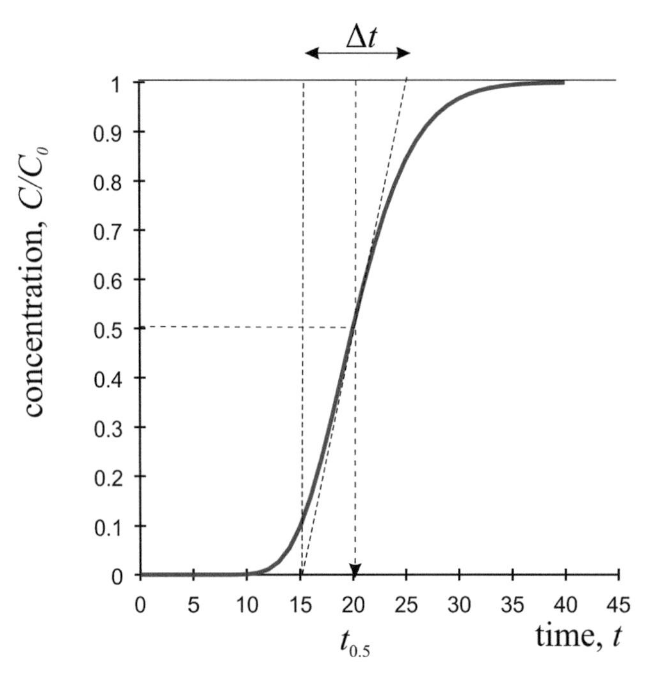

Figure 8.4 A breakthrough curve at x due to the release of a step input tracer source at $x = 0$.

Details of the procedure deriving Eq. (8.2.20) is given below:

$$\frac{dC/C_0}{dt} = \frac{x}{\sqrt{4\pi D t^3}} \exp\left[-\frac{(x-ut)^2}{4Dt}\right]. \tag{8.2.21}$$

If $x = ut$, the slope of the line is

$$\frac{1}{\Delta t} = \frac{dC/C_0}{dt}\Big|_{x=ut} = \frac{x}{\sqrt{4\pi D t_{0.5}^3}} = \frac{x}{\sqrt{4\pi \alpha u t_{0.5}^3}} = \frac{x}{\sqrt{4\pi \alpha x t_{0.5}^2}}. \tag{8.2.22}$$

Therefore,

$$\Delta t = \sqrt{\frac{4\pi\alpha x}{u^2}}. \tag{8.2.23}$$

Then, we have

$$\alpha_L = \frac{1}{4\pi}\frac{(u\Delta t)^2}{x} = \frac{1}{4t}\frac{u^2\Delta t^2}{u t_{0.5}} = \frac{u t_{0.5}}{4\pi}\frac{\Delta t^2}{t_{0.5}^2} = \frac{x}{4\pi}\left(\frac{\Delta t}{t_{0.5}}\right)^2. \tag{8.2.24}$$

This graphical method is also limited to the tracer breakthrough at long distances from the source when the breakthrough curve exhibits a cumulative normal distribution. Again, this limitation has nothing to do with non-Fickian transport. The arrival of the tracer at the observation location is a combined effect of advection and dispersion.

8.3 Spatial Moments

Another approach uses the snapshot or breakthrough curve's spatial or temporal moments (Aris, 1958).

8.3.1 The First Spatial Moment (Spatial Mean Position)

The mean position of a plume μ_x is defined as

$$\mu_x = \int_{-\infty}^{\infty} xf(x)dx = \int_{-\infty}^{\infty} xC(x,t)dx \Big/ \int_{-\infty}^{\infty} C(x,t)dx \tag{8.3.1}$$

where $f(x)$ is the probability density function, and this function is not restricted to a normal distribution and is the snapshot $C(x)$ at a given time t.

Mathematical Proof First, we assume a slug of a chemically inert tracer is released at $x = 0$. We further assume that the flow is one-dimensional with infinite lateral boundaries, and solute migration has reached the Fickian regime. With these assumptions, the governing solute transport equation is the ADE,

$$\frac{\partial C}{\partial t} + u\frac{\partial C}{\partial x} = D\frac{\partial^2 C}{\partial x^2} + \frac{M}{\rho A}\delta(x)\delta(t) \tag{8.3.2}$$

where $\delta(x)$ and $\delta(t)$ are Dirac delta functions in x and t, respectively. Now, we integrate Eq. (8.3.2) over x from $-\infty$ to $+\infty$. Thus,

$$\frac{\partial}{\partial t}\int_{-\infty}^{\infty} Cdx + u\int_{-\infty}^{\infty} \frac{\partial C}{\partial x}dx = D\int_{-\infty}^{\infty} \frac{\partial^2 C}{\partial x^2}dx + \frac{M}{\rho A}\delta(t). \tag{8.3.3}$$

Eq. (8.3.3) can be simplified as

$$\frac{\partial}{\partial t}\int_{-\infty}^{\infty} Cdx + uC|_{-\infty}^{\infty} = D\frac{\partial C}{\partial x}\bigg|_{-\infty}^{\infty} + \frac{M}{\rho A}\delta(t). \tag{8.3.4}$$

Notice that the second term on the left of Eq. (8.3.4) and the first term on the right of the equation are zero because the concentration and its gradient are zero at $x = \pm\infty$ (i.e., unbounded media). Next, integrating Eq. (8.3.4) for t from $-\infty$ to $+\infty$ leads to

$$\int_{-\infty}^{\infty} C dx = \frac{M}{\rho A} \tag{8.3.5}$$

which is the area under the concentration distribution and is equal to $M/\rho A$.

Next, we multiply Eq. (8.3.2) with x and integrate the result over x from $-\infty$ to $+\infty$.

$$\frac{\partial}{\partial t}\int_{-\infty}^{\infty} xCdx + u\int_{-\infty}^{\infty} x\frac{\partial C}{\partial x}cdx = D\int_{-\infty}^{\infty} x\frac{\partial^2 C}{\partial x^2}cdx + \frac{M}{\rho A}\int_{-\infty}^{\infty} x\delta(x)\delta(t)dx. \tag{8.3.6}$$

Integrating the second term on the left-hand side of Eq. (8.3.6) by parts yields

$$u\left(Cx\Big|_{-\infty}^{\infty} - \int_{-\infty}^{\infty} Cdx\right) = -u\int_{-\infty}^{\infty} Cdx. \tag{8.3.7}$$

The first term in the bracket in Eq. (8.3.7) is zero due to the unbounded domain assumption. Integration of the first term on the right-hand side of Eq. (8.3.6) yields

$$xD\frac{\partial C}{\partial x}\Big|_{-\infty}^{\infty} - \int_{-\infty}^{\infty} \frac{\partial C}{\partial x}dx = 0 \tag{8.3.8}$$

due to the unbounded domain assumption (i.e., zero concentrations at infinite distances). Then, the integral involving the input mass term in Eq. (8.3.6) vanishes because of the symmetry of concentration in x. Subsequently, we have

$$\frac{\partial}{\partial t}\int_{-\infty}^{\infty} xCdx - u\int_{-\infty}^{\infty} Cdx = 0. \tag{8.3.9}$$

Then, Eq. (8.3.6) can be rearranged to

$$\frac{\partial}{\partial t}\left[\int_{-\infty}^{\infty} xCdx \Big/ \int_{-\infty}^{\infty} Cdx\right] = u \qquad \frac{\partial}{\partial t}\left[\int_{-\infty}^{\infty} xf(x)dx\right] = \frac{\partial \mu_x}{\partial t} = u \tag{8.3.10}$$

Eq. (8.3.10) proves that the rate of change in the solute mass center in the x-direction μ_x is the velocity, u. At least two snapshots of a moving tracer plume are required or one snapshot if the location of the source of the tracer is known to determine the velocity. Note that the location of the center of mass, u_x, is the first moment. Notice that $f(x)$ is not required to be a normal distribution.

8.3.2 The Second Spatial Moment (Spatial Variance)

The second spatial moment is the spatial variance of a snapshot of a tracer plume at a given time t, σ_x^2, which is a statistical measure of the spread of the plume:

$$\sigma_x^2(t) = \int_{-\infty}^{\infty} (x-\mu_x)^2 f(x)dx = \int_{-\infty}^{\infty} (x-\mu_x)^2 C(x, t)dx \Big/ \int_{-\infty}^{\infty} C(x, t)dx \tag{8.3.11}$$

Let's define a new variable $\xi = x - ut = x - \mu_x$; it is the distance between x and the center of the plume and is the advective coordinate or the spread of a plume with respect to the center location of the plume. After this transformation, Eq. (8.3.11) becomes

$$\sigma_\xi^2(t) = \int_{-\infty}^{\infty} \xi^2 C(\xi, t)d\xi \Big/ \int_{-\infty}^{\infty} C(\xi, t)d\xi. \tag{8.3.12}$$

Likewise, using the advective coordinate system, the original ADE Eq. (8.3.2) without the source term is reduced to a dispersion equation:

$$\frac{\partial C}{\partial t} = D\frac{\partial^2 C}{\partial \xi^2}. \tag{8.3.13}$$

Multiplying both sides of Eq. (8.3.13) with the advective coordinates ξ, we have

$$\int_{-\infty}^{\infty} \frac{\partial C}{\partial t} \xi^2 C d\xi = D\int_{-\infty}^{\infty} \xi^2 \frac{\partial^2 C}{\partial \xi^2} d\xi. \tag{8.3.14}$$

Integration of the right-hand side of Eq. (8.3.14) leads to

$$D\left(\xi^2 \frac{\partial C}{\partial \xi}\Big|_{-\infty}^{\infty} - \int_{-\infty}^{\infty} 2\xi \frac{\partial C}{\partial \xi} d\xi \right) = -2D\left(C\xi|_{-\infty}^{\infty} - \int_{-\infty}^{\infty} C d\xi \right) = 2D\left(\int_{-\infty}^{\infty} C d\xi \right). \tag{8.3.15}$$

Substituting Eq. (8.3.15) into the right-hand side of Eq. (8.3.14) and rearrangement, Eq. (8.3.14) takes a new form

$$\frac{\partial}{\partial t}\left(\int_{-\infty}^{\infty} \xi^2 C d\xi \Big/ \int_{-\infty}^{\infty} C d\xi \right) = 2D. \tag{8.3.16}$$

From Eqs. (8.3.12) and (8.3.16), we find that

$$\frac{\partial \sigma_\xi^2}{\partial t} = 2D = 2u\alpha_L. \tag{8.3.17}$$

That is, the rate of the spatial spread of the plume is two times the dispersion coefficient if the ADE describes the solute movement (i.e., Fick's law is valid). Specifically, relating the rate of change in the spatial variance (spread) to the dispersion requires that the concentration distribution obeys the ADE. Otherwise, the spatial moment approach only approximates the dispersivity or dispersion coefficient in ADE.

For discrete cases, the spatial moments can be written as follows. The center of the plume at a given time, t, is

$$\bar{x} = \left(\sum_{i=1}^{n} x_i C_i\right) \Big/ \left(\sum_{i=1}^{n} C_i\right). \tag{8.3.18}$$

The spatial variance of the plume at a given time, t, is

$$\sigma_\xi^2 = \left(\sum_{i=1}^{n} (x_i - \bar{x})^2 C_i\right) \Big/ \left(\sum_{i=1}^{n} C_i\right) = \left(\sum_{i=1}^{n} x_i^2 C_i\right) \Big/ \left(\sum_{i=1}^{n} C_i\right) - \bar{x}^2 \tag{8.3.19}$$

where C_i is the concentration at the location x_i at a given time, t; n is the total number of sample locations in x.

For example, we have three snapshots of a plume at time = 1, 4, and 8 (Fig. 8.5), which was released at $x = -10$ at time zero. They are created by solving Eq. (8.3.2)

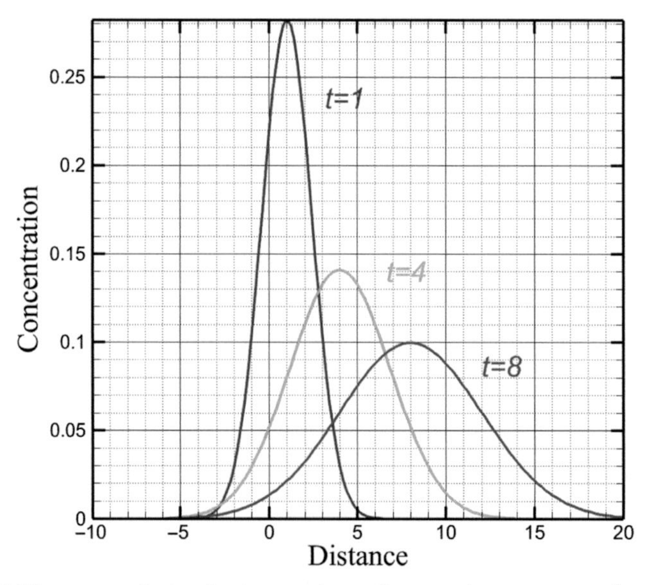

Figure 8.5 Three snapshots of a tracer plume from an instantaneous input.

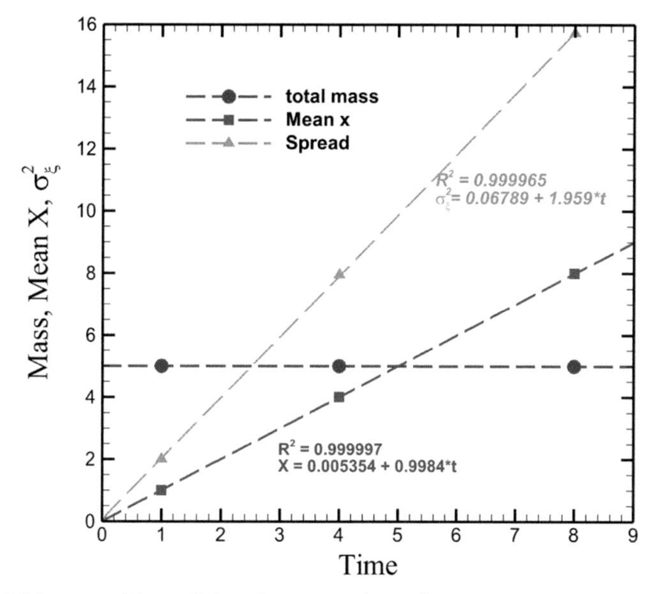

Figure 8.6 Mean position of the plume vs. time plot.

with $\alpha = 1$, and $u = 1$. Using Eq. (8.3.18), we calculate the total mass and the locations of the center of the mass at the three times, and we plot them as a function of time in Fig. 8.6 (red circle with a line and blue square with a line, respectively). It shows that the mass under the three snapshots remains the same (conservation of mass); the locations at the three times form a straight line with a slope of 1, consistent with $u = 1$. Similarly, we use Eq. (8.3.19) to determine the spatial variance of the three snapshots and they are shown in Fig. 8.6, as a function of time as orange triangles. The three spatial variances form a straight line, indicating that the plume behaves according to Fick's law. The slope of this line is twice the dispersion coefficient (D = 1), and dispersivity $\alpha = 1$ is consistent with $D = \alpha u$.

We emphasize that the spatial moment calculation is a method for finding the location of the mass center and the extent of the spread of the plume, independent of any assumption of the concentration distribution. However, as we apply the results of the moment analysis to determine dispersivity, we inevitably assume the validity of Fick's law and the normality of the distribution.

8.4 Temporal Moments

Collecting snapshots of a plume, even in a soil column, is challenging in many situations. Collecting breakthroughs at the end of a soil column is relatively straightforward and common practice for estimating solute transport parameters. The temporal moment has become a valuable tool for analyzing velocity and dispersivity. In effect, time moment analysis is a standard chemical engineering procedure for determining dispersive and rate parameters for packed bed reactors (Suzuki and Smith, 1971; Turner, 1972; Fahim and Wakao, 1982).

8.4.1 First Temporal Moment (Temporal Mean)

Consider concentration breakthrough data observed at a location, x. Using the first temporal moment, we can determine $\bar{t}$ or μ_t (i.e., the mean arrival time of the BTC at the observation location x). The first temporal moment formula is given below.

$$\bar{t} = \mu_t = \int_{-\infty}^{\infty} tf(t)dt = \left[\int_{-\infty}^{\infty} tC(x,\ t)dt\right] \Big/ \left[\int_{-\infty}^{\infty} C(x,\ t)dt\right]. \tag{8.4.1}$$

The numerator is the sum of the product of the arrival time of a solute concentration and the solute concentration. The denominator, the total mass of solute in a breakthrough curve (or total mass), then normalizes the numerator to obtain the first moment (the mean or average arrival time). If dispersion or diffusion is absent, all solutes from an impulse input arrive at x simultaneously due to uniform advection velocity. In this case, the mean arrival time $\bar{t}$ is identical to the arrival time of the peak concentration, t_{peak}. If the source location is known and the flow is one-dimensional, the advection velocity due to the flow can be determined $(u = x/\bar{t})$ or $(u = x/t_{\text{peak}})$.

On the other hand, if diffusion or dispersion exists, some solutes will arrive at x earlier or later, creating an asymmetric distribution. Consequently, the arrival time of the peak of the BTC of an impulse input differs from the arrival time of solute due to advection only $(t = x/u)$. Specifically, the arrival time of the peak of a BTC should not be used to estimate the advection velocity, u.

Fig. 8.7 shows the simulated BTC with $\underline{u} = 1$ and $\alpha = 1$ at $x = 5$, 10, and 20. It also displays the peak arrival, advection arrival, and mean arrival time of the BTC at $x = 10$ as indicated by the black dashed dot, blue solid, and orange long dashed lines, respectively. The peak arrives earlier than the advection only because it reflects the effects of advection and dispersion. The mean arrival time is the last.

Table 4 in Kreft and Zuber (1978) shows an analytical formula for the first temporal moment of a solute BTC from an impulse input described by the ADE. The mean arrival time based on Eq. (8.4.1) is

$$\bar{t} = \frac{x}{u}\left(1 + 2\frac{D}{ux}\right) = \frac{x}{u}\left(1 + 2\frac{\alpha}{x}\right). \tag{8.4.2}$$

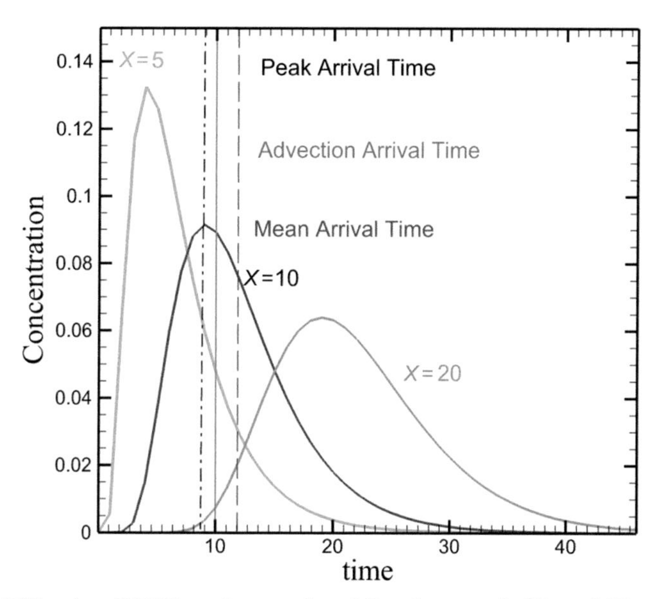

Figure 8.7 Simulated BTCs, using $u = 1$ and $D = 1$, at $x = 5$, 10, and 20 to illustrate the differences in the peak, advection, and mean arrival times.

In Eq. (8.4.3), the advection velocity is u and x/u is the arrival time of the BTC due to advection only. This equation shows that the mean arrival time of an asymmetric BTC $\bar{t}$ is not x/u. As remarked in Chapter 4, the arrival of the BTC at a fixed location includes the effect of dispersion, leading to the early rapid rise of the peak and long tailing (i.e., asymmetric distribution). As evident from Eq. (8.4.3), the mean arrival time $\bar{t}$ approaches x/u if the BTC is observed at a large distance x downstream when the dispersion term $D/(ux)$ in Eq. (8.4.2) becomes negligible or the dispersion and diffusion are negligible, or zero.

The above discussion suggests that neither u nor D can be estimated from the mean arrival time of a BTC unless an independently measured u is available (e.g., a flow experiment). We notice that $u\bar{t} = x + 2\alpha$ in Eq. (8.4.2). This relationship suggests that BTCs at two distances would determine u and α values for a soil column tracer experiment by solving a system of equations.

8.4.2 The Second Temporal Moment (Temporal Variance)

Eq. (8.4.3) below is the formula to calculate the second temporal moment of the BTC:

$$\sigma_\tau^2(x_1) = \int_{-\infty}^{\infty} (t-\mu_t)^2 f(t)dt = \left[\int_{-\infty}^{\infty} (t-\mu_t)^2 C(x_1, t)dt\right] \Big/ \left[\int_{-\infty}^{\infty} C(x_1,\ t)dt\right]. \tag{8.4.3}$$

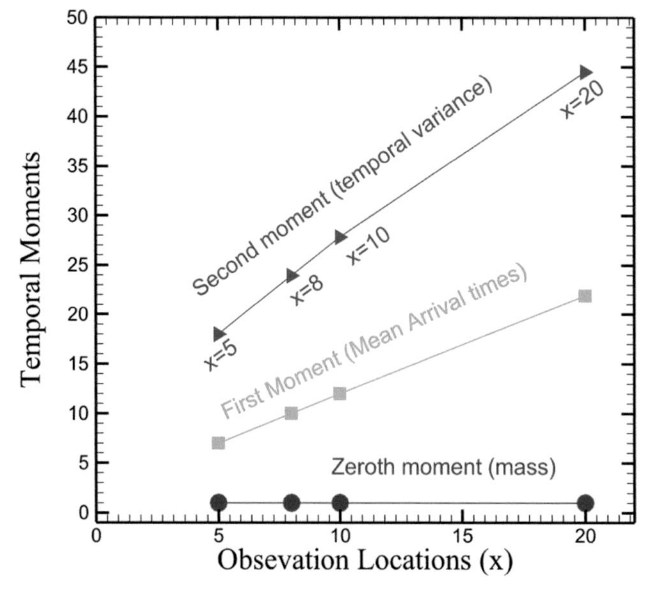

Figure 8.8 The zeroth, first, and second moments of the BTCs at x = 5, 10, and 20 as a function of time, as illustrated in Fig. 8.7. The BTC at x = 8 is omitted.

If the concentration conforms to the solution of the ADE, Kreft and Zuber (1978) showed that its second temporal moment is

$$\sigma_t^2 = \left(\frac{x}{u}\right)^2 \left(2\frac{D}{ux} + 8\left(\frac{D}{ux}\right)^2\right) = \left(\frac{x}{u}\right)^2 \left(2\frac{\alpha}{x} + 8\left(\frac{\alpha}{x}\right)^2\right). \tag{8.4.4}$$

Even though the ADE prescribes the solute transport process, the temporal variance is not linearly proportional to the dispersion coefficient, as is the spatial variance. Again, if the advection arrival time at a given distance x, x/u, is known from an independent measure of u, one can solve the nonlinear Eq. (8.4.4) to derive the dispersivity. Second temporal moments of BTCs from different locations will be helpful.

For discrete cases, the temporal moments can be estimated as follows:

$$\bar{t} = \sum_{i=1}^{n} (t_i C_i) \Big/ \left(\sum_{i=1}^{n} C_i\right). \tag{8.4.5}$$

$$\sigma_T^2 = \left(\sum_{-\infty}^{\infty} (t_i - \bar{t})^2 C_i\right) \Big/ \left(\sum_{i=1}^{n} C_i\right) = \left(\sum_{i}^{n} t_i^2 C_i\right) \Big/ \left(\sum_{i=1}^{n} C_i\right) - \bar{t}^2. \tag{8.4.6}$$

The results of applying Eqs. (8.4.5) and (8.4.6) to the BTCs in Fig. 8.7 are illustrated in Fig. 8.8. The zeroth moments (mass) are constant over the observation distances. The mean arrival times multiplied with the constant velocity (u) are linear with the distance (x) with an intercept (2α), consistent with Eq. (8.4.2). The nonlinear relationship between the temporal variance and the distance is evident and consistent with Eq. (8.4.4).

In the case where the input is a step input, taking the derivative of concentration to x (the snapshots) or t (the BTC) leads to the snapshot or BTC of an impulse input. Then, the approaches presented above for an impulse input should apply.

These temporal moment methods have been applied to estimating chemical reaction parameters from soil column experiments, as presented in Valocchi (1985), Pan et al. (2003), and many others. Because it involves complicated mathematics, we will refer readers to these references. Nonetheless, the above rudimentary explanation and examples elucidate that estimating both u and α using the temporal moment method uniquely would require at BTCs at least at two different locations in a soil column experiment, in addition to the ergodicity requirement.

We reiterate that spatial or temporal moment methods for estimating velocity and dispersivity require the snapshot or BTC to obey the ADE. If this requirement is met, the estimates are optimal. Otherwise, they are the best we can get. Nonetheless, they are a quantitative means to characterize the evolution of a plume in time.

8.5 Method of Moments for Multidimensional Plumes

The moment approach has been explored in the previous two sections for estimating the velocity and dispersivity during solute transport under one-dimensional steady-state flow. In real-world aquifers, the flow is multidimensional, and so is the solute migration. Consequently, multidimensional spatial moments must be considered.

8.5.1 Three-Dimensional Plumes

Aris (1958) quantified the migration and spread of the plume at different times, using the spatial moments of $C(x,y,z,t)$

$$M_{ijk}(t) = \int_{-\infty}^{+\infty}\int_{-\infty}^{+\infty}\int_{-\infty}^{+\infty} \Delta C(x, y, z, t) x^i y^j z^k dxdydz \tag{8.5.1}$$

where ΔC denotes the concentration after subtracting the background. The zeroth, first and second spatial moments correspond to the sum of either subscripts or superscripts $(i + j + k)$, equal to 0, 1, and 2, respectively. The zeroth moment (M_{000}) represents the changes in the solute concentration within the domain under investigation. The subscripts denote i, j, and k.

The normalized first spatial moments are

$$x_c = M_{100}/M_{000},\ y_c = M_{010}/M_{000},\ \text{and}\ z_c = M_{001}/M_{000}. \tag{8.5.2}$$

They represent the location (x_c, y_c, z_c) of the mass center of the plume at a given time. The spread of the plume about its center is described by the symmetric second spatial variance tensor:

$$\mathbf{\Sigma}^2 = \begin{bmatrix} \sigma_{xx}^2 & \sigma_{xy}^2 & \sigma_{xz}^2 \\ \sigma_{yx}^2 & \sigma_{yy}^2 & \sigma_{yz}^2 \\ \sigma_{zx}^2 & \sigma_{zy}^2 & \sigma_{zz}^2 \end{bmatrix} \tag{8.5.3}$$

where

$$\sigma_{xx}^2 = \frac{M_{200}}{M_{000}} - x_c^2,\ \sigma_{yy}^2 = \frac{M_{020}}{M_{000}} - y_c^2,\ \sigma_{zz}^2 = \frac{M_{002}}{M_{000}} - z_c^2 \tag{8.5.4}$$

$$\sigma_{xy}^2 = \sigma_{yx}^2 = \frac{M_{110}}{M_{000}} - x_c y_c \quad \sigma_{xz}^2 = \sigma_{zx}^2 = \frac{M_{101}}{M_{000}} - x_c z_c \quad \sigma_{yz}^2 = \sigma_{zy}^2 = \frac{M_{011}}{M_{000}} - y_c z_c \tag{8.5.5}$$

Note that the symmetry of the off-diagonal terms assumes the plume's ergodicity–the plume must encounter sufficient heterogeneity at different scales and reach the Fickian regime. The ergodicity of a solute plume in real-world aquifers may not hold (see Chapter 10).

8.5.2 Two-Dimensional Plumes

Suppose a two-dimensional domain is discretized into m nodes ($m = n \times k$ grid points). The first spatial moment (center of mass) of a tracer plume at a given time can be determined by

$$\begin{aligned} \overline{x} &= \left(\sum_{i=1}^{m} x_i C_i\right) \bigg/ \left(\sum_{i=1}^{m} C_i\right) \\ \overline{y} &= \left(\sum_{i=1}^{m} y_i C_i\right) \bigg/ \left(\sum_{i=1}^{m} C_i\right) \end{aligned}. \tag{8.5.6}$$

C_i is the concentration of the tracer at node i. The second spatial moments (spread of mass) of a tracer plume can be determined by

$$\sigma_{xx}^2 = \left(\sum_{i=1}^{m} x_i^2 C_i\right) \bigg/ \left(\sum_{i=1}^{m} C_i\right) - \overline{x}^2 \tag{8.5.7}$$

$$\sigma_{yy}^2 = \left(\sum_{i=1}^{m} y_i^2 C_i\right) \Big/ \left(\sum_{i=1}^{m} C_i\right) - \bar{y}^2 \tag{8.5.8}$$

$$\sigma_{xy}^2 = \sigma_{yx}^2 = \left(\sum_{i=1}^{m} x_i y_i C_i\right) \Big/ \left(\sum_{i=1}^{m} C_i\right) - \overline{xy}. \tag{8.5.9}$$

These moment approaches have been used in the analysis of solute plumes at Borden (Sudicky, 1986), Cap Cod (Garabedian et al., 1991), and Mississippi experiment sites (Adams and Gelhar, 1992). The moment analysis results reflect only the overall plume behavior, not the detailed shape of the plume. Furthermore, suppose the estimating average velocity and dispersivity are the objectives. The tracer observations' x, y, and z coordinates should align with the mean flow direction. Otherwise, a coordinate transformation to the flow directions must be undertaken. Because of the ergodicity assumption ingrained in the moment analysis, it should not surprise that the off-diagonal terms may not completely vanish. Finally, applications of the moment methods to field multidimension problems are presented in Chapters 10 and 11.

8.6 Least Square Nonlinear Regression Approach

Lastly, the most convenient but least intuitive approach for estimating the parameters is the least-squares regression. It uses an automatic algorithm to search for the parameter values that best fit the analytical or numerical solutions of the ADE to the snapshot or BTC data. Specifically, it minimizes the differences between the data and the solutions in the least-squares sense. That is,

$$\sum_{i=1}^{N} \left(C_i - \widehat{C}_i\right)^2 = \text{minimal}. \tag{8.5.10}$$

In the equation C_i and $\widehat{C}_i$ represent the numerical or analytical solutions and the observed data, respectively, at different time (BTC) or space (snapshot) intervals (from $i = 1$, to N, the total number of intervals). The nonlinear regression algorithm is available in EXCEL (solve package) as introduced in Homework 1.2 or other commercially available software (e.g., Mathlab or Mathematica). Cautions must be taken when applying this approach to a problem. The approach may lead to non-unique or unrealistic parameters, as manifested in Eqs. (8.4.2) and (8.4.4) showing the mutual dependence of velocity and dispersivity, even if the porosity value is known for the velocity determination. A set of initial guess values close to their true values could avoid this problem. The simple graphic methods discussed previously serve this purpose. Predetermining the specific discharge, porosity, and velocity values from hydraulic tests and then estimating dispersivity should also minimize this problem. Likewise, estimating the velocity and dispersivity using a conservative tracer and then assessing the chemical reaction parameters using reactive tracers would be a rational approach.

8.7 Remarks

- All the methods presented assume one-dimensional flow and solute transport and the validity of ADE or Fick's law. They are, nevertheless, valuable tools for estimating dispersivity, chemical reaction parameters, and velocity in soil column experiments. In the field situations, where flow is three-dimensional and the solute migration has not reached the Fickina regime or ergodicity, applications of these tools require cautions.
- The graphic approach is less mathematics intensive and is a first-cut tool. Although it is an approximation, it yields a better understanding of the physical meanings of the snapshot or the BTC.
- The calculations of the spatial or temporal moments are not affected by the distribution of the tracer plume. It is a tool for estimating parameters, but it is also a tool for quantifying the movement and spread of a solute plume. However, interpreting the results for velocity or dispersivity requires the normality, ergodicity, and Fickian assumptions.

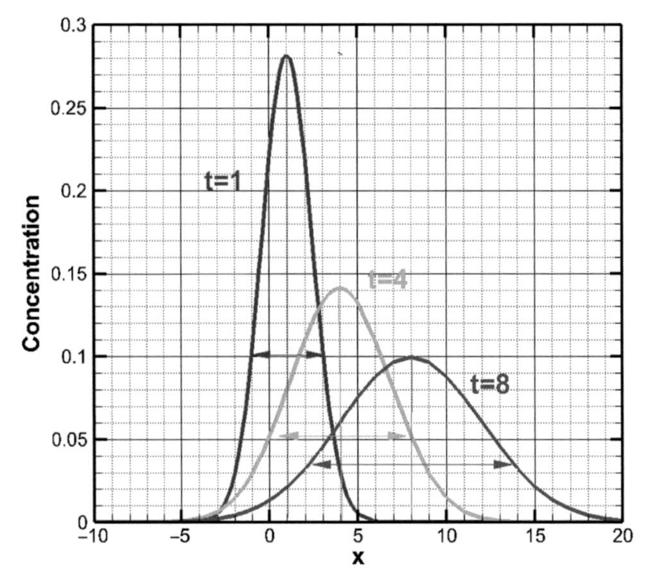

Figure 8.9 Concentration snapshots of an instantaneous input for problem 1.

- All methods presented in this chapter adopt the ADE – an ensemble mean equation – for describing solute transport in soil columns, which may not fully explain the solute transport in one realization (the outcome from one soil column). Therefore, a perfect match between the observed data and the ADE solution is unlikely despite measurement errors or noises.

8.8 Homework

1. Based on Fig. 8.9 and assuming one-dimensional flow and transport, determine the velocity and dispersivity.
2. If the concentration profiles of a step input are available, apply the spatial moment approach to determine the velocity and dispersivity (Hint, find the slope of the profile and plot it).

9

Solute Transport in Field-Scale Aquifers

9.1 Introduction

Diffusion from molecular-scale velocity variation (i.e., random motion) is explicated in Chapter 4. Chapter 6 attributes the dispersal of a tracer plume in a pipe to fluid-dynamic- and molecular-scale velocity variations. Afterward, Chapter 7 exposes the dominant role of the pore-scale velocity variation in spreading a tracer plume in a soil column in addition to the molecular-, and fluid-dynamic-scale velocity variations. As we analyze solute transport at field scales, Velocity variations due to hydraulic heterogeneity at CVs larger than molecules, fluids, and pores likely dictate the spread of the solute. Such heterogeneity may include cracks, laminations, and aggregates of clays, silts, sands, pebbles, and boulders exhibiting as layers, stratifications, and formations in addition to fissures, fractures, folds, and faults. Specifically, multiscale hydraulic property variability would play a significant role in transporting solutes in field-scale aquifers. Moreover, our observation, interest, and model scales are much smaller than the dominant heterogeneity in field-scale aquifers. For these reasons, quantifying the multiscale hydraulic heterogeneity in aquifers and their effects on solute transport becomes the primary task of this chapter.

Using spatial statistics, based on the statistics presented in Chapter 1, we first explain how to quantify spatial variability of hydraulic properties or parameters in the aquifer. Afterward, we present a highly parameterized heterogeneous media (HPHM) approach for simulating flow and solute transport in aquifers with spatially varying hydraulic properties to meet our interest and observation scale. Our limited ability to collect the needed hydraulic properties and information for this approach promotes alternatives such as zonation and equivalent homogeneous media (EHM) with macrodispersion approaches. Notably, this chapter will detail the EHM with macordispersion concept since it is pragmatic and has been widely used in many field-scale tracer experiments, as discussed in Chapter 10.

9.2 Stochastic Representation of Heterogeneity

Analysis of solute transport in field-scale aquifers or groundwater basins differs from in soil columns. The flow in soil columns is often regarded as one-dimensional and uniform because of our model, interest, and observation scales. On the other hand, flow in a groundwater basin at the scale of our interest and observation and model is multidimensional and nonuniform.

These differences demand a spatially distributed multidimensional flow model, conceptualizing a basin as a collection of many soil columns (or blocks) in a three-dimensional space. Under this conceptualization, each block may have a different hydraulic property. Blocks of similar hydraulic properties may connect, aggregate, and distribute over various basin parts, manifesting as laminations, layers, stratifications, formations, faults, or fold zones. In other words, this conceptualization views an aquifer or a basin as a pool of spatially varying hydraulic properties at multiple scales.

The complexity of the heterogeneity in aquifers and our limited sampling ability compels us to adopt the stochastic concept presented in Chapter 1 to describe and characterize them. We refer readers unfamiliar with the statistics, random variables and stochastic terminology to Chapter 1. Specifically, the spatial distribution of the hydraulic property (such as hydraulic conductivity K or specific storage S_s) is presented as a stochastic or random field as a function of the spatial coordinates, $\mathbf{x}$, characterized by the JPDs. To illustrate the stochastic representation approach, we use a K field as an example. We first convert the K to its natural logarithm (ln K), and then express it as the sum of its mean and perturbation.

$$\ln K(\mathbf{x}, \omega) = F(\mathbf{x}, \omega) = \overline{F} + f(\mathbf{x}, \omega) \tag{9.2.1}$$

in which $\mathbf{x}$ is the location vector (x, y, z), ω is the realization index, ranging from 1 to ∞. In Eq. (9.2.1), $\overline{F}$ is the ensemble mean, a constant in space and the ensemble. The perturbation is $f(\mathbf{x}, \omega)$ with a zero mean and is characterized by a normal JPD with given covariance (i.e., variance and autocorrelation function, Chapter 1), although normality is unnecessary. From now on, we will drop $(\mathbf{x}, \omega)$ and use f for convenience. Note that using the natural logarithm of the parameter avoids negative values of the parameters. Nevertheless, this book uses K for hydraulic conductivity for presentations unless it is necessary to specify it as ln$K$.

Extensive coverage is available in Yeh et al. (2015) on the stochastic representation of hydraulic heterogeneity and the flow analysis in multidimensional, heterogeneous groundwater basins. The following sections focus on the principles and physical meanings of the spatial random field's autocorrelation function, relevant to understanding the solute transport in heterogeneous groundwater basins.

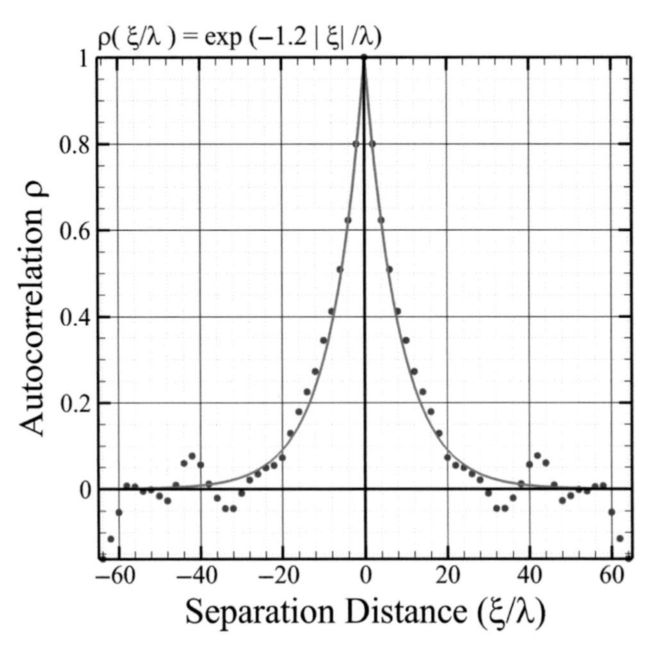

Figure 9.1 Autocorrelation function or autocorrelogram for the random field $K(x)$. The red dots are sample autocorrelation values, and the solid line represents a theoretical autocorrelogram. They all are symmetrical about $|\xi/\lambda| = 0$.

9.2.1 Autocorrelation Function

Recall that the autocorrelation analysis in Chapter 1 reveals periodicity in a time series. We may ask what it reveals as we apply the same analysis to a spatial series (e.g., the hydraulic conductivity K in heterogeneous geologic media). Specifically, suppose we plot the correlation coefficients of a K spatial series as a function lag (ξ) (the distance between two data points in space) and find the autocorrelation function or autocorrelogram behave in a manner as shown in Fig. 9.1. The physical meaning carried by autocorrelogram about the spatial variability of K becomes the question. Before answering this equation, we explore the property of the autocorrelation function.

All autocorrelation functions have the following properties:

(1) $\rho(0) = 1$(when nugget effect is not considered)
(2) $|\rho(\xi)| \leq 1$
(3) If $K(x)$ is a real value process, then $\rho(-\xi) = \rho(\xi)$, for all ξ. That is $\rho(\xi)$ is always an even function or symmetric about $\xi = 0$.

The hydraulic conductivity's spatial autocorrelation function generally exhibits an exponential decay as the separation distance increases in a seemly homogeneous formation. As such, the mathematical exponential model can quantify its behavior:

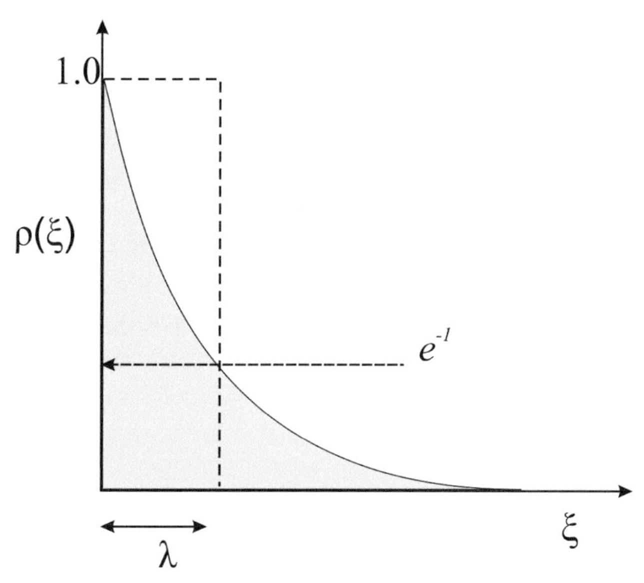

Figure 9.2 The exponential autocorrelation function of a hydraulic conductivity field and an integral scale for the exponential function are defined.

$$\rho(\xi) = \exp\left[-\frac{|\xi|}{\lambda}\right] \tag{9.2.2}$$

where ξ denotes the separation distance, and λ is the correlation scale (or integral scale). In the following, we will use a one-dimensional model as an example to define correlation and integral scales and discuss their meanings.

A **correlation scale** is a characteristic length (or the separation distance) beyond which the correlation between data is insignificant, or the correlation value becomes zero.

An integral scale (λ) is the yellow area under an autocorrelation function if the area is positive and nonzero (Fig. 9.2). For example, if an exponential autocorrelation function is considered (Fig. 9.2), the integral scale is given as

$$\int_0^\infty \rho(\xi)d\xi = \int_0^\infty e^{-\frac{|\xi|}{\lambda}}d\xi = -\lambda e^{-\frac{|\xi|}{\lambda}}\Big|_0^\infty = \lambda \tag{9.2.3}$$

when $\xi = \lambda$, the autocorrelation is

$$\rho(\lambda) = e^{-1} = 0.3678. \tag{9.2.4}$$

The integral scale means that if the spacing between any two samples is greater than the integral scale (i.e., $\xi > \lambda$), then the correlation of the two samples drops below 0.3678, and the two data sets are statistically uncorrelated (or independent). Notice that if the correlation scale definition (i.e., the separation distance at which the correlation is zero) is applied to the exponential correlation model, the correlation scale for the exponential model will be infinite, implying that the space series is correlated everywhere. As a result, the correlation scale definition is futile.

Since the significance of the correlation is a loosely defined concept, any arbitrary value (say, $\rho = 0.1$) can be used as the cutoff value for significance. While the integral scale provides a definite cutoff value, it may be zero or negative (e.g., autocorrelation functions with many negative values). For these reasons, we adopt 0.368 as the cutoff value for the correlation significance even if the integral scale does not exist. We then use the "correlation scale" to describe the distance beyond which any two samples are uncorrelated in the rest of the book.

To illustrate the practical implications of the correlation scale, consider a wall made of many large rocks of different sizes held together by cement at various spacing. Suppose we are blindfolded and asked to determine the average size of rocks. To do so, we touch the entire wall with two hands separated by a given distance. We then determine the number of times our two hands touch the same material (rock or cement) and record the number of times we encounter the same material. We then repeat this procedure with different separation distances. The recorded data will show that the number of the same materials decreases as the separation distance increases. In particular, if the separation distance is smaller than the average size of the rock, the number of times we touch the same material is high. The number will be low if the space between the two hands is larger than the size of the rock. Then, we graph the number of times we touched the same material as a function of the separation distance. From this graph, we can determine the average dimension of the rocks in the wall if a cutoff value for the number of times we encounter the same material is defined. This example is an analogy to the autocorrelation analysis discussed earlier. Thus, the correlation scale is a statistical measure of the average dimension of the heterogeneity in a basin.

Since geologic heterogeneity is always three-dimensional, a three-dimensional random field representation is most desirable, and a three-dimensional correlation function is necessary. For example, the following three-dimensional exponential autocorrelation function has been widely adopted for this purpose:

$$\rho(\boldsymbol{\xi}) = \exp\left\{-\left[\left|\frac{\xi_1}{\lambda_1}\right|^2 + \left|\frac{\xi_2}{\lambda_2}\right|^2 + \left|\frac{\xi_3}{\lambda_3}\right|^2\right]^{\frac{1}{2}}\right\} \tag{9.2.5}$$

where $\boldsymbol{\xi}$ is the separation vector with components ξ_1, ξ_2, and ξ_3 is the separation vector and λ_1, λ_2, λ_3, and are correlation scales in x, y, and z directions, respectively.

Synthetic two-dimensional random fields in Figs. 9.7–9.9 with different correlation scales in x and y directions demonstrate the correlation scale concept. These random fields mimic different shapes of geologic heterogeneity in outcrops. Note that the directions of the correlation scales in the figures align with the x and y coordinate system. Otherwise, coordinate rotation is needed.

Thinking Points:

- Is it critical to use different autocorrelation functions other than the exponential for the characterization? (Hint, it is a statistical description).

9.2.2 Variogram Analysis

Another popular method to characterize the spatial variability of geologic media statistically is the variogram analysis. The variogram analysis relies on an intrinsic hypothesis, which is less stringent than the second-order stationary assumption (i.e., mean and covariance are spatially invariant, Section 1.4.3). The intrinsic hypothesis assumes that even if the variance of a hydraulic conductivity field along a transect, $K(x)$, is not finite (i.e., the field is nonstationary), the variance of the first-order increments of K, $[K(x+\xi)-K(x)]$, where ξ is the separation distance, will be finite. Specifically, instead of treating the K ($\mathbf{x}$) as a stochastic process, the intrinsic hypothesis considers the difference between the pair of K separated by ξ as a new stochastic process (say, $Z(\xi)=K(x+\xi)-K(x)$). The increments $Z(\xi)$ will be a second-order stationary random field, satisfying the following requirements:

$$E[K(x+\xi)-K(x)]=m(\xi) \tag{9.2.6}$$

$$\text{var}[K(x+\xi)-K(x)]=2\gamma(\xi) \tag{9.2.7}$$

In other words, the difference between pairs of K's, separated by a given distance, ξ, will have a mean, $m(\xi)$, and variance, $2\gamma(\xi)$. This statement implies that the K field may not be a second-order stationary process, but the increment of the K field is. Specifically, the mean and variance of the increment of the K field are functions of ξ only, not x.

According to Eq. (9.2.7), a variogram or semi-variogram is

$$\gamma(\xi)=\frac{1}{2}\text{var}[K(x+\xi)-K(x)]. \tag{9.2.8}$$

This variogram should have the following properties:

(1) The value of the variogram at the origin is zero, $\gamma(0)=0$; (nugget)
(2) The values of the variogram are positive, $\gamma(\xi)>0$ when $\xi>0$, and
(3) The variogram is an even function $\gamma(\xi)=\gamma(-\xi)$.

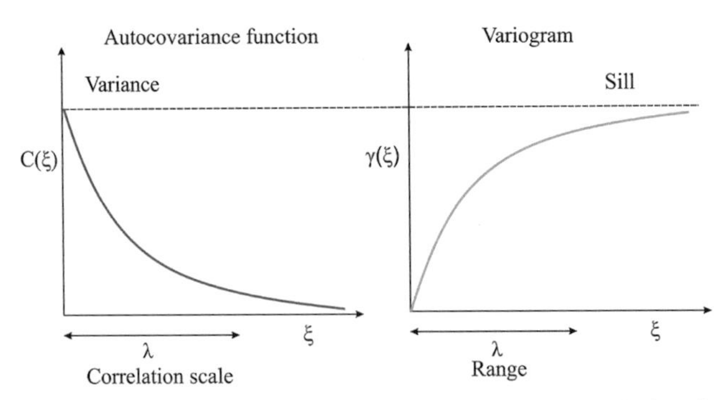

Figure 9.3 An illustration of the relation between the autocorrelation function and the variogram for the case where the stochastic process is second-order stationary (Yeh et al., 2015).

Now, if we assume

$$E[K(x+\xi) - K(x)] = m(\xi) = 0 \tag{9.2.9}$$

(i.e., the process K has a constant mean), then

$$\begin{aligned}\gamma(\xi) &= \frac{1}{2}E\left\{[K(x+\xi) - K(x)]^2\right\} \\ &= \frac{1}{2}\left[E\left\{[K(x+\xi)]^2\right\} - 2E[K(x+\xi)K(x)] + E\left\{[K(x)]^2\right\}\right].\end{aligned} \tag{9.2.10}$$

Furthermore, if the second moment of the K process is stationary, the variogram can be related to the covariance using the following equation.

$$\gamma(\xi) = C(0) - C(\xi).$$

Therefore, the variogram of a second-order stationary process mirrors the autocovariance function, as illustrated in Fig. 9.3. This figure also shows that the variance of a covariance function is equivalent to the sill of the variogram; the correlation scale in the covariance function is identical to the range in the variogram analysis.

9.2.2.1 Nested Variogram and Nugget Effect

A nested variogram showing multiple ranges and sills is a typical variogram of geologic media's multiscale heterogeneity. To illustrate the nested variogram, we present a spatial series with a correlation scale of 20 m, representing small-scale heterogeneity in Fig. 9.4a. Similarly, we use Fig. 9.4b to show a series with a correlation scale of 200 m, manifesting a large-scale spatial variability. The combination of the two series, representing a random field with two-scale spatial variability, is depicted in Fig. 9.4c. The corresponding variograms and autocorrelation functions of the three spatial series are plotted in Fig. 9.5a and b, respectively.

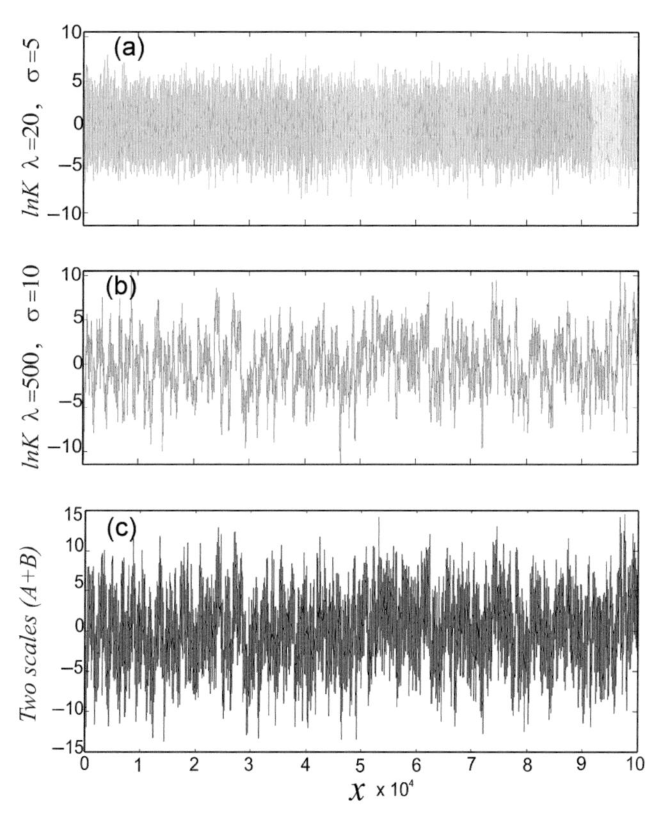

Figure 9.4 (a) The space series of the small-scale variability with a correlation scale of 20 m, (b) the space series of the large-scale heterogeneity with a correlation of 200 m. (c) the space series of the two-scale heterogeneity.

The variogram and autocorrelation function of the small-scale series are the green lines, the large-scale series are the red lines, and the combined series are the blue lines. The green variogram increases drastically after a few separation distances (20 m) and reaches a sill, while the red variogram reaches the sill after the separation distance is about 200 m. The variogram of the two-scale heterogeneity (the blue line) increases drastically at a small separation distance (20 m). It then increases parallel to the red line to reach a sill, the sum of the sills associated with the small- and large-scale series. Thus, the blue line represents the two-scale heterogeneity's nested variogram, containing the two-scale heterogeneity's spatial statistics.

As evident in Fig. 9.5a and b, an autocorrelation or variogram analysis can detect the correlation scales and sills for small- and large-scale heterogeneity of a space series with two-scale heterogeneity. Small-scale heterogeneity has a small correlation scale, while large-scale heterogeneity possesses a large correlation scale. The variance or sill of the two-scale heterogeneity is the sum of both small- and large-scale heterogeneity.

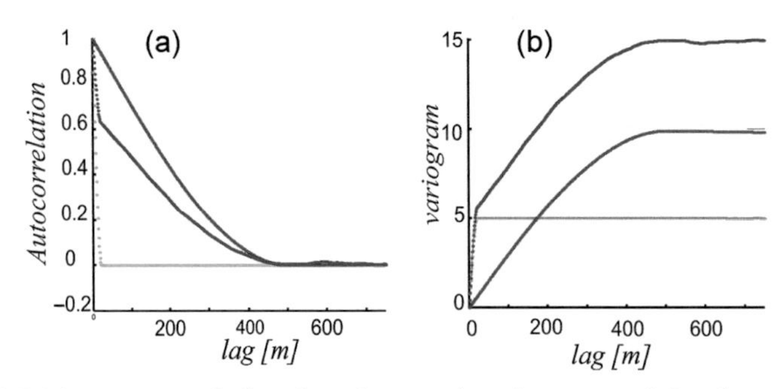

Figure 9.5 The autocorrelation functions and variograms of the three series in Fig. 9.4 (a, b, and c).

Note that the data collection spacing must be smaller than the correlation of the small-scale heterogeneity to detect the nested variogram. Otherwise, the resulting sample variogram exhibits a continuous rising of the variogram from the origin. This sudden jump is called the nugget effect. Nugget is a constant variogram (the green line) in Fig. 9.5, often employed to explain a rapid change in the value of a variogram near its origin. The nugget represents (1) measurement errors, which often are assumed independent or (2) the small-scale variability omitted by large sampling spacing.

9.2.2.2 Anisotropic Variogram

Like the autocorrelation function of hydraulic properties, the properties' variogram can have different ranges in different directions. This type of variogram is called an anisotropic variogram. Theoretically, the variograms in different directions have the same sill because the sill or variance is always a scalar.

If the variograms of a multidimensional data set have the same sill values but different range values in different directions, the data set has *geometric anisotropy*. In contrast, a data set has *zonal anisotropy* if its variogram has different sills and ranges in different directions. The geometric and zonal anisotropic variograms are illustrated in Fig. 9.6.

Zonal anisotropy results from an extensive layering structure of the data set, and the horizontal dimension of the data sampling area is much smaller than the horizontal correlation scale of the data. For example, suppose a geologic formation consists of layers of different porosity values; on average, the layers are 500 m long and 5 m thick. If porosity samples are collected from only a 50 m $\times$ 50 m area of the formation, the porosity variogram will exhibit zonal anisotropy. In this case, the vertical dimension of the sample area is ten times the vertical correlation scale (thickness). Taking the sample at a 1 m interval over the 50 m in the vertical direction, one likely collects all possible heterogeneity in the vertical.

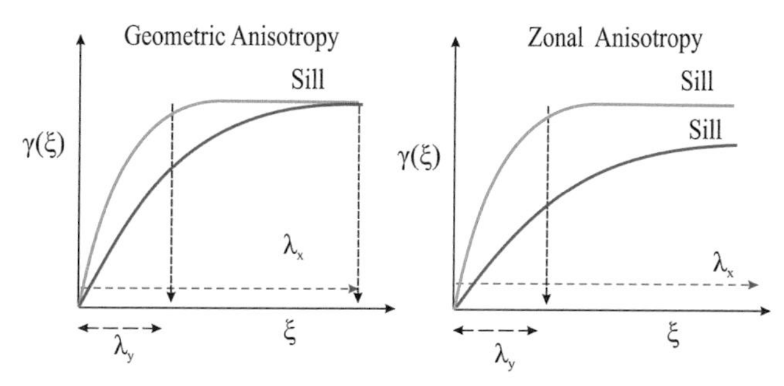

Figure 9.6 Schematic illustration of the two variograms showing the geometric and zonal anisotropies (Yeh et al., 2015).

On the other hand, the horizontal dimension of the sample area is only one-tenth of the horizontal correlation scale. Consequently, the horizontal sampling only covers correlated heterogeneity and does not include all possible heterogeneities in the horizontal direction. Specifically, the zonal anisotropy can be avoided if the sampling area extends over an area much larger than the correlation scale in all directions. Under this condition, the sill of the horizontal variogram will reach the same sill as the vertical variogram. The zonal anisotropy, in other words, is an artifact due to an improper sampling network.

9.3 Classification of Random Fields

Using the aforementioned spatial statistics concept, we can classify an aquifer as a statistically isotropic or anisotropic random field. Note that the statistical isotropy and anisotropy are different from hydraulic conductivity isotropy or anisotropy. The statistical isotropy or anisotropy defines the shape or dimensions of heterogeneity in geologic media. In contrast, hydraulic conductivity anisotropy defines the preferred flow directions of fluids in geologic media. Nonetheless, these types of isotropy and anisotropy are related. Fig. 9.14 in Section 9.8.2 demonstrates the linkage between hydraulic conductivity anisotropy and statistical anisotropy.

9.3.1 Statistically Isotropic Fields

A random field with the same correlation scales in all directions $\lambda_1 = \lambda_2 = \lambda_3$ is statistically isotropic. Suppose an aquifer is said to have a statistically isotropic random field. We may visualize that this aquifer is made of spherical inclusions (for instance, a conglomerate formation or a sandbox packed with uniform sand) of an average diameter equal to the correlation length or λ. Of course, these spherical inclusions could have different shapes and properties. Fig. 9.7 illustrates a

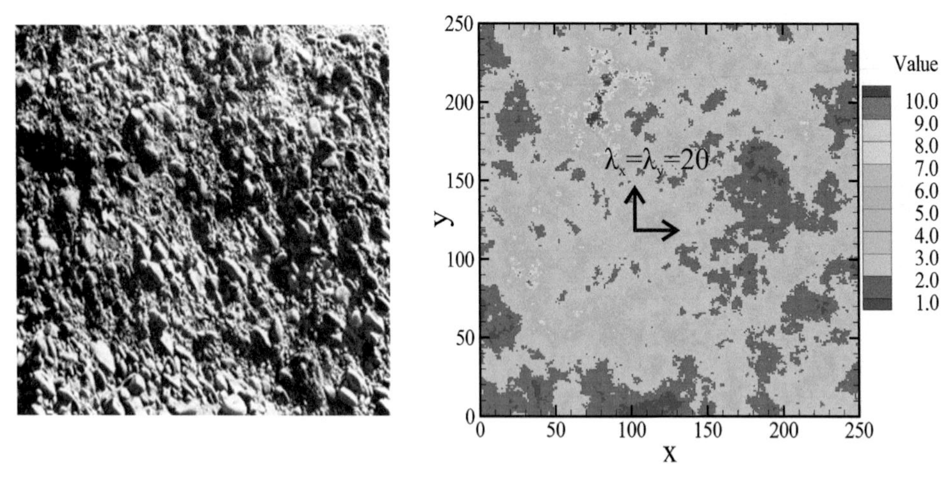

Figure 9.7 An outcrop of a conglomerate formation and a numerically generated statistically isotropic hydraulic conductivity field with a correlation of 20 units in both x and y directions (Yeh et al., 2015).

conglomerate outcrop and a generated two-dimensional random field (one realization) using an isotropic exponential correlation function with a correlation scale of 20 units in both x and y directions.

9.3.2 Statistically Anisotropic Fields

Suppose the correlation scales of a random field in x, y, *and* z directions are different (i.e., $\lambda_1 \neq \lambda_2 \neq \lambda_3$). This field is statistically anisotropic comprising of many heterogeneities of elliptic shapes. Because of geologic deposition processes, geologic materials often spread over larger horizontal extents than the vertical. The statistical anisotropy is a typical spatial structure of geologic heterogeneity. An outcrop showing geologic heterogeneity with statistical anisotropy is displayed in Fig. 9.8, along with a single realization of a generated random field with correlation scales 100 units in the x-direction and 20 units in the y-direction. This realization is generated with the same random seed number (required for a random field generator), mean, and covariance as in Fig. 9.7. Comparing these figures, we observe that the two fields have the same high and low values distribution, except that high and low values zones are greatly horizontally stretched in Fig. 9.8.

A perfectly stratified aquifer is an extreme case of the statistically anisotropic random field. It is made of many disc-shaped lenses (long horizontal dimensions and small vertical one), characterized by $\lambda_1 = \lambda_2 \gg \lambda_3$. Fig. 9.9 illustrates a perfectly stratified geologic outcrop, and a single realization of the generated random field with an exponential covariance function with a correlation scale of

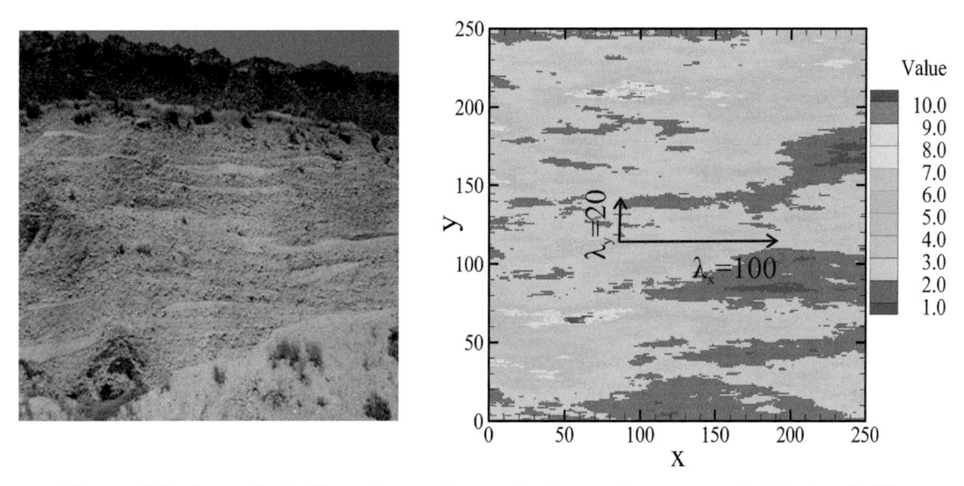

Figure 9.8 A statistically anisotropic geologic medium at a field site (left). A synthesized statistically anisotropic hydraulic conductivity field (right) with a correlation length of 100 units in the *x*-direction and 20 units in the *y*-direction (Yeh et al., 2015).

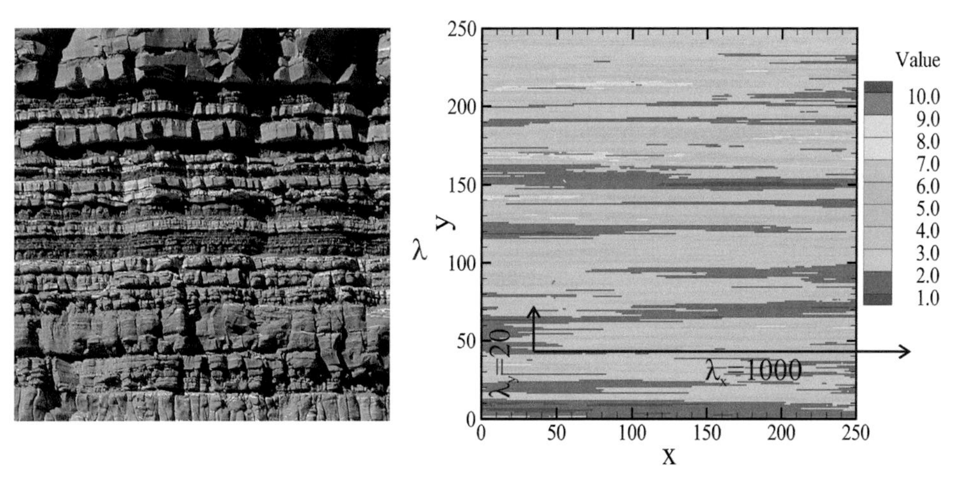

Figure 9.9 A perfectly stratified geologic media (left) and a synthetic, statistically anisotropic hydraulic conductivity field correlated over 1000 units in the *x*-direction and 20 units in the *y*-direction (Yeh et al., 2015).

1000 units in the *x*-direction is and 20 units in the *y*-direction. Again, this realization has the same seed number, mean, and covariance as those in Figs. 9.7 and 9.8. These three figures demonstrate how statistical anisotropy affects the distribution of the synthetic geologic heterogeneity and how it can describe general geologic fabrics within an aquifer.

We emphasize that the generated random field resembles the actual field of a geologic medium only in terms of the statistic properties (i.e., mean, and covariance). Moreover, a generated random field is only one realization of the ensemble. Its spatial distribution, mean, and covariance will equal the specified ensemble statistics if the generated field satisfies the ergodicity (Chapter 1) – sufficiently large to encompass all possible heterogeneities in the ensemble. Otherwise, many realizations are necessary to derive the same spatial statistics.

Also, the statistical anisotropy discussed above assumes that the maximum and minimum correlation scales align with our chosen coordinates system (i.e., horizontal and vertical). The directions of the maximum and minimum correlation scales are the principal directions of the statistical anisotropy. They are similar to the principal directions of hydraulic conductivity anisotropy. In many situations, the orientations of geologic structures may not align with our chosen coordinates. These cases require a coordinate transformation to determine the principal directions of the statistical anisotropy. A question for the readers: How should the statistical approach characterize a highly folded geologic formation?

9.3.3 Statistically Homogeneous Fields

Statistical homogeneity is the same as stochastic stationarity discussed in Chapter 1. This concept also resembles the definition of hydraulic homogeneity. While the latter means that hydraulic property defined over some CV does not vary with locations, the former states that the JPD of a random field is independent of the location. In other words, a heterogeneous random field does not have any large-scale trends. A typical statistically homogeneous field is hydraulic conductivity variation within a sand deposit.

9.3.4 Statistically Heterogeneous Fields

A random field that is not statistically homogeneous is a statistically heterogeneous or nonstationary random field. For example, an outcrop consists of thick layers of sandstone, schist, dolomite, and conglomerate. If we focus on K variability at a small CV within each layer, we likely find the K values of each layer vary but have a constant mean and variance (statistically homogeneous). However, as we sample the entire outcrop, we observe different means and variances for the K values for different layers. The K field of the entire outcrop, thus, is statistically heterogeneous. This example illuminates that both the stationary and nonstationary definitions depend on the scale of the sample area.

Furthermore, stationarity and nonstationarity are ensemble concepts in theory. Strictly, we cannot classify a field (one realization of the ensemble) as stationary or nonstationary unless sufficiently large samples are available. The ergodicity, however, is never verifiable in reality. As a result, one can always conceptualize a heterogeneous field as stationary despite the trend and trend-dependent variance, as many have done for field problems. In other words, this nonstationary field is just one realization of the ensemble, which consists of realizations with different trends.

In one realization (the outcrop discussed above), the trend of the nonstationary field can be removed first, and the residual is examined, as demonstrated in Rehfeldt et al. (1992). Similarly, we can examine the spatial statistics of each unit separately to describe its heterogeneity and geo-fabrics (or spatial structures) in each unit, as demonstrated in Ye et al. (2005). Alternatively, we can treat the entire formation as one unit, ignore the trend and consider the variability of the property over the whole formation as a stationary field with a constant mean. This approach is tantamount to adopting a CV as large as the entire aquifer to homogenize the aquifer as homogeneous (i.e., Ensemble REV, Chapter 1). If this is the case, the spatial statistics include the trend and the variability around the trend. Thus, the trend variation dominates the spatial statics due to its greater variability than the residual. A nested variogram (Section 9.3.1) is a means to describe such a multiscale or nonstationary variability.

Natural aquifers always are heterogeneous at multiple scales. Accordingly, they could be classified as statistically heterogeneous and isotropic or anisotropic, or they could be viewed as statistically homogeneous and isotropic or anisotropic

9.4 Field Example

After the above presentation of the stochastic representation of heterogeneity, this section examines its application to some field aquifers. In particular, Sudicky (1986) investigated the hydraulic conductivity field's spatial structure at the Borden tracer test site (Chapter 10) by extracting thirty-two 2 m-long cores of the aquifer material along two transect lines (Fig. 10.1 in Chapter 10). Each spaced at 1 m horizontal intervals, twenty cores were obtained along line A–A' (19 meters long). The transverse core line B–B' comprises 13 cores with 1 m horizontal spacing (12 meters long). The line B–B' bisected the line A–A' perpendicularly at its 10th core location. Each 2 m-long core was obtained from depths between about 2.5 and 4.5 m below the ground surface (Fig. 9.11).

Hydraulic conductivity profiles along each of the 32 cores were derived using the following procedures. First, each 2 m core was subdivided into 0.05 m equal length subsamples and allowed to dry. Because of small 0.05 m intervals, the material within each segment was relatively homogeneous in texture. The K value of each subsample was measured using the standard falling head procedure. Twelve hundred and seventy-nine K measurements were made along the two transect lines.

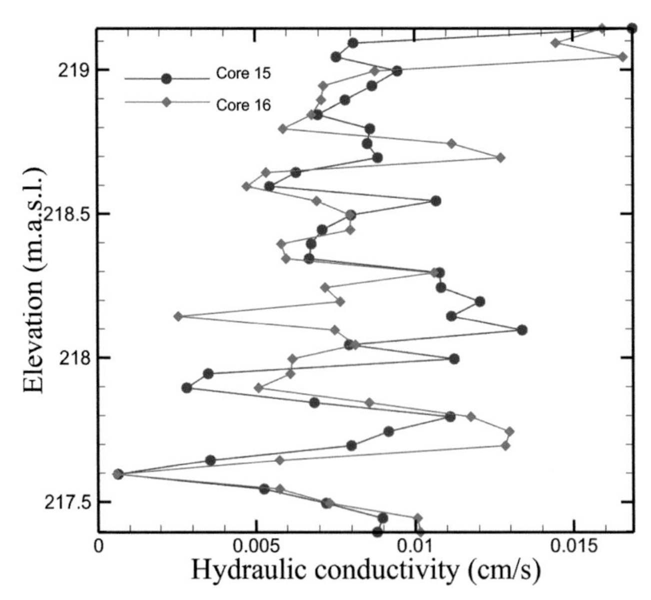

Figure 9.10 Hydraulic conductivity profiles from two vertical boreholes at the Borden site (Sudicky, 1986).

Figure 9.10 compares the measured K profiles for core locations 15 and 16, separated by a horizontal distance of 1 m, of the transect line $A-A'$, showing the type of depth variability at the site. The K value ranges between about 6×10^{-4} and 2×10^{-2} cm/s, exhibiting a contrast in values of slightly more than a factor of 30. The lowest hydraulic conductivity zones for both cores 15 and 16 occur as two individual bands near the bottom of the 2 m depth, with the thicknesses of these bands being on the order of 0.1–0.2 m. Except for these zones, the hydraulic conductivities range over about a factor of 5. The two depth profiles for locations 15 and 16 suggest no evidence for a hydraulic conductivity trend with depth. The aquifer, thus, is a statistically homogeneous random field.

Contour plots of the measured K field for the two vertical cross-sections A–A' and B–B' are shown in Fig. 9.11a and b, respectively. The negative of ln$K$ was contoured to avoid domination by the extreme values in the contour maps because the range of ln$K$ is less than K. Several distinct but thin lenses of low hydraulic conductivity are visible along with A–A' (Fig. 9.11a) at distances between 10 and 19 m. These low hydraulic conductivity zones' presence leads to more variability between distances about 10 and 19 m than over the remainder of the region along the transect A–A'. However, several exploratory cores taken at various locations throughout the aquifer indicate no larger-scale trend, confirming that the heterogeneity in this plot is statistically homogeneous.

The frequency distributions of ln$K$ and K values along the two transects are shown in Fig. 9.12a and b. The distribution of ln$K$ is approximately normal, while the K value distribution has a distribution skewed to the right. The mean

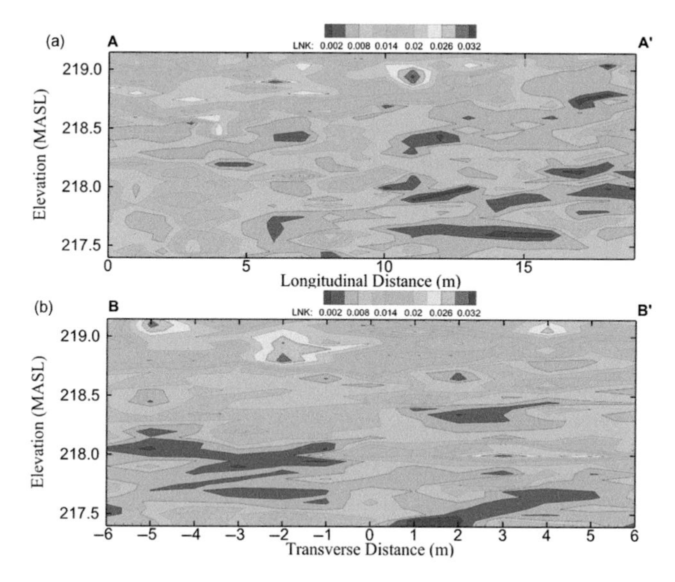

Figure 9.11 Hydraulic conductivity sampling network for the two cross-sections in the Borden site and their ln$K$ distributions.

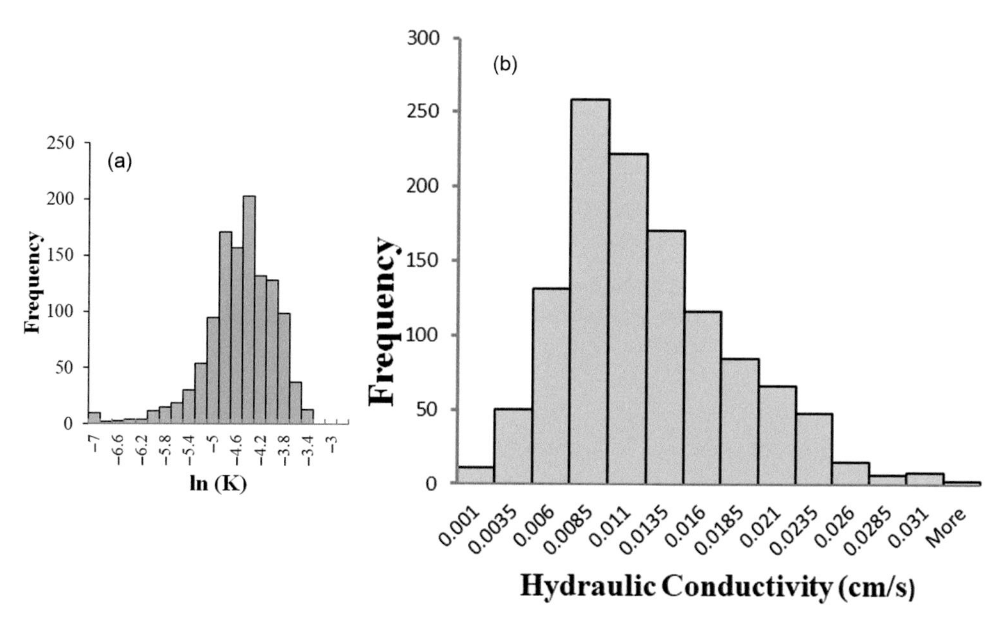

Figure 9.12 The histograms for (a) ln K and (b) K of the two cross-sections at the Borden site.

and variance ($\overline{F}$ and σ_f^2) are -4.63 and 0.38, respectively. The corresponding geometric mean for K is 9.75×10^{-3} cm/s.

Horizontal and vertical autocorrelation functions of ln$K$ estimated from the equally spaced measurements along A–A' or B–B' using Eq. 1.4.12 Chapter 1

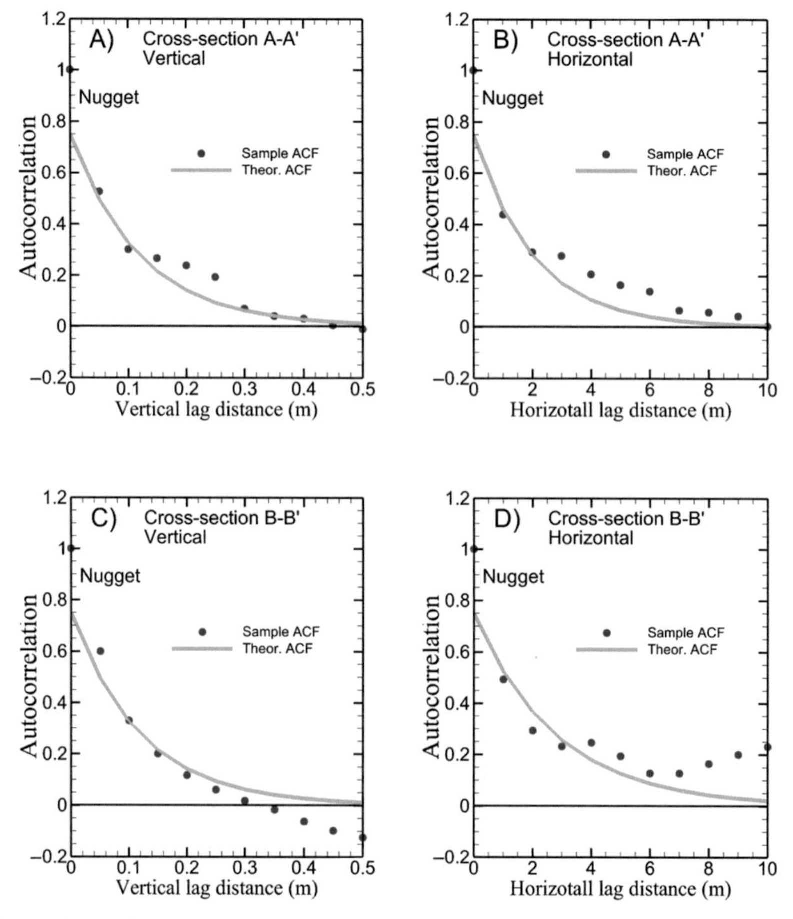

Figure 9.13 The autocorrelograms in the x- and z-directions of the hydraulic conductivity in the two cross-sections at the Borden site. They show the nugget effects and statistical anisotropy.

are illustrated in Fig. 9.13, compared to an exponential model of the form $\rho(s_j) = \eta \exp(-s_j/\lambda_j)$. This autocorrelation function considers the nugget effect (Section 9.2.3.1) using a correction factor $\eta = \left(\sigma_f^2 - \sigma_0^2\right)/\sigma_f^2$. The term σ_0^2 is the variance representing uncorrelated variations below the scale of the measurement CV for K or the variability not detected from the sample interval. In other words, the factor was self-imposed.

Figure 9.13 shows that the sample autocorrelation function drops continuously from 1 at the origin as the lag (separation distance) increases – the absence of the nugget effect. With a guess nugget effect, the exponential model with a vertical correlation length scale $\lambda_3 = 0.12$ m and $\eta = 0.75$ still closely approximates the estimated vertical autocorrelation function for B–B' and provides an excellent fit to that estimated from A–A'. In the transverse horizontal direction along B–B', the use of a correlation scale $\lambda_2 = 2.8$ m and $\eta = 0.75$ also provides good agreement with the estimated function, as does use of $\eta = 0.75$ and $\lambda_1 = \lambda_2 = 2.8$ m in obtaining the fit with the estimated horizontal correlation function in the direction A–A'. This result suggests statistical isotropy in the horizontal plane ($\lambda_1 = \lambda_2$). With $\eta = 0.75$, variability smaller than the measurement CV (or the variability beyond the sampling interval), and the measurement error σ_0^2 is about 0.09.

Note that the analysis assumed that the horizontal and vertical directions follow the principal correlation axes. Some of the lenses in Fig. 9.11 are at a slight angle about 2° or 3° below horizontal. Since the correlation scale concept describes the overall heterogeneity structure, such a minor difference plays only a minor role in the analysis. Neglecting this minor difference, $\lambda_1 = \lambda_2 = 2.8$ m was obtained in the horizontal direction along A–A' and B–B' and $\lambda_3 = 0.12$ m was derived in both profiles. In addition, the mean and variance of ln$K$ obtained from A–A' and B–B' did not differ significantly. As a result, measurements along both sample lines were lumped together to estimate a single horizontal and a single vertical autocorrelation function. The resulting directional autocorrelation functions are very close to the average of the individual horizontal and vertical estimates shown in Fig. 9.13. They can be approximated reasonably well by the exponential autocorrelation function with $\lambda_1 = \lambda_2 = 2.8$ and $\lambda_3 = 0.12$ m. Woodbury and Sudicky (1991) revisited this dataset using a variogram analysis (Section 9.3) and derived similar results with minor differences.

Sudicky (1986) demonstrated that the stochastic representation approach describes the spatial variability of ln$K$ at the Borden site as a statistically homogeneous and anisotropic random field with a mean and variance ($\overline{F}$ and σ_f^2) of -4.63 and 0.38, respectively. Moreover, the spatial variability at the site has a spatial structure, quantified by the spatial covariance:

$$Cov(\boldsymbol{\xi}) = \sigma_f^2 \rho(\boldsymbol{\xi}) = \sigma_f^2 \exp\left\{-\left[\left|\frac{\xi_x}{\lambda_x}\right|^2 + \left|\frac{\xi_y}{\lambda_y}\right|^2 + \left|\frac{\xi_z}{\lambda_z}\right|^2\right]^{\frac{1}{2}}\right\} \tag{9.4.1}$$

where σ_f^2 is the variance and $\boldsymbol{\xi}$ is the separation vector between any two points in space. Of course, a Gaussian joint pdf is implicitly assumed, but it is not essential for practical purposes. At this moment, one must recognize that a stochastic characterization is a quantitative tool that provides information about the overall heterogeneity at a field site. Because it quantifies the general variability, the precision of the analysis may not be critical. For example, an accurate stochastic representation of the Borden site's heterogeneity does not precisely reveal how the ln$K$ contours in Fig. 9.11 should be. Nevertheless, the approximate statistics are helpful for comparison of the heterogeneity at different aquifers. For example, the small variance ($0.38 \ll 1$) of the Borden site indicates that the site is relatively homogeneous compared to other sites (e.g., MADE site, which has a variance of 4.5, Rehfeldt et al., 1992).

Suppose we imagine the Borden site aquifer as a large homogeneous soil column. The CV size of sampled K and concentration becomes tantamount to a pore in the soil column. How could the aforementioned stochastic representation of heterogeneity help us understand the migration and spread of a tracer cloud at this site? Would the solute plume behave like in a soil column presented in Chapter 7? We intend to answer these scale issues in the following sections and Chapters.

Thinking points:

- Is it worth deriving these estimates of the spatial statistics using such an intensive sampling effort?
- Are the statistics from the samples from the two transect representative of the entire three-dimensional variability in the area?
- How dense is the sampling network needed to obtain representative spatial statistics?
- Is the normal distribution assumption necessary? Hint: does anyone even have a sufficient number of samples to prove its ensemble distribution?
- How would the spatial correlation affect the JPD?

9.5 Highly Parameterized Heterogeneous Model (HPHM)

During the past few decades, various approaches have emerged to address a fundamental question: how to predict water and solute movement over hundreds and thousands of meters in an aquifer using theories from soil column-scale results (Chapter 7). Aquifers are known to consist of hydraulic heterogeneity at various scales (e.g., variations in the pore size, laminations, cross-beddings, aggregates, layers, fractures, folds, faults, stratifications, and facies changes). Furthermore, our observation and interest scales are frequently smaller than the scales of the most heterogeneity. High-resolution flow and transport models capable of capturing heterogeneity effects at various scales become essential to our interests.

As explained in Chapter 7, laboratory soil column experiments have proven that the flow and transport model based on Darcy's and Fick's laws is sufficient at the scale of our observation and interest for these experiments. Accordingly, flow and solute migration through a heterogeneous groundwater basin can be viewed as flow and transport through a collection of interconnecting soil columns of different properties in three dimensions. Such conceptualization leads to the highly parameterized heterogeneous model (HPHM) concept. In other words, HPHM discretizes a heterogeneous aquifer or groundwater basin into many small **local-scale CVs**, similar to soil columns. However, there is no clear definition of the local-scale CV's size. Ideally, soil columns in laboratory experiments are the local-scale CV. However, when dealing with aquifers or groundwater basins, the local-scale CV may refer to a volume "much" smaller than the domain under investigation, containing relatively uniform geologic material with insignificant variations in hydraulic properties at our interest and observation scales.

Flow and solute transport in these interlaced local-scale CVs is then depicted by continuous partial differential equations (PDE) governing the flow and transport processes over the assembly of soil columns. In particular, the PDE for flow is

$$\frac{\partial}{\partial x_i}\left(K(\mathbf{x})\frac{\partial H}{\partial x_j}\right) = S_s(\mathbf{x})\frac{\partial H}{\partial t}. \tag{9.5.1}$$

In Eq. (9.5.1), i and j = 1, 2, and 3. $x_1 = x$, $x_2 = y$, and $x_3 = z$. $K(\mathbf{x})$ is the saturated hydraulic conductivity [L/T], and $S_s(\mathbf{x})$ is the specific storage [1/L] of a local-scale CV, varying with the location $\mathbf{x}$ (x, y, z). The conductivity is generally isotropic in HPHM, assuming the effects of variation in $K(\mathbf{x})$ is more profound than its anisotropy at the CV scale. With the specification of $K(\mathbf{x})$ and $S_s(\mathbf{x})$ (e.g., their values in each finite-difference block, finite element cell, or local-scale CV), Eq. (9.5.1) then can be solved with the initial conditions:

$$H(\mathbf{x}, 0) = H^*(\mathbf{x}, 0),$$

where 0 denotes time = 0, and $H^*(\mathbf{x}, 0)$ represents the prescribed heads everywhere in the aquifer, and with the boundary conditions specified:

$$H(\mathbf{x})|_{\Gamma_D} = H_D \text{ and } K\nabla H{\cdot}\mathbf{n}|_{\Gamma_N} = q_N \tag{9.5.2}$$

where H is the hydraulic head [L]. The head at the Dirichlet (prescribed head) boundary Γ_D is H_D. q_N is the specific flux [L/T] at the Neumann (prescribed flux) boundary Γ_N. $\mathbf{n}$ is the unit vector normal to the boundary Γ_N. These boundary conditions are specified at the edges of the aquifer.

Once Eq. (9.5.1) is solved, the local-scale groundwater flux (or specific discharge) $q_i(\mathbf{x}, t)$ at a location $\mathbf{x}$ and a given time is determined from the Darcy law:

$$q_i(\mathbf{x}, t) = -K(\mathbf{x})\frac{\partial H(\mathbf{x}, t)}{\partial x_i}. \tag{9.5.3}$$

The flux in Eq. (9.5.3) is an average over the CV, at a given time, omitting effects of the velocity variation within this CV on the solute movement. Eqs (9.5.1) to (9.5.3) are called the **local-scale governing flow equations,** in which K and S_s are **the local-scale parameters**, and H, and q are **local-scale state variables**.

Analogous to the analysis in Chapter 7, we must include solute mass flux at scales smaller than the local-scale CV and omitted in the local-scale flux in HPHM. For this reason, we must employ the local-scale advection-dispersion equation (Chapter 7) in addition to Eq. (9.5.3):

$$n\frac{\partial C}{\partial t} = \frac{\partial}{\partial x_i}\left(d\frac{\partial C}{\partial x_j}\right) - \frac{\partial q_i C}{\partial x_i}. \tag{9.5.4}$$

The term $C(\mathbf{x}, t)$ is the **local-scale solute concentration**, and n **is** the **local-scale porosity**, and d is **the local-scale dispersion coefficient**. This **local-scale dispersion coefficient** d is commonly defined as the product of the **local-scale dispersivity** α, [L], and the local-scale flux magnitude (Bear, 1979, p235).

$$d = \alpha|q| = \alpha\left(\sqrt{q_x^2 + q_y^2 + q_z^2}\right). \tag{9.5.5}$$

This dispersivity is generally considered isotropic. Its anisotropic effect is usually minor compared to the variability of local-scale hydraulic parameters over the entire aquifer. Eqs. (9.5.4) and (9.5.5) are the **local-scale solute transport equations**.

Therefore, modeling solute transport first solves Eqs. (9.5.1) through (9.5.3) to obtain the spatial distribution of the local-scale fluxes at different times. Then, with the known local-scale flux distribution, dispersivity, and porosity values, initial and boundary conditions, Eqs. (9.5.4) and (9.5.5) are solved to include the flux variation below the local scales (i.e., pore-scale dispersion and molecular diffusion). HPHM approach considers the effects of multiscale heterogeneity on solute transport in a natural groundwater basin. This approach, nevertheless, generally considers the local-scale porosity and dispersivity uniform over the entire basin because their variability is small compared to the variability of hydraulic parameters (see Chapter 11).

Similar to the analysis of solute migration in the pipe (Chapter 6) and soil column (Chapter 7), using the HPHM, we, in theory, could predict the solution transport in detail (at our interest and observation scales). This conjecture should be valid so long as we can simulate the precise velocity field at the local scale CV over the entire aquifer. How to characterize the aquifer in such detail becomes the issue, which is the subject of high-resolution aquifer characterization technology to be discussed in Chapter 12. Without such technology, an alternative is the zonation modeling approach discussed next.

9.6 Zonation Conceptual Model

Instead of using HPHM to predict detailed solute distributions, one often adopts a straightforward geologic zonation approach, only capturing the effects of the large-scale trend of a statistically heterogeneous aquifer on solute distribution. This approach assumes that the aquifer consists of several large-scale CVs, mimicking geologic zones or facies. Each CV possesses unique homogeneous hydraulic properties, invoking the ensemble REV concept, regardless of the absence or existence of spatial REV. Then, Eqs. (9.5.1) through (9.5.5) are solved to simulate flow and transport processes in the aquifer. Because of the homogenization of the complex heterogeneity in each zone, the predicted H and C fields are the ensemble-average responses over the ensemble REV. Thus, at best, they may reproduce the general trend of the actual responses but could be significantly different from point observations in each zone.

Nevertheless, the zonation approach faces the same dilemma as the HPHM. First, the locations and geometries of the zones are unknown, although geologic maps, GIS information, geophysical surveys, and well logs could be helpful. Furthermore, appropriate values of the hydraulic parameters for each zone remain undetermined. Of course, calibrating the zonation model to some observed aquifer responses is an option, as discussed in Chapter 10. Are there any alternatives if the geologic information or measurements of heads, concentrations and other aquifer responses are unavailable? The possible solutions are Monte Carlo simulation (MCS) and equivalent homogeneous model with effective hydraulic conductivity and macrodispersion approaches to be discussed next.

9.7 Ensemble Mean Based on Monte Carlo Simulation (MCS)

As remarked in Section 9.1, under a given spatial statistics of $K(\mathbf{x})$, and $S_s(\mathbf{x})$, many possible realizations of random fields $K(\mathbf{x})$ exist and could be created by a random field generator. With these generated fields in Eqs. (9.5.1)–(9.5.5), MC experiments could simulate many corresponding $H(\mathbf{x})$, $q_x(\mathbf{x})$, $q_y(\mathbf{x})$, $q_z(\mathbf{x})$ and $C(\mathbf{x}, t)$ fields. Statistical analysis of these fields yields their ensemble means and variances. The ensemble means are the most likely local-scale aquifer response (e.g., head, flux, or concentration) at any location $\mathbf{x}$ of all possible realizations with the given spatial variability. On the other hand, their variances represent the deviation of the local-scale aquifer responses at $\mathbf{x}$ from their mean values due to the unknown variability of the hydraulic parameter fields.

Overall, MCS is a straightforward but computationally expensive and brute-force approach. Further, the statistically best predictions without conditioning (or constraining) with observations may depart significantly from the actual aquifer responses. Fig. 12.27 of Chapter 12 provides an example of MCS of solute transport in the riverbed.

9.8 Equivalent Homogeneous Media (EHM) Approach

Besides the MCS, the EHM approach is a widely used and pragmatic approach to predicting the most likely concentration distribution without knowing the exact spatial distribution of local-scale properties. This approach embraces a large-scale CV, much larger than many local-scale CVs (e.g., soil columns and blocks) but smaller than the entire basin. Specifically, this CV includes variations in pore characteristics and local-scale K and Ss variability due to laminations, bedding, clay lenses, or clusters of gravels or layers. If the K defined over this large-scale CV is the same regardless of the CV's location. This heterogeneous field is a statistically homogeneous random field. The entire basin can be perceived as homogeneous with a constant K at this large-scale CV (i.e., a large-scale spatial REV).

On the other hand, different geologic facies, formations, folds, faults, or other large-scale geologic features makes the basin's K and S_s fields nonstationary (or statistically heterogeneous), and the spatial REV for the field does not exist. Despite this, many have treated such a basin as homogeneous for convenience. In this case, they implicitly invoke an ensemble REV (Chapter 1) to homogenize the entire basin. As such, the uniform K for the entire basin represents the average value over all possible realizations of nonstationary random K fields within this basin. Because this average includes all possible large-scale trends, the variability of local-scale K in the basin is much more significant than in an aquifer or a basin without large-scale trends.

EHM, with a homogeneous K value over the large-scale spatial or ensemble REV, aims to derive the mean head $\overline{H}(\mathbf{x})$ and the mean fluxes $\overline{q}_i(\mathbf{x})$. An ensemble-mean ADE, in turn, derives the most likely solute concentration distributions with these mean fluxes without any local-scale information. In this EHM approach, local-scale flux variation $q_i'(\mathbf{x})$ omitted by the mean $\overline{q}_i(\mathbf{x})$ is subsequently included in the ensemble-mean ADE as the dispersion to determine the corresponding $\overline{C}(\mathbf{x}, t)$. This rationale follows the dispersion concepts in shear flow and soil columns, except the scale of the investigation is a basin or a large aquifer, consisting of heterogeneity of a wide range of scales. However, applications of this approach to aquifers forget that our interest and observation scale problem is as small as a point in a pipe or a soil column experiment.

Accepting the above rationale and ignoring the scales of our interest and observation, we substitute the means and the perturbations of the flux and concentration into Eq. (9.5.4) to yield

$$n\frac{\partial \overline{C} + C'}{\partial t} = d\frac{\partial^2 \overline{C} + C'}{\partial {x_i}^2} - \frac{\partial \left(\overline{q}_i + q_i'\right)\left(\overline{C} + C'\right)}{\partial x_i} \tag{9.8.1}$$

Since we seek the most likely concentration distribution, we take the expectation of Eq. (9.8.1), and we have

$$n\frac{\partial \overline{C}}{\partial t} = d\frac{\partial^2 \overline{C}}{\partial {x_i}^2} - \overline{q}_i\frac{\partial \overline{C}}{\partial x_i} - \frac{\partial E\langle q_i' C'\rangle}{\partial x_i}. \tag{9.8.2}$$

E< > denotes the expectation (the average over the ensemble). It is an ensemble operator, independent of the time or spatial derivative operator. These two operators' positions are, thus, exchangeable. Also, notice that the expected value of the product of the mean and perturbation of a variable is always zero; that of the product of perturbations of two different variables is the cross-covariance, which could be zero if the variables are mutually independent.

The last term on the right-hand side of Eq. (9.8.2) is the cross-covariance, representing the statistical relationship between q_i' and C', reflecting the effects of perturbations q_i' and C', omitted by $\overline{q}_i$ and $\overline{C}$. Their mutual dependence makes this cross-covariance term nonzero. Many have shown that this cross-covariance, including Gelhar (1993) and Dagan (1984 and 1987), is related to the q_i' covariance. Further, it is related to the mean concentration gradient similar to Fick's law for molecular diffusion or shear flow dispersion after the tracer has traveled long distances or reached Fickian's regime. That is,

$$D_{ij}\frac{\partial \overline{C}}{\partial x_i} = E\langle q_i' C'\rangle \tag{9.8.3}$$

where D_{ij} is the large-scale dispersion coefficient, a tensor, denoting the rate of spreading of a solute plume due to the variation of the local-scale velocity. Substitution of Eq. (9.8.3) into Eq. (9.8.2) leads to the large-scale ensemble-mean ADE:

$$\begin{aligned} n\frac{\partial \overline{C}}{\partial t} &= \frac{\partial}{\partial x_i}\left(\left(D_{ij} + d\right)\frac{\partial \overline{C}}{\partial x_j}\right) - \overline{q}_i\frac{\partial \overline{C}}{\partial x_i} \\ &= \frac{\partial}{\partial x_i}\left(E_{ij}\frac{\partial \overline{C}}{\partial x_j}\right) - \overline{q}_i\frac{\partial \overline{C}}{\partial x_i}. \end{aligned} \tag{9.8.4}$$

The term E_{ij} (not the expectation) is the macrodispersion coefficient, which includes D_{ij} and d. The dispersion coefficient D_{ij} accounts for influences of the unknown variation of the local-scale $q_i'(\mathbf{x})$ from the unexplored variability of local-scale K and S_s fields. This variability includes any heterogeneity larger than pore-scale variability. On the other hand, d accommodates the unknown flux variations below the local-scale heterogeneity (such as pore variability and random motion of molecules). Thus, E_{ij} makes up the effects of unknown multiscale flux perturbations on the mean solute transport process.

The macrodispersion coefficient assumes a linear relationship with the mean velocity, as does the dispersion coefficient in the soil column experiments:

$$E_{ij} = A_{ij}U \tag{9.8.5}$$

The macrodispersivity is A_{ij} $[L]$ and is a tensor, and U is the mean velocity, equal to the mean specific discharge divided by the mean porosity, often assumed spatially invariant.

The ensemble–mean concept of Eq. (9.8.4) and its derivation are analogous to the analysis of shear flow dispersion in pipes (Taylor, 1953), as presented in Fischer et al. (1979), and have been elucidated in Chapters 4, 6, and 7 of this book. We reiterate that the linear relationship between the $E\langle q_i'C'\rangle$ (cross-covariance between q' and C') and mean concentration gradient in Eq. (9.8.3) is valid only if the solute plume has experienced sufficient heterogeneity. In other words, the ergodicity embedded in the cross-covariance is satisfied. To satisfy this requirement, the size of the initial plume must be larger than many correlation scales of the heterogeneity, or the solute plume has to travel over a long distance in a statistically homogeneous (stationary) random field. Unless this condition is satisfied, Fick's law is inappropriate. The macrodispersivity will exhibit scale-dependent behavior, and the observed plume distribution will deviate from the predicted. Many have reported this deviation as anomalous dispersion.

Note that the local-scale ADE (Eq. 9.5.4) adopts the ensemble concept, except that it is applied to a local-scale CV, where the subscale flux variation results from variations in pore sizes and pore throats, sorting of grains, and the torturous nature of pore channels. Because of the small size of the local-scale CV (in turn, small variation within the CV) and reasons discussed in Chapter 7, we often forget the ensemble-mean nature of $C(\mathbf{x}, t)$ in Eq. (9.8.4) and as well that in Fick's law for molecular diffusion.

Once accepting the ensemble-mean ADE for the aquifer, we face the issue of determining the parameters for the ADE. A possible solution is to conduct a flow and tracer experiment at the basin scale. The equivalent homogeneous hydraulic properties and dispersivity are then estimated from observed tracer information (see Freyberg, 1986 in Chapter 10). Another is to derive these parameters from spatial statistics of hydraulic property values measured at local-scale CVs (much smaller than the entire basin or aquifer), as Sudicky (1986) did so (Sections 9.1–9.5). This approach is, in effect, the stochastic effective parameter (upscaling) theories advocated by Dagan (1982), Gelhar and Axness (1983), or Gelhar (1993).

9.8.1 Effective Hydraulic Conductivity

Most macrodispersion theories focus on solute transport under steady-state, uniform flow conditions, and so does our discussion. Evaluation of Eq. (9.8.4) requires the ensemble mean velocity, which depends on the ensemble homogeneous hydraulic conductivity and the mean hydraulic gradient. Generally, the mean hydraulic gradient can be estimated from the available contour map of the observed groundwater levels. The effective hydraulic conductivity for the basin is derived from the stochastic theories that relate the spatial statistics of local-scale hydraulic conductivity to the effective hydraulic conductivity for the entire basin. Because of their complexity, we will present the concepts and results by Gelhar and Axness's (1983), and we refer interested readers to the details of the theories in the book by Gelhar (1993).

The theory considers the local-scale hydraulic conductivity a spatial random field characterized by its mean and covariance (Sections 9.1–9.5). The approach then implicitly adopts a highly parameterized steady-state flow model,

$$\frac{\partial}{\partial x_i}\left(K(\mathbf{x})\frac{\partial H(\mathbf{x})}{\partial x_j}\right) = 0 \tag{9.8.6}$$

with specified boundary conditions to calculate the hydraulic head field first. The $K(\mathbf{x})$ is the random field with many possible realizations with the given spatial statistics. Instead of solving the ensemble mean head field using Monte Carlo simulation, the theory solves it via an analytical approach. Specifically, the $K(\mathbf{x})$ is expressed as the sum of its mean and perturbation: $K(\mathbf{x}) = \overline{K} + K'(\mathbf{x})$ and likewise, the hydraulic head is $H(\mathbf{x}) = \overline{H} + H'(\mathbf{x})$, where the overbar denotes the mean and the prime represents the perturbation. Substituting these expressions into Eq. (9.8.6), we have

$$\frac{\partial}{\partial x_i}\left[\left(\overline{K} + K'(\mathbf{x})\right)\frac{\partial\left(\overline{H} + H'(\mathbf{x})\right)}{\partial x_j}\right] = 0. \tag{9.8.7}$$

After expanding this equation and dropping $\mathbf{x}$ notation, we have

$$\frac{\partial}{\partial x_i}\left[\left(\overline{K} + K'\right)\frac{\partial \overline{H}}{\partial x_j} + \left(\overline{K} + K'\right)\frac{\partial H'}{\partial x_j}\right] = \frac{\partial}{\partial x_i}\left[\overline{K}\frac{\partial \overline{H}}{\partial x_j} + K'\frac{\partial \overline{H}}{\partial x_j} + \overline{K}\frac{\partial H'}{\partial x_j} + K'\frac{\partial H'}{\partial x_j}\right] = 0. \tag{9.8.8}$$

Then, we take the expected value of this equation while recognizing the expected value ($E\langle\rangle$) of the products of the mean and perturbation are zero, i.e.,

$$\frac{\partial}{\partial x_i}\left[E\left\langle \bar{K}\frac{\partial \bar{H}}{\partial x_j}\right\rangle + \cancel{E\left\langle K'\frac{\partial \bar{H}}{\partial x_j}\right\rangle} + \cancel{E\left\langle \bar{K}\frac{\partial H'}{\partial x_j}\right\rangle} + E\left\langle K'\frac{\partial H'}{\partial x_j}\right\rangle\right] = 0. \tag{9.8.9}$$

The resulting ensemble mean flow equation becomes

$$\frac{\partial}{\partial x_i}\left[\bar{K}\frac{\partial \bar{H}}{\partial x_j} + E\left\langle K'\frac{\partial H'}{\partial x_j}\right\rangle\right] = 0. \tag{9.8.10}$$

Rearranging the second term as follow,

$$\frac{\partial}{\partial x_i}\left[\left(\bar{K} + E\left\langle K'\frac{\partial H'}{\partial x_j}\right\rangle\left(\frac{\partial \bar{H}}{\partial x_j}\right)^{-1}\right)\frac{\partial \bar{H}}{\partial x_j}\right] = 0. \tag{9.8.11}$$

The superscript -1 indicates the inverse of the gradient vector. The ensemble mean equation becomes

$$\frac{\partial}{\partial x_i}\left[\hat{K}_{ij}\frac{\partial \bar{H}}{\partial x_j}\right] = 0 \qquad \hat{K}_{ij} = \left(\bar{K} + E\left\langle K'\frac{\partial H'}{\partial x_j}\right\rangle\left(\frac{\partial \bar{H}}{\partial x_j}\right)^{-1}\right) \tag{9.8.12}$$

This equation has the same form as the local-scale Eq. (9.8.6). However, it uses the ensemble effective hydraulic conductivity tensor $\widehat{K}_{ij}$, and its solution is the ensemble mean head $\overline{H}$ under given boundary conditions. Note that the effective hydraulic conductivity $\widehat{K}_{ij}$ differs from the statistical mean hydraulic conductivity $\overline{K}$ by an additional term. Physically, it includes the effects of the variability in the head and heterogeneity omitted in the mean head and hydraulic conductivity. For this reason, it is called an effective hydraulic conductivity tensor in the ensemble sense.

Similarly, if flow in each local-scale CV is laminar, which Darcy's law is valid, we can derive the ensemble mean specific discharge of the heterogeneous basin. Considering the K and H at the local scale are random fields, the specific discharge would be a random field, dictated by

$$q_i(\mathbf{x}) = -K(\mathbf{x})\frac{\partial H(\mathbf{x})}{\partial x_i}. \tag{9.8.13}$$

Expressing each random variable in terms of its mean and perturbation, we can rewrite Eq. (9.8.13) as

$$\overline{q_i} + q_i'(\mathbf{x}) = -\left(\overline{K} + K'(\mathbf{x})\right)\frac{\partial\left(\overline{H} + H'(\mathbf{x})\right)}{\partial x_i}. \tag{9.8.14}$$

After taking the expected value on both sides of the equation, we have

$$\overline{q_i} = -\left(\bar{K}\frac{\partial \bar{H}}{\partial x_i} + E\left\langle K'(\mathbf{x})\frac{\partial H'}{\partial x_i}\right\rangle\right) = -\left(\bar{K} + E\left\langle K'(\mathbf{x})\frac{\partial H'}{\partial x_i}\right\rangle\left(\frac{\partial \bar{H}}{\partial x_i}\right)^{-1}\right)\frac{\partial \bar{H}}{\partial x_j} \tag{9.8.15}$$

The ensemble mean discharge vector in the ith direction becomes

$$\overline{q_i} = -\widehat{K}_{ij}\frac{\partial \overline{H}}{\partial x_j} \tag{9.8.16}$$

Notice that the effective hydraulic conductivity $\widehat{K}_{ij}$ could be a tensor or scalar and is constant under uniform mean flow.

Gelhar and Anexx (1983) developed a formula relating the effective hydraulic conductivity to the geometric mean K_g, variance of the local-scale K in terms of the natural logarithm of K, ln$K$, designated as σ_f^2, correlation scales of the local-scale hydraulic conductivity λ_x,λ_y,λ_z, and the mean flow gradient J_x,J_y,J_z:

$$\widehat{K}_{ij} = F\left(K_g, \sigma_f^2, \lambda_x, \lambda_y, \lambda_z, J_x, J_y, J_z\right). \tag{9.8.17}$$

The use of ln$K$ is merely for mathematical convenience. Details of the formula are available in Gelhar (1993). Figure 9.9 presents the approximate effective hydraulic conductivity for uniform mean gradient conditions to illustrate the practical utility of the theory. It shows the effective hydraulic conductivities in x, y, and z directions, normalized by K_g (geometric mean of K), as a function of σ_f^2, and the aspect ratio ($\eta = \lambda_x/\lambda_x$), under uniform, mean gradient flows. The graphical results also assume that the correlation scales in the $x-y$ plane are identical (i.e., statistically isotropic in the x–y plane). The green line in the figure represents that the random field is statistically isotropic in x, y, and z directions. It indicates the effective hydraulic conductivity is slightly greater than K_g. The red lines above the green line represent the normalized hydraulic conductivity in the direction parallel to the mean hydraulic gradient. As the aspect ratio approaches infinity (i.e., the medium is perfectly stratified), the effective hydraulic conductivity becomes the arithmetic mean of the local-scale hydraulic conductivity values.

Meanwhile, the blue dash-dot-dot lines below the green line represent the effective hydraulic conductivity values perpendicular to the mean flow direction. They approach the harmonic mean. These results are consistent with the deterministically derived volume-average K of an equivalent homogenous medium of perfectly layered geologic media (e.g., Yeh et al., 2015).

Sudicky (1986) demonstrates the applicability of this theory in a field tracer test, discussed in Chapter 10. Suppose samples of the local-scale hydraulic conductivity

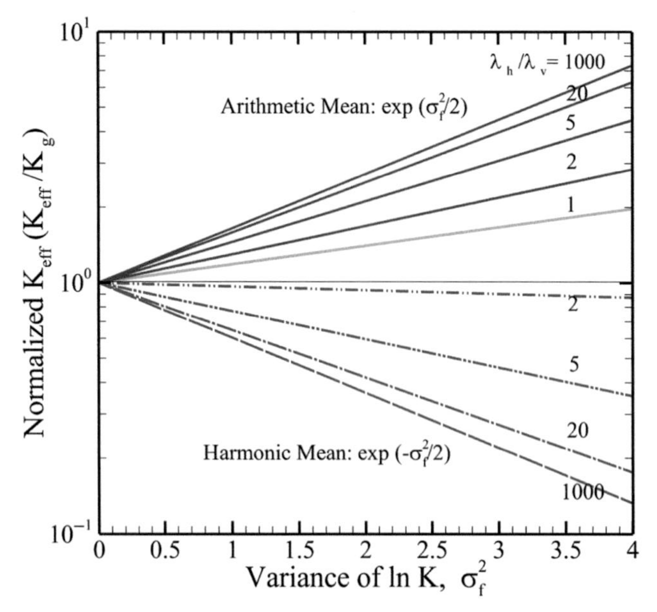

Figure 9.14 Normalized effective hydraulic conductivity as a function of the variance of local-scale hydraulic conductivity under different aspect ratios and mean flow directions.

are available, from which the geometric mean and their variance can be determined. Variogram or autocovariance analysis then yields the correlation scales if spatial data density is sufficient. Otherwise, visual estimations of the correlation scales could be the alternative. With this information and knowledge of the mean gradient direction, one can directly estimate the effective hydraulic conductivities from this graph (Fig. 9.14).

9.8.2 Macrodispersion and Macrodispersivity

With the effective hydraulic conductivity and the mean hydraulic gradient, Eq. (9.8.16) determines the specific discharge vector. Such a specific discharge field is analogous to the averaged velocity over the pipe's cross-sectional area or the average flow in the soil column experiment. The effects of velocity variations around the mean velocity on the spreading of the solute is subsequently represented by "**macrodispersion**" concept. The word "macro" distinguishes itself from molecular diffusion, fluid-dynamics dispersion, and pore scales (local-scale) dispersion, as presented in Chapters 4, 6, and 7. The macrodispersion accounts for the effects of velocity variations at the aquifer- or basin-scale CVs, much larger than molecular, fluid-dynamic, pore-scale CVs, without any detailed characterization of the CV. It, therefore, makes up the effects of velocity variations due to dominant or large-scale heterogeneous over the aquifer or groundwater basin.

The mathematics development of macrodispersion theory, Gelhar (1993), follows similar steps in Taylor's shear-flow dispersion analysis. This section avoids its detailed formulation and present simple formulas that illustrate the logic behind the macrodispersion. Overall, the procedures are similar to those presented in Section 5.3, but the velocity field is a stochastic random field, and the ensemble average replaces the cross-sectional area average. For instance, $\overline{C}$ and $\overline{u'C'}$ terms in Eq. (6.3.14) in Chapter 6 are the cross-sectional area mean concentration and the cross-sectional area average of the product of random velocity and concentration perturbations. They are the ensemble mean concentration and the cross-covariance of the velocity and concentration perturbations in the macrodispersion analysis. Similarly, while the dispersion coefficient in Eq. (6.3.21) in Chapter 6 is related to the cross-sectional area average of the products of the velocity variations, the macrodispersion uses the velocity covariance in the ensemble sense. The significant scientific breakthrough of the macrodispersion theory expresses the velocity covariance in terms of the ensemble mean and covariance of the hydraulic conductivity and mean hydraulic gradient in aquifers.

For the case of statistically anisotropic and heterogeneous geologic media ($\lambda_x = \lambda_y \neq \lambda_z$) and with x coordinate parallel to the mean flow direction, Gelhar and Axness (1983) showed that

$$\begin{aligned} A_{xx} &= \sigma_f^2 \lambda_x \mu / \left(\gamma^2 \zeta\right) \\ A_{yy} &= \sigma_f^2 \lambda_x \mu J_z^2 / \left[2(1+\zeta)^2 \gamma^2 J_x^2\right] \\ A_{zz} &= \sigma_f^2 \lambda_x J_z^2 (1+2\zeta) / \left[2(1+\zeta)^2 \gamma^2 J_x^2\right] \\ A_{xz} &= A_{zx} = \sigma_f^2 \lambda_x \mu J_z / \left[(1+\zeta)^2 \gamma^2 J_x\right]. \end{aligned} \tag{9.8.18}$$

In these formulas, $\mu = \lambda_z/\lambda_x$. The mean gradient's x component is J_x, and the z component is J_z. $\zeta = \sqrt{(\sin\theta)^2 + \mu^2(\cos\theta)^2}$ and θ is the inclined vertical angle of the mean flow direction from the horizontal correlation scale direction. The flow factor $\gamma = q_x/\left(K_g J_x\right)$, where q_x is the x component of the mean discharge. The detailed theory and formulation are available in Gelhar (1993). The applications of the theorem to field-scale tracer experiments are presented in Chapter 10.

Besides Gelhar and Axness's macordispersivity theory, several stochastic transport studies have investigated the behavior of the ensemble covariance of the solute plume concentration as a function of time for spatially variable local-scale velocity fields. The studies by Dagan (1982, 1984, 1987) are of particular interest here. Using a linearized, first-order approximation and an isotropic exponential covariance function for ln*K*, Dagan derived expressions for the evolution of spatial covariance for the solute plume under multidimensional steady flow in an infinite aquifer. The theory is beyond the scope of this book but is available in Dagan (1987). Briefly, his two-dimensional model is present here for discussion in the next chapter:

$$\sigma^2_{x'x'}(t) = 1.48\sigma_f^2\lambda^2\left\{\frac{3}{4} - \frac{3}{2}E + \frac{L}{\lambda} + \frac{3}{2}\left[Ei\left(-\frac{L}{\lambda}\right) - \ln\left(\frac{L}{\lambda}\right) + \frac{\lambda}{L}\exp\left(-\frac{L}{\lambda}\right)\left(1+\frac{\lambda}{L}\right) - \frac{\lambda^2}{L^2}\right]\right\} \tag{9.8.19}$$

$$\sigma^2_{y'y'}(t) = 1.48\sigma_f^2\lambda^2\left\{\frac{3}{2}\left[\left(\frac{\lambda}{L}\right)^2 - \frac{\lambda}{L}\left(1+\frac{\lambda}{L}\right)\exp\left(-\frac{L}{\lambda}\right)\right] - \frac{1}{2}\left[Ei\left(-\frac{L}{\lambda}\right) - \ln\left(\frac{L}{\lambda}\right)\right] - \frac{3}{4} + \frac{1}{2}E\right\} \tag{9.8.20}$$

where σ_f^2 is the variance of ln*K*; $\lambda = \lambda_x = \lambda_y$ is the correlation scale of the *K*; $E = 0.5772\cdots$ is Euler's constant; *Ei* is the exponential integral; and $L = \bar{u}t$. These equations are for the migration of the solute plume under a two-dimensional flow, representing vertically averaged conditions.

After all, Eqs. (9.8.19) and (9.8.20) are valuable and practical tools for predicting the evolution of a tracer plume's spatial variances (lateral spreading or size) under steady-state mean uniform flow conditions.

These stochastic theories extended previous molecular diffusion, pipe dispersion, and atmospheric diffusion theories to dispersion in geologic media. They are a breakthrough in hydrogeology. They not only elucidate factors controlling macrodispersion processes in large-scale aquifers but also developed tools for directly estimating either the macrodispersivity or the evolution of the plume's size from limited samples of hydraulic conductivity in a field site. Most importantly, their works and others (such as Matheron and de Marsily, 1980 and Naff et al., 1988) on stochastic modeling of macrodispersion provide a better understanding of the philosophy and limitations of ADE for predicting solute movements in inherently heterogeneous groundwater basins. Their application to a field tracer experiment is discussed in Chapter 10.

9.9 Homework

1. Is detrending a field K dataset necessary when the macrodispersion theories are applied to this field site? Hint: these theories are ensemble mean approaches.
2. Could an autocovariance function have negative values, even though the variance must be positive? What does the negative autocovariance mean?
3. The local-scale Ks at the Borden site were collected along two-dimensional cross-sections (i.e., 36 in the vertical and 20 in the horizontal in A–A' section). How could the data set be analyzed using the one-dimensional formula in Eq. (1.4.12)? Of course, we assume that the principal directions of the correlation scales align with the directions of the x and y coordinates of the sampling network.

10

Field-Scale Solute Transport Experiments under Natural Gradient

10.1 Introduction

This chapter reviews two well-controlled field tracer experiments – Borden aquifer in Canada and the Macrodispersion Experiment (MADE) Site, Columbus Air Force Base in northeastern Mississippi. Both experiments injected tracers in aquifers and monitored their movements over fields at 10–100 meters under natural gradient conditions. The behaviors of the tracer plumes at these two sites are distinctly different because the Borden aquifer is statistically homogeneous, and the aquifer of MADE is statistically heterogeneous. As a result, the validity of the classical ADE and non-Fickian dual-domain models for solute transport in aquifers becomes a contentious debate in the groundwater hydrology community. This controversy deserves articulating the differences between the theories based on the ensemble mean nature of the ADE and the observations in one realization.

These two field experiments, in essence, provided opportunities for understanding solute transport theories and mathematical models developed from soil-column experiments under real-world scenarios where the subsurface is heterogeneous at multi-scales, and groundwater flow varies spatiotemporally. More importantly, the domain scale of heterogeneity is larger than the observation, model, and interest scales. We explore and discuss the strengths and weaknesses of the theories and models. The complication in comprehensive mapping of multidimensional solute distribution over aquifers is further exposed.

10.2 Borden Site, Canada

The tracer experiment was conducted in the shallow water table aquifer at Canadian Forces Base Borden, Ontario (Freyberg, 1986; Mackay et al., 1986; Roberts et al., 1986; Sudicky, 1986). The goal was to produce a detailed database describing conservative and selected halogenated organic tracers' transport, transformation, and fate in the saturated zone. Such a database was necessary to quantitatively test the hypothesized models for describing the dominant fundamental processes (Mackay et al., 1986).

10.2.1 Site Description and Experiments

The aquifer primarily consists of horizontal, discontinuous lenses of medium-grained, fine-grained, and silty fine-grained sand. Infrequent silt, silty-clay, and coarse sand layers also exist. The water table's fluctuation at the tracer site varies seasonally within about a 1.0 m interval below the ground surface.

In late August 1982, 2 m^3 of a combined chloride and bromide tracer solution and five halogenated organic solutes were injected into the aquifer and allowed to migrate under natural gradient conditions. Following injection of the tracer solution into a zone between 2.0 and 3.6 m below the ground surface, an array of 275 multilevel sampling devices monitored the spatiotemporal evolution of the tracer plume (Fig. 10.1). Each device extracted water samples from 14 to 18 discrete points over depth, with the vertical interval varying between 0.2 and 0.3 m.

In addition, an array of piezometers at various depths in the aquifer collected the hydraulic heads at the site. Contours of the water table position measured on August 10, 1984, and November 15, 1984, are also shown in Fig. 10.1. The hydraulic head contours indicate that the head field at the tracer site was relatively smooth and uniform but flow meandered horizontally. The smooth head contours suggest that the aquifer is statistically homogeneous (Chapter 9), without apparent large-scale heterogeneity. The average vertical gradient is downward according to the vertical movement of the tracer plume.

As of May 31, 1984, over 18,700 groundwater samples were collected for chemical analyses, capturing the 3-D spatial distribution of the tracer cloud for 12 sampling sessions. Fig. 10.2 presents contours of vertically averaged solute concentration for chloride at 1, 85, 462, and 647 days and carbon tetrachloride at 16, 380, and 633 days after injection. Initially, the plumes were nearly rectangular in plan view. They then moved at an angle to the x, y, and z coordinate system defined in Fig. 10.1 and became progressively more ellipsoidal. The chloride plume moved at a constant velocity, yet a distinct bimodality developed during the first 85 days and remained visible after 647 days. Significant spreading in the longitudinal direction and dilution were apparent for both the inorganic and organic plumes. Relatively minor horizontal transverse spreading was observed.

Fig. 10.2 also shows that the carbon tetrachloride mobility is significantly less than the chloride, supporting retardation due to chemical reactions. The retardation of the other organic solutes was even more significant (Roberts et al., 1986), generally in accord with their hydrophobicity.

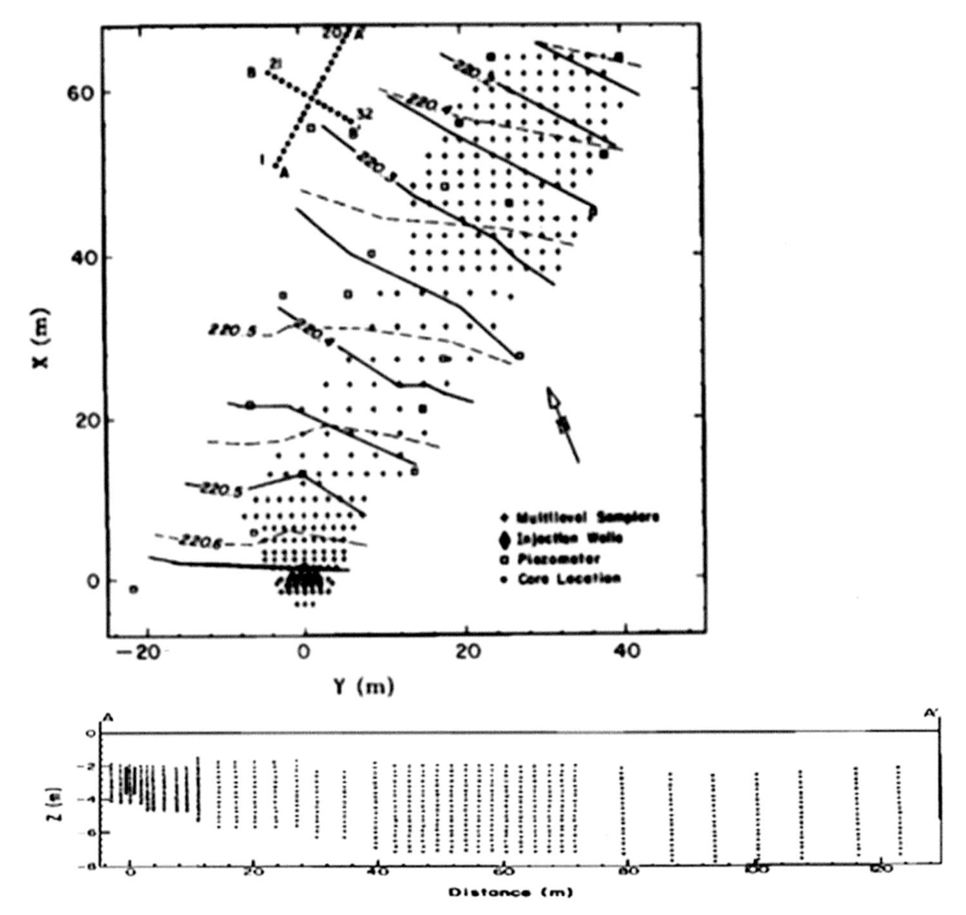

Figure 10.1 The plan view and cross-sectional view of the tracer and water level sampling array. It also shows the water level contours at two times, indicating the general flow conditions. Sudicky (1986).

The carbon tetrachloride mobility decreased somewhat with time; the plume had not traveled as far after 633 days as expected based on its position after 380 days and a constant velocity. Roberts et al. (1986) reported the evidence of the decreased mobility, which increased retardation of the organic solutes over time.

Concentration distributions of the chloride plume in vertical sections (1 and 462 days after injection) are illustrated in Fig. 10.3. It shows that the vertical component of the chloride plume movement is small, and its spreading in the horizontal direction is greater than the vertical.

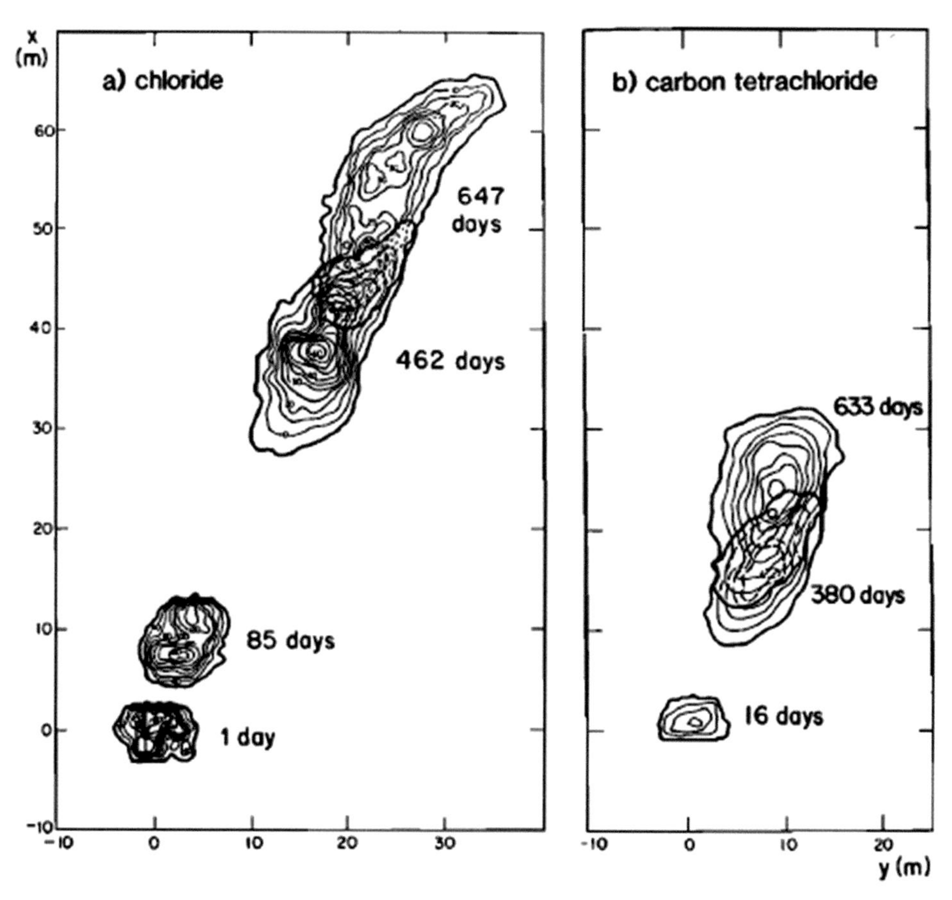

Figure 10.2 The plan views of the spatiotemporal evolution of the chloride plume (left) and the carbon tetrachloride plume (right). Mackay et al. (1986).

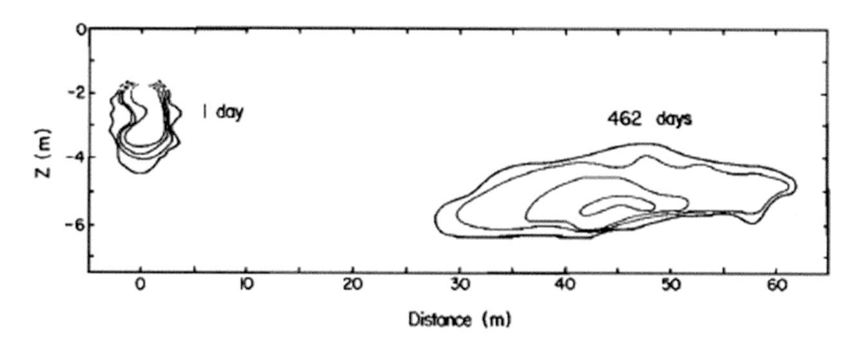

Figure 10.3 The cross-section view of the migration of the chloride plume. Mackay et al. (1986).

10.2.2 Quantitative Description of the Tracer Plume

The tracer spatial moments analysis quantifies the plume distributions. It can also evaluate the plume's advective and dispersive characteristics, as presented in Chapter 8. Freyberg (1986) estimated the spatial moments for the 3-D chloride plume distribution at time t using the formula by Aris (1956).

$$M_{ijk} = \int_{-\infty}^{\infty}\int\int x^i y^j z^k nC(x, y, z, t)dxdydz. \tag{10.2.1}$$

where $i,j,k = 0$, 1, and 2, and n is the mean porosity of the site. The field x–y coordinate system is shown in Fig. 10.1, with z being vertically downward from the land surface. Notice that this moment method applies to any distributions.

10.2.2.1 The Zeroth and First Moments

Using Eq. (10.2.1) with that i, j, and k are zero, the plume's zeroth moment represents the tracer's total mass at time t. When one of the indexes is 1 and the others are zero, Eq. (10.2.1) yields the first moment of the plume, which can determine the coordinate of the center of the plume (not necessarily the peak of the plume):

$$\begin{aligned} x_c &= M_{100}/M_{000} = x \text{ coordinate of the center of mass at time } t. \\ y_c &= M_{010}/M_{000} = y \text{ coordinate of the center of mass at time } t. \\ z_c &= M_{001}/M_{000} = z \text{ coordinate of the center of mass at time } t. \end{aligned} \tag{10.2.2}$$

If several plume snapshots (at least two) are available, the velocity components of the center of the plume at time t can be determined by

$$u_x = \frac{dx_c}{dt}, \quad u_y = \frac{dy_c}{dt}, \quad u_z = \frac{dz_c}{dt}. \tag{10.2.3}$$

The zeroth moments for bromide and chloride in Fig.10.4 show that both tracers' total mass normalized by their known input mass fluctuates with time. The consistent underestimate of the input mass of both tracers at the early time in Fig. 10.4 likely indicates that the sampling network near the injection location was not sufficiently dense to capture the actual plumes or interpolation errors. On the other hand, the plumes at the late time may have spread out more extensively than the area of the sampling network and led to the underestimation. The fluctuations in the middle time could be due to measurement errors or the variability of the porosity used in calculating the moments, which was assumed uniform for the moment calculation.

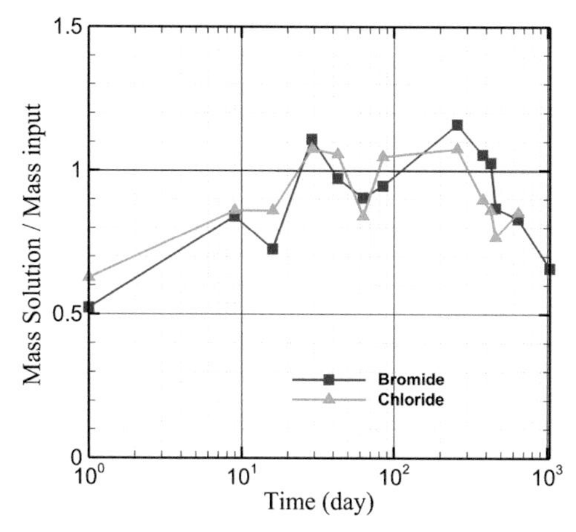

Figure 10.4 Estimated mass in chloride and bromide solutions relative to their input masses as a function of time. Modified from Freyberg (1986).

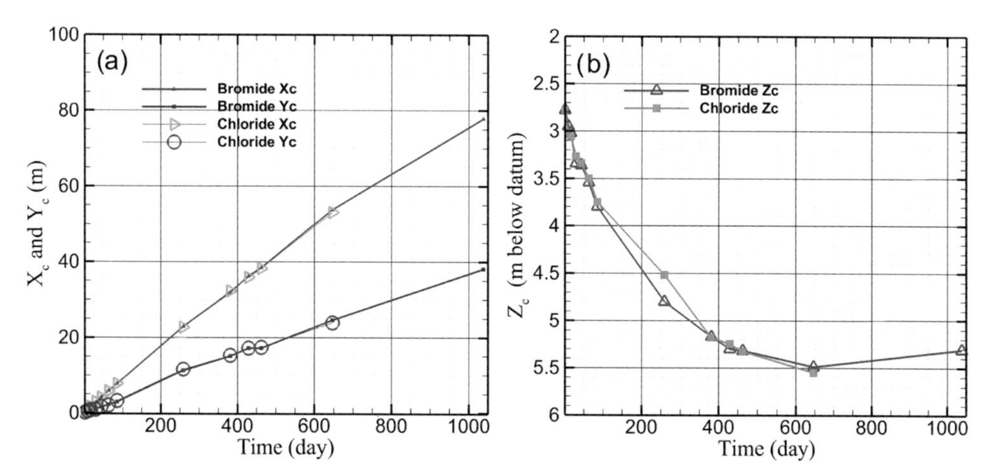

Figure 10.5 (a) The plot of the center of mass x and y coordinates as a function of time and (b) the center of mass z coordinate as a function of time.

The estimated coordinates x_c, y_c, and z_c of the chlorite plumes' center of mass are plotted as a function of time in Fig. 10.5. Fig. 10.5a indicates that the center of mass's movement rate in the horizontal x-y plane is nearly constant, with velocity components $u_x = 0.082$ m/day and $u_y = 0.038$ m/day (Fryberg, 1986). However, slight changes in the x- and y-directions are likely due to regional flow changes. Overall, the plume center moved at 0.09 m/day along a N47° trajectory (25° clockwise to the field x-axis).

The plot of z_c vs. time (Fig. 10.5b) reveals that the plume moved downward at a decreasing rate during the first 200–300 days. The rate is relatively constant at larger times with a velocity of about 10^{-3} m/day. This velocity is nearly two orders of magnitude less than the horizontal velocities. The tracer's density likely caused the plume's rapid downward movement during the early transport stages, although a small vertical gradient existed. Then, the plume spread and gradually was diluted, so the effects of density became small.

10.2.2.2 The Second Moments

The shape or the spread of the plume in 3-D is defined by

$$\begin{aligned}
\sigma_{xx}^2 &= M_{200}/M_{000} - x_c^2 = \text{spatial variance (or spread) in } x \text{ direction} \\
\sigma_{yy}^2 &= M_{020}/M_{000} - y_c^2 = \text{spatial variance (or spread) in } y \text{ direction} \\
\sigma_{zz}^2 &= M_{002}/M_{000} - z_c^2 = \text{spatial variance (or spread) in } z \text{ direction} \\
\sigma_{xy}^2 &= M_{110}/M_{000} - x_c y_c = \sigma_{yx}^2 = \text{spatial variance (or spread) in } x - y \text{ direction} \\
\sigma_{xz}^2 &= M_{101}/M_{000} - x_c z_c = \sigma_{zx}^2 = \text{spatial variance (or spread) in } x - z \text{ direction} \\
\sigma_{yz}^2 &= M_{011}/M_{000} - y_c z_c = \sigma_{zy}^2 = \text{spatial variance (or spread) in } y - z \text{ direction}
\end{aligned} \tag{10.2.4}$$

and the spatial variance matrix (tensor) thus becomes

$$\sigma_{ij}^2 = \begin{bmatrix} \sigma_{xx}^2 & \sigma_{xy}^2 & \sigma_{xz}^2 \\ \sigma_{yx}^2 & \sigma_{yy}^2 & \sigma_{yz}^2 \\ \sigma_{zx}^2 & \sigma_{zy}^2 & \sigma_{zz}^2 \end{bmatrix}. \tag{10.2.5}$$

If the x, y, z coordinate system of the plume are rotated to a new coordinate system (x', y', and z') that coincides with the mean advection directions based on the analysis of the first moment (e.g., Fig. 10.5), the spatial variance tensor becomes

$$\sigma_{ij}^2 = \begin{bmatrix} \sigma_{x'x'}^2 & 0 & 0 \\ 0 & \sigma_{y'y'}^2 & 0 \\ 0 & 0 & \sigma_{z'z'}^2 \end{bmatrix}. \tag{10.2.6}$$

The off-diagonal terms vanish. Again, the symmetry of the diagonal terms in Eq. (10.2.5) rests upon the ensemble mean concept, and so does their disappearance. Further, they may not hold in one single realization of heterogeneous aquifer unless ergodicity is met. Besides, the mean advection direction under field conditions likely changes over time, as shown in Fig. 10.5.

After evaluating the moments, following the diffusion concept in Chapters 4–6, if we assume the spread of the plume reaches the Fickian regime, the rate of spreading of a solute plume can be related to the dispersion coefficient tensor

$$D_{ij} = \frac{1}{2}\frac{d\sigma_{ij}^2}{dt}. \tag{10.2.7}$$

This coefficient is the macrodispersion coefficient, representing the effects of velocity variation at a scale larger than those at molecular, fluid dynamic, and pore scales. Suppose that $D_{ij} = A_{ij}|U|$ is true as is in soil column experiments (Chapter 5). The A_{ij} is the macrodispersivity and $|U|$ is the magnitude of the velocity in the mean flow.

Instead of examining the 3-D spatial variances, Freyberg (1986) averaged the concentration vertically across a 6.0-m-depth interval containing the tracer plume and investigated spatial variance tensor components in the horizontal plane.

The spatial variances $\sigma_{x'x'}^2\sigma_{y'y'}^2$, and $\sigma_{x'y'}^2$, after rotating to the new coordinate system x' and y' that is aligned with the mean velocity vector, as functions of time are shown in Fig. 10.6a, and b, respectively. This figure includes both chloride and bromide spatial variances. Notice that in Fig. 10.6b, the off-diagonal term $\sigma_{x'y'}^2$ does not vanish after the coordinate transformation. This non-zero term may indicate the effect of assumed unchanged mean flow directions, deviating from Fig. 10.5a. Alternatively, it reflects that the plume did not reach the ergodic condition that the ensemble nature of the coordinate transformation concept is applicable. Specifically, the sample size must be sufficiently large such that the ensemble theory (vanish of the off-diagonal terms) fits the data from one realization. Of course, measurement and other errors may play some roles.

Despite these issues, Freyberg used a linear relationship to fit all the data points except the last point (Fig. 10.6). He excluded the last one because the mean velocity field changed during the last sampling period of the experiment. He then calculated the macrodispersivities from the slopes of the lines, assuming that the spread of the plume has reached the Fickian regime (Eq. 10.2.7) and obtained $A_{x'x'} = 0.36$ m, $A_{y'y'} = 0.039$ m, and $A_{x'y'} = 0.023$ m.

Besides, he used the following formula to estimate the macrodispersivity values over successive periods:

$$A_{x'x'}(t) = \frac{1}{2|u|}\frac{\left(\sigma_{x'x'}^2(t) - \sigma_{x'x'}^2(1)\right)}{(t-1)} \quad A_{y'y'}(t) = \frac{1}{2|u|}\frac{\left(\sigma_{y'y'}^2(t) - \sigma_{y'y'}^2(1)\right)}{(t-1)} \tag{10.2.8}$$

where the values of $\sigma_{x'x'}^2(t)$, $\sigma_{y'y'}^2(t)$, $\sigma_{x'x'}^2(1)$ and $\sigma_{y'y'}^2(1)$ are the spatial variances at a given t and t = 1 day, respectively (see Table 10.1). The denominator $|u|$ is the mean velocity over the time interval between $t = 1$ to a given t.

The average values of the macrodispersivity $A_{x'x'}(t)$ and $A_{y'y'}(t)$ of the two tracers are shown in Fig. 10.7a and b, respectively. These figures reveal that macrodispersivity values scatter, but the longitudinal macrodispersivity $A_{x'x'}(t)$

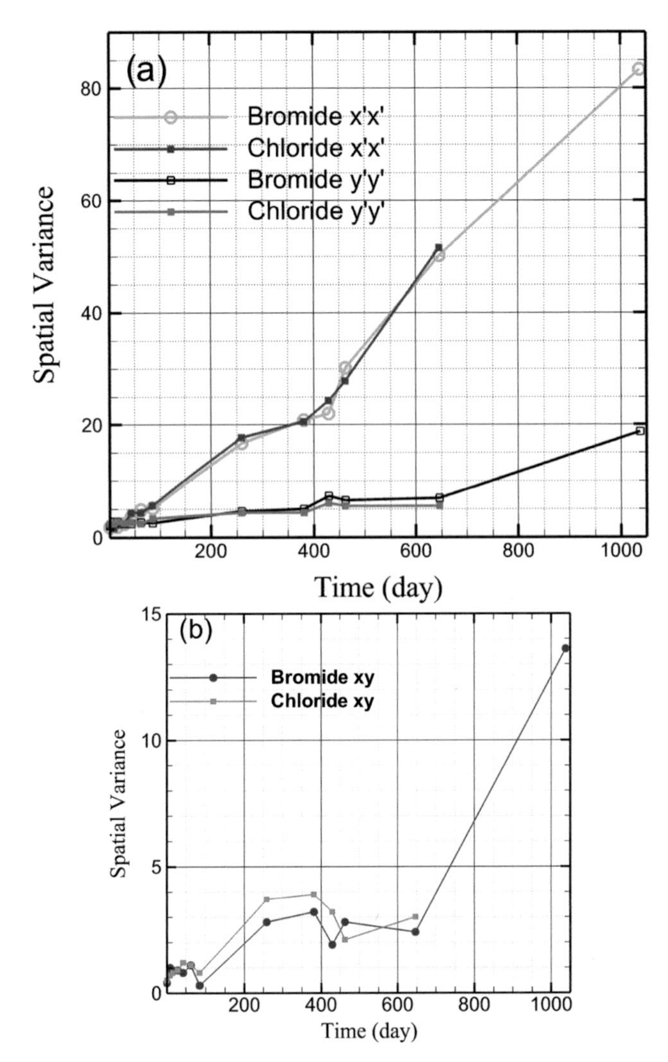

Figure 10.6 The estimated spatial covariance components (a) $\sigma^2_{x'x'}$ $\sigma^2_{y'y'}$ and (b) $\sigma^2_{x'y'}$ (given in Table 10.1) of the depth-averaged concentration plume at a function of time. Freyberg (1986).

increases with time, exhibiting a scale-dependent dispersive behavior (Fig. 10.7a). The value of $A_{x'x'}(t)$ increases from 0.06 to 0.43 m throughout the experiment. The $A_{y'y'}(t)$ is large at early times but shows no significant increase in magnitude with time.

Freyberg (1986) and Sudicky (1986) noticed that the magnitudes of the longitudinal dispersivity are at the low end of the range of dispersivities reported by Anderson (1979). They are, however, much larger than values commonly obtained by tracer tests from a laboratory column containing homogeneous sand.

Table 10.1 *Estimates of mass in solution, location of center of mass, and spatial covariance for the bromide and chloride plumes*

			Center of mass			Spatial covariance		
Date	Elapsed Time, Days	Mass in Solution, kg	x_c	y_c	z_c	$\sigma^2_{x'x'}$	$\sigma^2_{y'y'}$	$\sigma^2_{x'y'}$
Bromide								
Aug. 24, 1982	I	2.02	0.3	−0.2	2.78	1.8	2.0	0.4
Sept. 1, 1982	9	3.25	0.6	0.4	2.95	2.3	2.8	1.0
Sept. 8, 1982	16	2.81	1.7	0.9	3.02	1.9	2.8	0.9
Sept. 21, 1982	29	4.29	2.7	1.0	3.34	2.4	2.5	0.9
Oct. 5, 1982	43	3.76	4.1	1.5	3.36	3.6	2.5	0.8
Oct. 25, 1982	63	3.50	5.8	2.2	3.54	4.9	2.7	1.1
Nov. 16, 1982	85	3.66	7.7	2.9	3.80	5.2	2.6	0.3
May 9, 1983	259	4.49	22.9	11.3	4.80	16.7	4.6	2.8
Sept. 8, 1983	381	4.08	31.9	15.1	5.17	20.8	5.0	3.2
Oct. 26, 1983	429	3.97	36.2	17.2	5.30	22.0	7.3	1.9
Nov. 28, 1983	462	3.36	38.4	17.2	5.32	30.1	6.6	2.8
May 31, 1984	647	3.21	53.8	24.6	5.49	50.1	6.9	2.4
June 26, 1985	1038	2.55	77.8	38.1	5.31	83.3	18.8	13.6
Chloride								
Aug. 24,1982	1	6.7	0.2	0.1	2.78	2.1	2.4	0.5
Sept. 1, 1982	9	9.2	0.7	0.4	3.02	1.7	2.4	0.7
Sept. 8, 1982	16	9.2	1.6	0.7	3.06	2.3	2.8	0.8
Sept. 21, 1982	29	11.5	2.9	0.9	3.27	2.5	2.6	0.9
Oct. 5, 1982	43	11.3	4.1	1.6	3.34	4.4	2.7	1.2
Oct. 25, 1982	63	9.0	5.7	2.0	3.50	4.4	2.4	1.1
Nov. 16,1982	85	11.2	7.7	3.2	3.75	5.7	3.3	0.8
May 9, 1983	259	11.5	22.7	11.6	4.52	17.8	4.4	3.7
Sept. 8, 1983	381	9.6	32.3	15.3	5.18	20.6	4.4	3.9
Oct. 26, 1983	429	9.2	35.9	17.2	5.25	24.3	6.0	3.2
Nov. 28,1983	462	8.2	38.2	17.4	5.33	27.8	5.5	2.1
May 31, 1984	647	9.1	53.1	23.9	5.55	51.5	5.5	3.0

The center of mass is given in the field coordinate system (x, y, and z).
The spatial covariance is described in the rotated coordinates: x' is parallel to the linear horizontal trajectory, y' is perpendicular to the linear horizontal trajectory (Freyberg, 1986).

Different from the analysis by Freyberg (1986), Sudicky (1986) carried out a detailed characterization of the spatial variability of K at a small plot adjacent to the Borden tracer test site(see Chapter 9) to obtain the spatial statistics of the plot. Assuming the statistics represent the entire tracer site, he applied Gelhar and

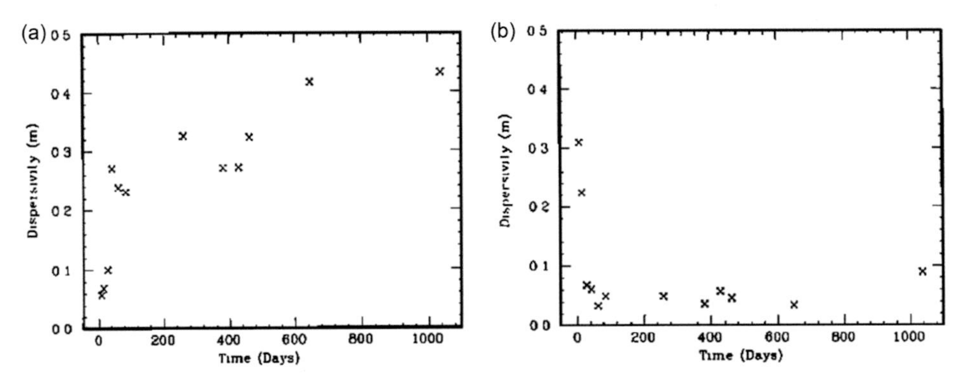

Figure 10.7 (a) The estimated longitudinal macrodispersivity and (b) the transverse macrodispersivity as a function of time. Freyberg (1986).

Axness's theory (1983) to determine the components of marcodispersivity for the entire tracer site from Eq. (9.5.18) in Chapter 9.

Evaluating the macrodispersivity formulas of Gelhar and Axness (1983) requires the value for the geometric mean hydraulic conductivity and correlation scales λ_x, λ_y, and λ_z to estimate the effective hydraulic conductivity tensor of the entire tracer field. From the spatial statistics from the characterization plot, Sudicky derived the effective hydraulic conductivity:

$$\begin{bmatrix} K_{xx} & & \\ & K_{yy} & \\ & & K_{zz} \end{bmatrix} = \begin{bmatrix} 8.20 \times 10^{-3} & & \\ & 8.20 \times 10^{-3} & \\ & & 6.33 \times 10^{-3} \end{bmatrix}. \quad (10.2.9)$$

These ensemble effective conductivity components are in cm/s and for a 10° C field temperature, indicating the hydraulic conductivity anisotropy (K_{xx}/K_{zz}) about 1.3. Note that this ensemble anisotropy is a result of the statistical anisotropy of the heterogeneous local-scale K field (i.e., disk-shaped heterogeneity, $\lambda_x = \lambda_y$ and $\lambda_x/\lambda_z = 23$.)

Next, he determined the value for the horizontal component of the groundwater velocity calculated using K_{xx}, the mean hydraulic gradient J = 4.3×10^{-3}, and a porosity equal to 0.34 and obtained a value of about 9.0 cm/day. This value is very close to the tracer velocity determined from the moment analysis (9.1 cm/day).

Using $J_z = n\, u_z/K_{zz}$, where $n = 0.34$ is the porosity, $u_z = 0.1$ cm/day (the vertical component of the tracer velocity estimated from the moment analysis (see Fig. 10.5b)), and K_{zz}, he obtained the estimated value for J_z about 6×10^{-5}. With this value for J_z, the following values for A_{ij} were derived from Eq. (9.5.18, Chapter 9):

$$\begin{bmatrix} A_{xx} & A_{xy} & A_{xz} \\ A_{yx} & A_{yy} & A_{yz} \\ A_{zx} & A_{zy} & A_{zz} \end{bmatrix} = \begin{bmatrix} 0.61m & \simeq 0 & \simeq 0 \\ \simeq 0 & \simeq 0 & \simeq 0 \\ \simeq 0 & \simeq 0 & \simeq 0 \end{bmatrix}. \quad (10.2.10)$$

Because the inclination angle θ and magnitude of J_z are small, the estimated values for other components in Eq. (10.2.10) are extremely small. Therefore, the macrodispersivities A_{ij} in Eq. (10.2.10) are in the principal directions. Such small A_{ij} values in transverse directions indicate that the column-scale dispersivity values are small, accounting for pore-scale velocity variation. The weak transverse dispersivity values are consistent with the observations (Figs. 10.2 and 10.3).

Rather than determining the macrodispersivity values for the ensemble ADE to evaluate ensemble mean concentration distributions, Sudicky also evaluated the spatial covariance of the plume in the x' -direction $\sigma^2_{x'x'}$ and in the y' -direction $\sigma^2_{y'y'}$ using Dagan's theory Eqs. (9.5.19) and (9.5.20) in Chapter 9 with estimated spatial statistics and the calculated v = 9.0 cm/day. The coordinates x' and y' align with those of the mean flow direction. The results are shown in Fig. 10.8, compared with actual values determined from the field test. The field estimates of $\sigma^2_{x'x'} = 1.8$ m^2 and $\sigma^2_{y'y'} = 2.6$ m^2 at t = 0 were added Eqs. (9.5.19) and (9.5.20), respectively, to account for the initial dimensions of the tracer slug following injection. From Fig. 10.8, we observe that the theory provides reasonable agreement with the actual spread of the tracer in the longitudinal direction throughout most of the duration of the experiment.

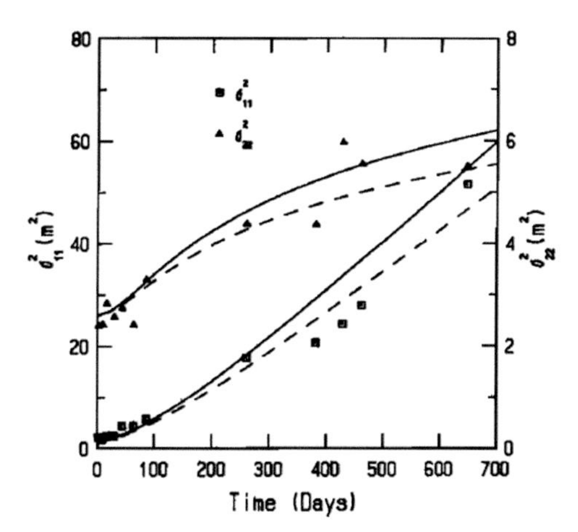

Figure 10.8 The spatial variances in the horizontal plane of observed plumes (symbols) at different times and those by the theory (solid lines). The left vertical axis and triangles are for $\sigma^2_{x'x'}$ and the right and square for $\sigma^2_{y'y'}$. Sudicky (1986).

The results from calibrating Eqs. (9.5.19) and (9.5.20) to the estimated spatial variances from observed data by Freyberg are shown as the dashed curves in Fig. 10.8. The best fit parameter values are: u = 0.091 m/day, $\sigma_f^2 = 0.24$, and $\lambda = \lambda x = \lambda y = 2.7$ m. These values are close to the values estimated from the observed ln$K$ variability (Sudicky, 1986). Considering the scatter of the moment data and the uncertainty of the statistical parameters describing the ln$K$ variability, each of the predictions shown in Fig. 10.8 is equally acceptable. Notice that the spatial moments describing the evolution of the plume do not invoke Fickian or the normal distribution assumption. In other words, calibrating an ADE based on EHM to fit the plume data may not be the same as using the moment analysis to calculate the moments and corresponding parameters (mass, velocity, and dispersion coefficient) unless the plume is approximately Gaussian.

10.3 Macrodispersion Experiment (MADE) Site, Mississippi

After the Borden site experiment, Boggs et al. (1992), Adams and Gelhar (1992), Rehfeldt et al. (1992), and Boggs and Adams (1992) conducted a large-scale natural-gradient tritium tracer test in an alluvial aquifer at Columbus Air Force Base in northeastern Mississippi. This experiment is commonly referred to as the Macrodispersion Experiment or MADE site. The average thickness of the aquifer formation is 11 m, and it is composed of poorly sorted to well-sorted sandy gravel and gravelly sand with minor amounts of silt and clay.

The hydraulic head field at the site exhibits complex temporal and spatial variability due to the heterogeneity of the aquifer, and it also displays large seasonal fluctuations in the water table. The potentiometric surfaces based on head measurements from the deep and shallow piezometer networks are shown in Fig. 10.9. The general direction of groundwater movement is northward. However, local differences in the magnitude and direction of the horizontal hydraulic gradient are evident, particularly in the far-field region of the site.

The V-shaped contours in the southern part of the site indicate converging groundwater flow toward a narrow zone of relatively high hydraulic conductivity aligned with the vertex of the V-shaped contours, leading to an increasing groundwater velocity. The potentiometric surface in Fig. 10.9 also reveals the relative K distribution at the test site. The widely spaced contours (corresponding to small hydraulic gradients) shown in the far-field region of the test site indicate a relatively high K zone. In contrast, the closely spaced contours (large hydraulic gradients) in the near-field indicate a relatively low K zone. Specifically, a large-scale trend of K variation exists, and the aquifer is statistically heterogeneous, quite different from the Borden site, which is statistically homogeneous (Fig. 10.1).

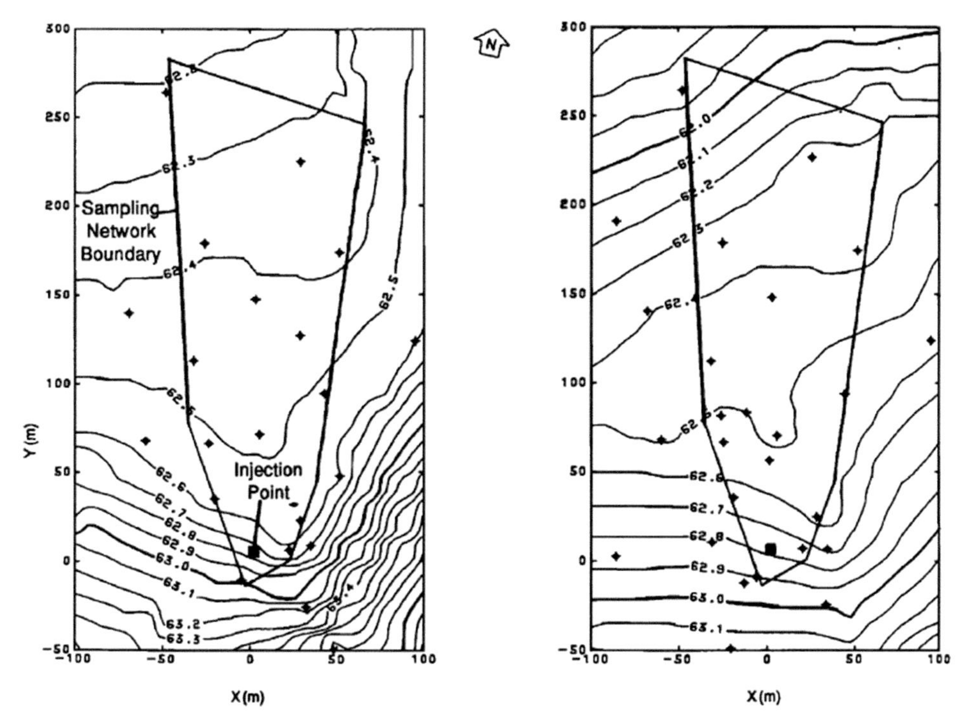

Figure 10.9 Potentiometric surface maps derived from October 11, 1989, head measurements in (left) shallow and (right) deep observation wells. Mean well screen elevations for shallow and deep well networks were 61.1 m and 56.3 m, respectively. From Bogg et al. (1992).

Rehfeldt et al. (1992) conducted 2187 borehole flow-meter tests in 49 wells at the site. They confirmed the statistically heterogeneous nature of this aquifer with large-scale trends and reported a variance of $\ln K = 4.5$ for the entire aquifer. This magnitude is significantly greater than any previously reported major tracer test sites. For example, the reported variance of $\ln K$ is 0.29 for the Borden site (Sudicky, 1986), 0.26 for the Cape Cod site (LeBlanc et al., 1991), and 0.031 for the Twin Lake site (Killey and Moltyaner, 1988). In addition, notice that the tracer injection point (black square in Fig. 10.9) is at the high gradient region (a low K zone), indicating that the solute migration in this experiment would behave differently from the Borden experiment.

Three natural-gradient tracer tests were conducted at this site. Zheng et al. (2011) provided an extensive overview of these experiments. In the first test (MADE-1, October 1986–June 1988), conservative tracers, including bromide, were injected into a linear array of wells over 48 hours (Boggs et al., 1990, 1992). The plume migration was then monitored through a network of multilevel samplers. Eight sampling events were conducted over the test period to obtain

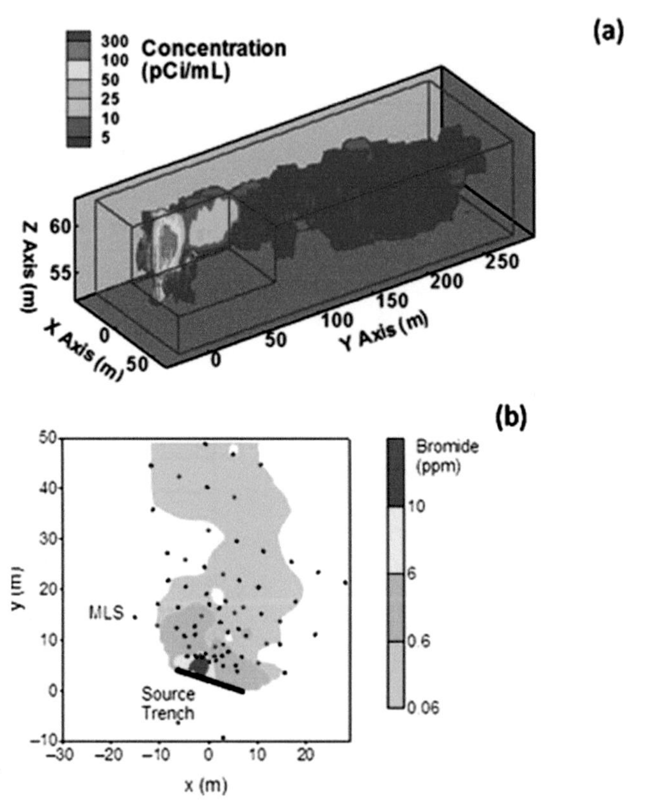

Figure 10.10 (a) The snapshot at 328 days after the injection in MADE-2 experiment. (b) the plane view of the plume 152 days after injection in MADE-3 experiment. Zheng et al. (2011).

snapshots of the plume. Snapshots of the plume in a cross-sectional view are shown in figs. 7, 8, and 9 of Boggs et al. (1992).

From June 1990 to September 1991, tritium was injected into the same array over 48 hours (Boggs et al., 1993; MacIntyre et al., 1993) and sampled in monitoring wells on five occasions. This experiment was named MADE-2. The snapshot at 328 days after the injection is shown in Fig. 10.10a.

The third experiment (MADE-3) started from December 1995 to September 1997, in which hydrocarbon- and bromide-coated sand was placed in a trench near the previous injection array (Boggs et al., 1995; Libelo et al., 1997; Stapleton et al., 2000; Brauner and Widdowson, 2001; Julian et al., 2001), and concentrations were sampled six times over 20 months. The snapshot at 152 days is shown in Fig. 10.10b.

All snapshots reveal that the bromide and tritium plumes spread out highly asymmetrically (a typical non-Fickian behavior). Specifically, the highest concentrations stayed close to the source, while the plume fronts spread downstream far away from the source in diluted concentrations. These non-Gaussian distributions are strikingly different from those observed at the Borden site and consistent with the observed hydraulic head field (Fig. 10.9). They deviate from the expected behaviors based on Fick's law. Such anomalous behaviors have fostered vigorous discussions and debates within the groundwater research community about the validity of the classical ADE and new theories for modeling solute transport (e.g., Harvey and Gorelick, 2000; Hill et al. 2006; Molz et al., 2006; Guan et al., 2008).

Llopts-Albert and Capilla (2009) and Zheng et al. (2011) reviewed many past modeling attempts to reproduce the observed non-Fickian behaviors at MADE site. They questioned the adequacy of the classic ADE based on Fick's law, developed for solute transport in pipes and soil columns per the scale of our model, observation, and interest. As such, they championed the adoption of the dual-domain (mobile/immobile) (Chapter 7) coupling HPHM (highly parameterized heterogeneous model) for velocity variation and a spatially uniform mass transfer mechanism for the effects of unaccounted heterogeneity in HPHM.

10.4 Summary and Discussions

Results of experiments at the Borden and MADE sites lead to several discussion points. Table 10.2 summarizes the differences in the two tracer tests prior to the discussion.

Table 10.2 *Results of experiments carried out at the Borden and MADE sites*

	Borden	MADE
size	120 m ×120 m	200 m × 200 m
Head field	Uniform, smooth	nonunifrom Diverging
Heterogeneity	$\sigma^2_{\ln k} = 0.25$, $\lambda_x = \lambda_y = 2.5\ m\ \lambda_z = 0.12\ m$ Statistically homogeneous	$\sigma^2_{\ln k} = 4.5$, $\lambda_x = \lambda_y = 12.8\ m\ \ \lambda_z = 1.6\ m$ Statistically heterogeneous
Snapshot	Approximately Gaussian	Highly skewed distribution
BTC	NA	NA

Borden Site.

- The head contour revealed no apparent large-scale heterogeneity pattern at the site–a statistically homogenous random field.
- A large number of local-scale K measurements at an off-site reported small variance, and small correlation scales, reflecting existing small/thin lenticular layering structures.
- The observed plume exhibited general Gaussian concentration distributions with some multi-peaks, suggesting that EHM with ADE approach is generally valid for one realization of the plume, although small-degree uncertainty exists.
- The spatial statistics from the off-site plot combined with stochastic solute transport theories reproduced the plume data well, confirming the statistical homogeneity of the site – statistics from a plot infers the entire site.

MADE site.

- The aquifer has a noticeable large-scale heterogeneity pattern (a statistically heterogeneous random field), different from the Borden site aquifer, as manifested in the head contours.
- Geostatistical analysis of many K measurements reported large variance and large-correlation scales for the site geological structures.
- The tracer injection point was located at the low permeable zone, retaining most of the injected tracer. The high permeable zone downstream from the injection point and the nearby converging and divergent flow field facilitated the rapid spread of the plume front over a long distance, creating non-Gaussian plume distributions.
- The dominant heterogeneity is much larger than the injections' size, ergodicity is unlikely to be met, and the ensemble mean approach leads to significant uncertainty.

The scale issues related to dispersion discussed in previous chapters should have explained the differences in the plume behaviors at these two sites. Specifically, the dispersion phenomenon stems from simplifying velocity variations at various scales below the scale of the velocity model and observation CV scale (e.g., pipe, soil column, finite difference or finite element block, or a groundwater basin). Dispersion theory is an ensemble-mean concept, which could be valid only when the solute has experienced sufficient heterogeneity to reach ergodicity. In uniformly packed soil column experiments, pore-scale variation is the dominant heterogeneity (i.e., a statistically homogeneous random field). Well-mixed tracers are uniformly injected over the column's cross-sectional area, much larger than many pores. As the tracer travels over a short distance, it encounters different pores and reaches the ergodic condition. Thus, the ensemble-mean Fick's law and ADE predict the BTC averaged over the cross-section area with reasonable satisfaction (Chapter 7).

The heterogeneous ln$K$ field at the Borden has no noticeable trends over the entire test area of 120 m × 120 m. It is a statistically homogeneous random field with small correlation scales (i.e., average dimension of heterogeneity) of about 2.5 m and a small variance of 0.25. Besides, the tracer injection area spanned over 8 m in width, larger than the correlation scale; the tracer plume thus experiences sufficient heterogeneity initially as it does in the soil column experiment. Further, the flow field was generally unidirectional. Hence, the plume behaves more or less according to Fick's law after migrating a short distance.

In contrast, the MADE site (200 m × 300 m) has large-scale high and low permeability zones, evident from the irregularly spaced piezometric surface contour maps (Fig. 10.9). Thus, the site is a statistically heterogeneous random field. Rehfeld et al. (1992) reported that the overall variance of ln$K$ is 4.5, with horizontal and vertical correlations 12.8 m and 1.6 m, respectively. However, these spatial statistics may not represent the tracer encountered since they were derived from the flow meter tests in the center part of the test site. In particular, the injection array spanned about 5m in a large low permeable zone of approximately 200 m × 50 m, where the high hydraulic gradients existed (Fig. 10.9). As a result, most of the injected tracers were trapped in this zone and slowly released to an adjacent high permeability zone. Then, the nearby converging flow facilitated the solute's rapid spread over a long distance. In other words, the tracer initially encountered heterogeneity in this low permeability zone only, which acts as a dead-end or immobile zone.

Due to the large-scale trend, an ensemble REV as large as the site is necessary to homogenize the entire aquifer as an EHM (equivalent homogeneous model). In this case, the ensemble mean plume from either macrodispersion ADE or dual-domain ADE model only agrees with the observed plume in the ensemble-mean sense. The actual plume likely deviates significantly from the prediction.

Although EHM with macrodisperison ADE and dual-domain approaches produce ensemble mean plume behaviors, the latter is conceptually more attractive and has additional adjustable parameters than the former. The effective K and macrodispersivity can be related to the spatial statistics, as shown in the Borden experiment. Nevertheless, the dual-domain model must fit the observed solute pattern to derive the mass transfer coefficient and the fraction of the immobile zone since these parameters are ensemble-averaged (or bulk) products of the aquifer. This fitting enables the dual-domain approach to reproduce closer bulk behaviors of the observed plume than the macrodispersion approach. Regardless, both approaches would yield equally uncertain plume at the scale of our interest (e.g., BTCs at wells).

Zonation models with the ADE for each zone are a step toward the HPHM with the classical ADE (Chapter 9). They adopt an ensemble REV for each zone. Each zone is assumed equivalently homogeneous even though a spatial REV with spatially invariant averaged property does not exist (Chapter 1). As a result, calibrating the model to the observed heads and concentrations is necessary to derive the flow and transport parameters. For example, by calibrating the effective *K*, uniform dispersivity, and recharge of only five geologic zones to the observed heads and concentrations observed at the MADE site, Barlebo et al. (2004) showed that this approach captured the general pattern of the observed plume. Their results suggest the dominance of the large-scale heterogeneity on plume migration at MADE site. Dispersivity (or heterogeneity) within each zone played a minor role.

HPHM is a higher-resolution ensemble mean model than both EHM and zonation models. Feehley et al. (2000) compared the performances of the HPHM with the ADE and HPHM with the dual-domain ADE for predicting the plume distributions at the MADE site. Using the measured *K* data from the flow meter tests, they created a single realization of the conditional random field and a kriged (conditional mean) field. A conditional realization is one of many generated random fields with the same value as the measured K values at sample locations but with random values at other locations. In contrast, the conditional mean field is the average of all possible conditional realizations. Thus, this field agrees with the measured *K* at the sampled location and possesses the averaged values of the realizations at other locations. In effect, this conditional mean field is a kriged field created by kriging. Kriging is a spatial interpolation and extrapolation tool, considering the site's spatial statistics and point measurements (see Yeh et al., 2015). Using the conditional realization and the kriged fields, the HPHM with ADE reasonably reproduced the observed tritium plume above a certain concentration limit. Both approaches failed to reproduce the extensive spreading of the tracer at diluted concentrations as observed in the field. Feehley et al. (2000) claimed that with the conditional random field, the HPHM with dual-domain mass transfer model produced the rapid, anomalous spreading significantly better while retaining high concentrations near the injection point. Nevertheless, they did not present comparisons of predicted and observed BTCs at various locations at the site.

The above comparison of different approaches is not rigorous since it did not compare point concentrations of the snapshots and BTCs at many observation locations. The results are expected because the dual-domain ADE has additional mass transfer mechanisms and parameters to adjust for uncharacterized velocity variations. These variations present the effects of unresolved or uncharacterized heterogeneity at locations where no direct measurement is available or from heterogeneity below the local-scale CV of their model (Chapter 7). The major weakness of the dual-domain model lies in the fact that the new hydraulic tests remain to be developed yet to characterize the volume fraction of mobile/immobile zones and mass transfer coefficients unless a tracer test is conducted. This issue also holds for dispersivity in the HPHM with ADE. Note that the local-scale CV (model block or element) becomes sufficiently small, the velocity is sufficiently characterized, the dispersion within the CV becomes less relevant, and so does the mass transfer mechanism.

Again, Fick's law and mass transfer mechanism are ensemble-mean equations, overlooking the velocity variation below the CV scale that defines Darcy-scale velocity. As we have reiterated in the book, the CV becomes small, the velocity variation within the CV becomes small, and the validity of Fick's law for dispersion and mass transfer mechanism becomes irrelevant unless our observation and interest scale are microscopic.

Without thoroughly conditioning the velocity field with observations, the prediction problem is underdetermined, and many possible solutions match the observed plume's bulk behavior. In other words, both approaches likely produce equally uncertain BTCs at every location and time at the site. Hence, developing high-resolution characterization technology is the future of subsurface hydrology discussed in Chapter 12.

10.5 Lessons Learned

- Our inability to characterize multi-scale velocity variations in aquifers promotes upscaling the principles for column-scale flow and solute transport to aquifers. This upscaling approach fosters the development of the unconditional ensemble-mean EHM with ADE models and, in turn, the effective hydraulic conductivity and macrodispersivity concepts. This approach predicts the statistically most likely flow and solute behaviors under a limited knowledge of aquifer heterogeneity.
- Stochastic macrodispersion theories recognize that spatial variability of local-scale K heterogeneity is the primary cause of local-scale velocity variation, which leads to the spreading of a solute plume in EHM. These theories relate the mean-ADE and macrodispersion processes to the effective K, the variance of the local-scale K, correlation scales, and mean hydraulic gradient to predict the most probable evolution of a trace plume in aquifers in a stochastic framework.
- These theories present a paradigm shift in research to understand the effects of heterogeneity and scale issues in subsurface hydrology.
- Borden site's heterogeneity is statistically stationary (i.e.,, spatial REVs exist) and mild. Its correlation scales of the dominant heterogeneity are much smaller than the distance the tracer plume has traveled. The solute plume, thus, reached approximately the ergodic condition and Fickian regime. The application of the macrodispersion theories led to relatively successful predictions, even though the site heterogeneity is just one realization of the ensemble.

- In contrast to the Borden site, MADE site is a statistically heterogeneous aquifer with a large-scale K trend. The spatial REV does not exist, and the variance of hydraulic conductivity is enormous. Besides, the size of the tracer released is also much smaller than the correlation scale of the dominant heterogeneity. As a result, the tracer was trapped in the large-scale low permeability zone, and the entire tracer could not experience the heterogeneity in the ensemble. Applications of ensemble REV as large as the field site for the EHM with macrodispersion ADE, thus, led to significant uncertainty (i.e., failure to capture the bulk distribution of the observed plume).
- The EHM with the dual-domain approach is also an ensemble mean ADE with an ensemble mean mass transfer mechanism acting as a sink or source to the ADE. This approach is more flexible than EHM with ADE. However, the immobile zone's location and its fraction in the dual-domain media are unknown. This approach performs the best if it is calibrated to observed tracer distributions or BTCs. Nonetheless, the calibrated parameters may change as the plume migrates or vary with the location of the source zone in the same aquifer. For these reasons, EHM with the dual-domain ADE may yield uncertain predictions as the EHM with the ADE.
- Instead of treating the entire domain as an ensemble REV, the zonation approach divides the domain into many zones based on geology. It regards each zone as the CV or ensemble REV with its effective hydraulic parameters and zonal macrodispersivity. In essence, the zonation approach first attempts to seize the large-scale heterogeneity (i.e., the spatial trend of the local-scale heterogeneity). After appropriately determining each zone's effective hydraulic parameter and macrodispersivity, this approach could mimic the large-scale velocity variation responsible for the observed plume's general behaviors. The effects of heterogeneity below the zones on the plume spread become less significant than in the EHM approach.
- A step toward a higher resolution than the zonation approach is the HPHM approach. It discretizes the domain into numerous small zones or blocks (local-scale CVs) that possess distinct parameters, apprehending greater detail velocity variations than the zonation approach. The effects of heterogeneity below the local-scale CV on solute transport become even less relevant than those below the zone-scale CV in the zonation approach. Hence, the discrepancies in predictions between HPHM with ADE and the dual-domain ADE should be indistinguishable if local-scale K distribution is sufficiently characterized at our interest and observation scale. Developing high-resolution aquifer characterization technology is undoubtedly the future of the subsurface sciences (see Chapter 12).

11

Forced Gradient Field-Scale Tracer Experiments

11.1 Introduction

Chapter 10 presents the field experiments under generally unidirectional mean flow conditions. Under many real-world situations, the mean flow varies spatiotemporally. This section presents field experiments that examine the role of velocity variation at the local and large scale in solute migration.

Laboratory studies have demonstrated that natural organic matter (NOM) can affect contaminant mobility by enhancing the transport of pollutants that bind to mobile NOM in groundwater (Magee et al., 1991; Dunnivant et al., 1992). Conversely, contaminants may be further retarded if the NOM introduced to an aquifer adsorbs to the solid phase, increasing its affinity to bind contaminants (Zsolnay, 1993; Murphy et al., 1994). NOM can also indirectly influence contaminant migration through its effect on the colloidal stability of inorganic colloids. Adsorption of NOM can promote the mobilization (Ryan and Gschwend, 1990) and stability (Amirbahman and Olson, 1993) of colloids within an aquifer and facilitate the transport of contaminants adsorbed on the mobile colloids. Although these laboratory studies improved our understanding of NOM's adsorption on model minerals and aquifer sediments, field-scale transport experiments are necessary to "ground truth" (McCarthy et al., 1996).

For the above purposes, three force gradient tracer experiments were conducted in an unconfined, sandy, coastal plain aquifer in Hobcaw Field at the Baruch Forest Science Institute, Georgetown, South Carolina (Williams and McCarthy, 1991; McCarthy et al., 1993; Mas-Pla, 1993; Yeh et al., 1995; McCarthy et al., 1996). The first experiment (May 1990 experiment) used the field set up in Fig. 11.1a for exploratory purposes. Afterward, the second experiment (May 1992) developed a new site (Fig. 11.1b) on the right of the May 1990 site. At this new site, three-dimensional characterization of the aquifer was carried out. Subsequently, a chloride tracer experiment tested the three-dimensional site characterization. The third

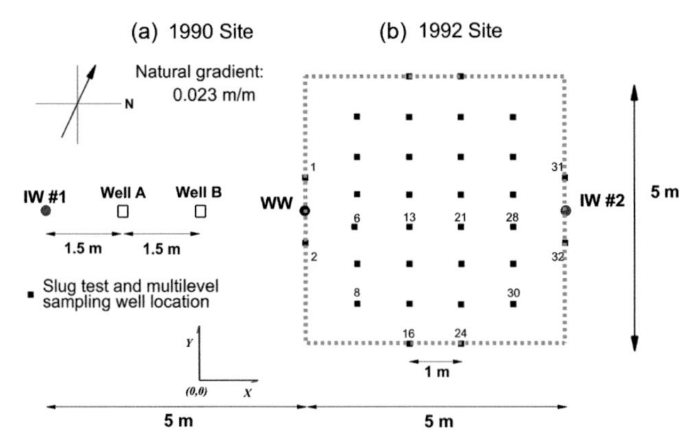

Figure 11.1 Well distribution pattern at the Georgetown site. (a) IW 1: injection well used in the May 1990 test; (b) IW 2: injection well used in the May 1992 and August–September 1992 tests; WW: withdrawal well (Yeh et al., 1995).

injection experiment took place in August–September 1992 to study NOM chemistry mechanisms (McCarthy et al., 1996).

11.2 May 1990 Two-Well Tracer Experiment

The aquifer is approximately 3 m thick, bounded by an impervious clay layer at the bottom, and exhibits distinct layering. The aquifer's upper part consists of a sand layer containing iron oxide and some clay (approximately 9% by weight). Below this layer exists a zone of gleyed sand with 4% clay content, ranging in color from gray to pale olive-gray. The deepest part of this aquifer is composed of a layer of coarse sand, ranging from 0.15 to 0.3 m thick, with a clay content of less than 2%.

The left-hand side of Fig. 11.1 illustrates the injection, withdrawal, and sampling wells for the experiment in May 1990. The distance between the injection and the withdrawal wells is 5 m; two multilevel sampling wells (wells A and B) were located at 1.5 and 3 m from the injection well, respectively. The injection and withdrawal wells were screened over the entire thickness of the saturated zone. Each sampling well had 11 sampling ports at 0.15 m intervals between 1.05 to 2.7 m below the land surface.

The forced gradient was established by recirculating water from the withdrawal well to the injection well at a steady flow rate of 3.7 L/min. After several weeks of recirculation, the flow appeared to reach a steady-state condition. The hydraulic head was measured at the injection, withdrawal, and monitoring wells; the head difference between the injection and withdrawal wells was approximately 0.9 m. A regional flow gradient, approximately 0.005, was observed perpendicular to the well dipole. This gradient was small compared to the overall gradient induced by the forced injection and pumping dipole.

A potassium chloride (KCl) solution was injected into the aquifer under the steady-state flow condition with an average concentration of 140 mg/L for the first 23 hours. Then, the injection concentration was reduced to 30 mg/L for the remainder of the test.

11.2.1 Analysis

The following analysis adopted a parsimonious approach to estimate the solute transport properties of an aquifer. The aquifer was divided into 11 layers of constant thickness, corresponding to the locations of the 11 sampling ports. Each layer was considered homogeneous and isotropic and had a uniform thickness. Because the injection and withdrawal wells were fully screened, flow within each layer was assumed to be two-dimensional horizontal flow, neglecting possible vertical flow components. The streamline and head patterns are presented in Fig.11.2a, showing that the injected tracer at the injection well diverges outward and converges at the withdrawal well. Because of the converging flow to the withdrawal well, the BTC collected at the withdrawal well should reflect the effects of large-scale velocity variations (i.e., tracer's arrival from different streamlines) and the pore-scale velocity variations due to pore-scale heterogeneity.

The aquifer is seemingly homogeneous at the scale of our observation or based on our common perception of a coastal sandy aquifer. While the dipole flow pattern is well-documented in many textbooks, the velocity along each streamline is unknown without knowing the K of each layer. To overcome this difficulty, Mas-Pla et al. (1992) assumed that under the steady flow, the hydraulic gradient, varying along the streamline, is constant over time and for each layer. Then, they estimated the flow rate for each layer at the injection or withdrawal well $Q(z)$ according to:

$$Q(z) = \left(\frac{K_r}{K_{\text{ave}}}\right) Q_{\text{aver}} \tag{11.2.1}$$

where K_r denotes the relative hydraulic conductivity, which is the actual K normalized to the highest K among all the layers (Pickens and Grisak, 1981). K_r was estimated by computing the relative arrival time of the peak concentration in each BTC conductivity, i.e., as $t_{\min}/t_i$, where t_i is the peak arrival time for layer i, and $t_{\min}$ is the minimum peak arrival time (highest permeability layer). The term K_{ave} is the mean relative hydraulic conductivity of all the layers and Q_{aver} is the mean flow rate per layer.

The vertical distribution of K_r values (Table 11.1) is remarkably similar at Wells A and B and consistent with the aquifer lithology. $Q(z)$ was calculated using the

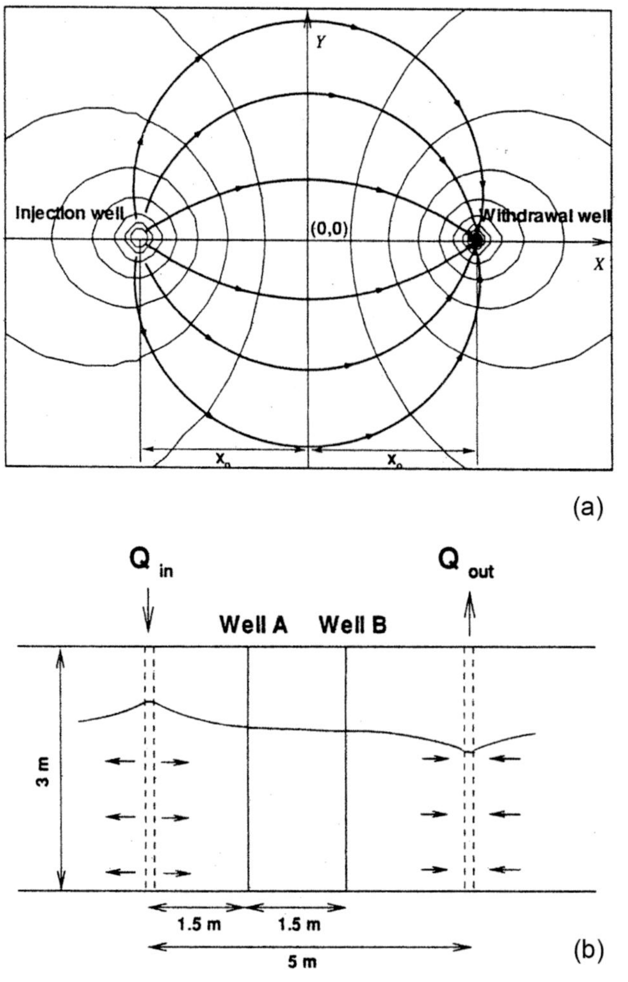

Figure 11.2 Schematic illustration of a plan (a) and a cross-section view (b) of the aquifer and flow pattern (Mas-Pla, 1993).

relative hydraulic conductivities estimated at Well A (Kaver = 0.53, and Qaver = 0.341/min). Once the $Q(z)$ values are known, the nonuniform seepage velocity components, u_x and u_y, was computed from (Huyakorn et al., 1986):

$$
\begin{aligned}
u_x &= -\frac{Q(z)}{2\pi bn}\left[\frac{x-x_o}{(x-x_o)^2+y^2}-\frac{x+x_o}{(x+x_o)^2+y^2}\right] \\
u_y &= -\frac{Q(z)}{2\pi bn}\left[\frac{y}{(x-x_o)^2+y^2}-\frac{y}{(x+x_o)^2+y^2}\right]
\end{aligned}
\tag{11.2.2}
$$

where $Q(z)$ is the flow rate of injection or pumping well; b is the aquifer thickness; n is the porosity, and x_0 is one-half the well spacing. The origin of the coordinate axes (x, y) is at the center of the line joining the withdrawal and injection wells. Considering

Table 11.1 *BTC analysis and relative hydraulic conductivity for Wells A and B. Time is the peak arrival time (hours), Conc. is the peak concentration (mg/L), Kr is the relative hydraulic conductivity (dimensionless), and Q(z) is the flow rate at the injection and the withdrawal wells at depth z (Mas-Pla, 1993).*

	Well A			Well B			
DEPTH	TIME	CONC	K_r	TIME	CONC	K_r	Q(z)
1.05	16	107	0.75	..	..	..	0.477
1.20	16	82	0.75	58	59	0.72	0.477
1.35	22	82	0.72	54	13	0.78	0.457
1.50	22	53	0.72	260	23	0.16	0.457
1.65	46	50	0.26	191	26	0.22	0.165
1.80	42	37	0.28	128	23	0.32	0.178
1.95	*38*	72	0.31	143	42	0.29	0.197
2.10	34	83	0.35	143	41	0.29	0.222
2.25	30	89	0.40	128	33	0.33	0.254
2.40	38	70	0.31	54	35	0.77	0.197
2.70	12	137	1.00	42	102	1.00	0.636

the streamline that connects Wells A and B, and the injection and withdrawal wells (streamline coordinate, s), the terms involving the y-coordinate were omitted.

The study then analyzed the pore-scale dispersion using a 1-D ADE to simulate the tracer's migration along the streamline between the injection and withdrawal wells. This model neglects the lateral mixing of tracer. The ADE is:

$$\frac{\partial}{\partial s}\left[D(s)\frac{\partial C}{\partial s}\right] - u(s)\frac{\partial C}{\partial s} = \frac{\partial C}{\partial t} \tag{11.2.3}$$

where C is the concentration in mass per unit volume; s is the distance along a streamline; $u(s)$ the seepage velocity; D is the longitudinal hydrodynamic dispersion coefficient along a streamline, and t represents time. Since the flow field in the dipole test involves diverging and converging streamlines, the velocity in Eq. (11.2.3) varies with distance, so does the longitudinal dispersion coefficient.

$$D(s) = \alpha u(s) \tag{11.2.4}$$

where α is the longitudinal dispersivity of each layer. By matching the calculated concentration using Eqs. (11.2.3) and (11.2.4) to the measured BTC, the dispersivity was determined with the velocity, estimated as follows.

A finite element model (Chapter 5) was employed to solve Eq. (11.2.3) based on the calculated velocity distribution. The Peclet and Courant number criteria (Chapter 5) were followed to avoid numerical oscillations and numerical dispersion. A prescribed concentration boundary condition was defined at the injection well, considering the chloride input variations. The model was subsequently used to estimate the dispersivity and porosity values by adjusting these parameters until it visually reproduced the observed breakthrough data at Well A.

11.2.2 Results and Discussions

The calibrated BTCs (solid lines) versus the observed (points) at the eleven depths at Well A are illustrated in Fig. 11.3. The observed BTCs exhibit multiple peaks, and the simulated captured only the general trends of the observed BTCs since the ADE is an ensemble mean equation for each heterogeneous layer. While the fit of the simulated BTCs to the observed data seems satisfactory, some of the calibrated porosity and dispersivity values (e.g., 0.95, 0.97, and 1.01) are physically impossible, indicating that the calibration problem is not well-defined, which has many possible solutions.

The calibration results suggest that (1) the goodness of fit of the observed data alone is insufficient for estimating physically correct values for the parameters. (2) From Eq. (11.2.4), we notice that the dispersion (the spread) depends on the dispersivity and velocity, and from Eq. (10.3.2), we see that velocity is inversely proportional to the porosity. Therefore, the tracer plume spread is proportional to the ratio of dispersivity and porosity. Since Q for each layer is constant, a large dispersivity or a small porosity value could lead to the same degree of spreading. Thus, one should expect non-unique or physically impossible estimates of the porosity and dispersivity if the tracer's spreading is the only fitting criterion. The results of this simple analysis suggest that matching the peak of the tracer plume, controlled by the porosity and Q, would be the logical first step, and then adjusting the dispersivity value to match the plume's spread.

This figure also shows some numerical oscillation in the simulated BTC at a depth of 2.7 m, where the peak concentration rises rapidly and much earlier than at the other depths. This behavior is consistent with coarse sand at the bottom of the aquifer, which yields large velocities, and the advection dominates the tracer movement. Further, the velocity is the largest near the injection and withdrawal wells, where the hydraulic gradient is high. As a result, the grid size and time step employed during the simulation likely violate the Courant and Peclect number criteria, leading to some oscillations in the BTC.

Although some estimated porosity and dispersivity values are physically impossible, they reasonably reproduced the observed BTCs. Because they reproduced the observed and the aquifer is a seemly homogeneous coastal sandy aquifer, it is logical to speculate if these parameter values could predict the BTCs at Well B, just 1.5 m away from Wells A (i.e., validating the usefulness of the parameters).

Fig. 11.4 shows the observed and predicted BTCs at ten depths of Well B, and the agreements are not as satisfactory as those at Well A. In particular, the

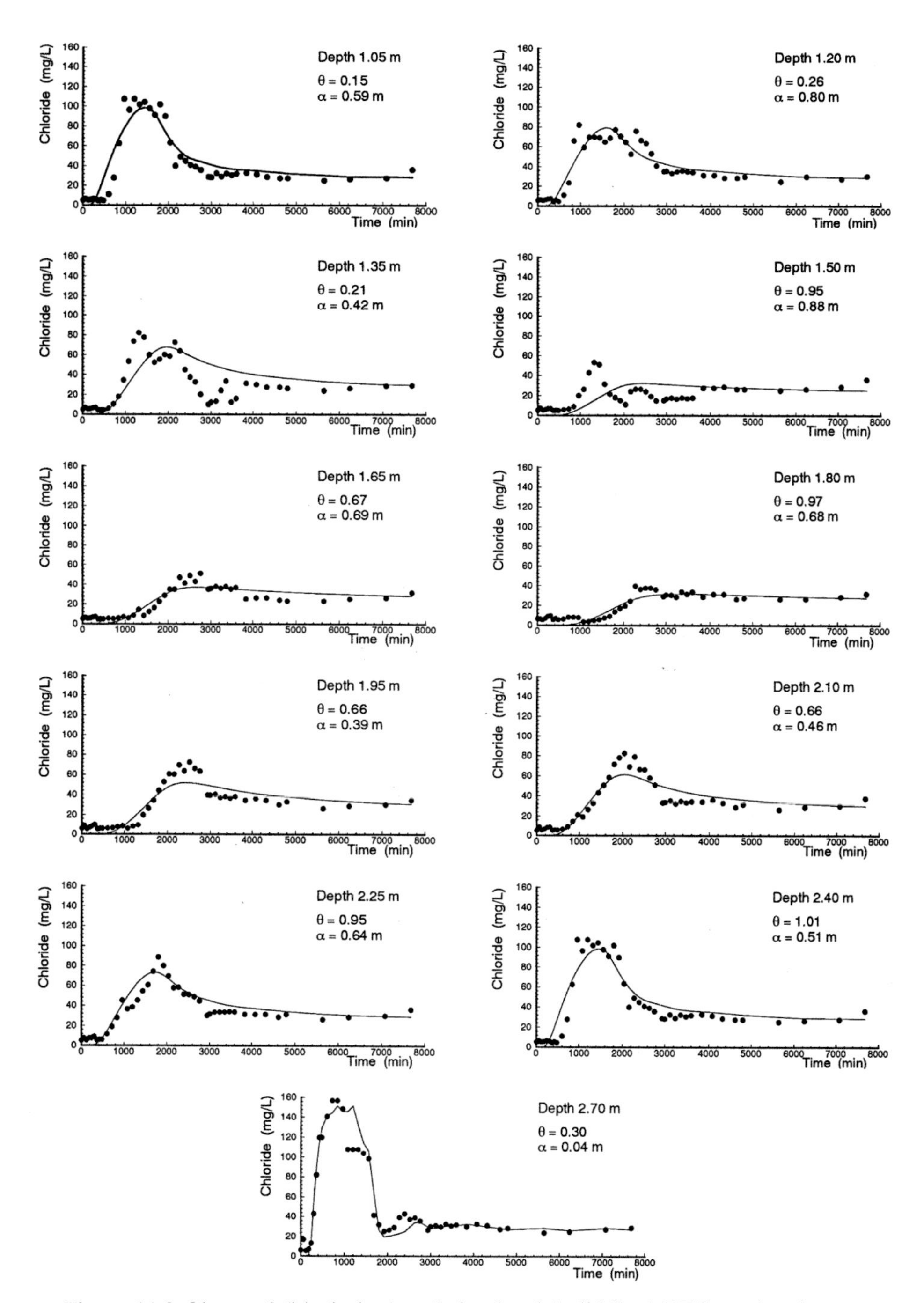

Figure 11.3 Observed (black dots) and simulated (solid line) BTCs at the eleven depths at Well A (Mas-Pla, 1993).

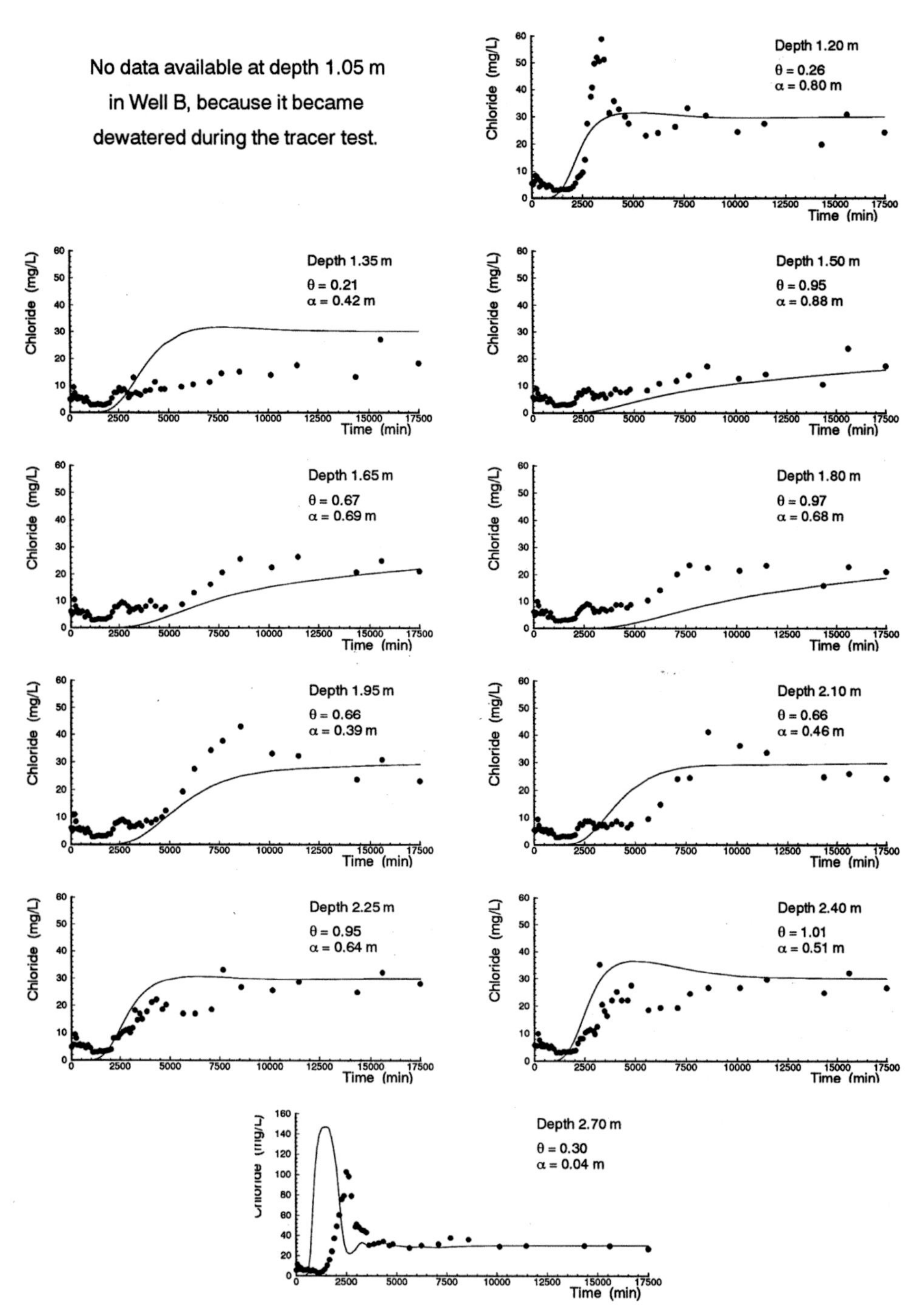

Figure 11.4 Observed (black dots) and simulated (solid line) BTCs at the 10 depths at Well B (Mas-Pla, 1993).

peak arrival time at a depth of 2.7 m is much sooner than observed. Such poor prediction raises questions about the cause of the discrepancy. Possibly, the results could be owing to (1) the spatial REV does not exist over the 5 m distance (i.e., statistically heterogeneous medium). Homogenization of the aquifer layers leads to significant uncertainty in predictions, (2) the three-dimensional flow field was assumed to be two-dimensional, and ADE is inherently an ensemble equation. These postulations promoted a three-dimensional characterization of the K distribution and modeling of the solute transport at the site, discussed next.

11.3 May–September, 1992 Field Experiments and Analysis

In searching for answers to problems associated with May 1990 experiments, thirty-two sampling wells were installed over a 5 $\times$ 5 m area between the withdrawal well (WW) and the injection well (IW2) (the right-hand side of Fig. 11.1). This new site is adjacent to the May 1990 experiment site, abandoned due to an accessibility issue. Each well was fully screened from 1 m below the surface to the aquifer's bottom. Falling head slug tests were conducted at eleven 15-cm depth intervals in each well. Because of some sampling problems, only 308 time-drawdown curves were collected and analyzed for hydraulic conductivity using Hvorslev (1951) and Cooper et al. (1967) methods. The former neglects the effects of aquifer storage (see Mas-Pla, 1993). Nonetheless, both analysis techniques assume horizontal radial flow, which deviates from the three-dimensional flow in our field slug tests. The K dataset from the former will be called the H conductivity and the latter the C conductivity.

Means and variances of ln$K$ derived from H and C conductivity are listed in Table 11.2. No autocorrelation or variogram analysis was conducted because of limited K measurement in either horizontal or vertical directions. The difference in the mean and variance of the two datasets is significant, creating a dilemma: which conductivity value set is the most representative of the aquifer? Viewing the significantly larger variance of the C conductivity, contradicting the characteristics of a coastal sandy aquifer, one may exclude them. However, the ability of the two K sets to reproduce the observed evolution of tracer distributions could provide a definitive answer.

A three-dimensional view of the ln$K$ distribution (based on the H conductivity) is shown in Fig. 11.5. According to the figure, ln$K$ values increase with depth, consistent with the geology, and exhibit several localized low permeability zones

Table 11.2 *Mean and variance of ln K (m/s) at each depth.*

			Hvorslev		Cooper et al	
Layer	Depth, m	No. of Data*	Mean	Variance	Mean	Variance
1	1.15	17 (11)	−11.75	0.41	−12.97	2.12
2	1.30	17 (13)	−11.83	0.44	−12.77	1.16
3	1.45	21 (18)	−11.88	0.44	−13.21	1.60
4	1.60	25 (22)	−11.87	0.61	−13.03	3.49
5	1.75	24 (22)	−11.69	0.73	−12.43	3.66
6	1.90	25 (23)	−11.32	1.01	−11.98	3.61
7	2.05	23 (24)	−11.28	0.81	−12.29	3.64
8	2.20	24	−10.56	0.96	−10.84	5.09
9	2.35	24	−10.24	0.92	−10.31	5.12
10	2.50	25	−9.72	0.59	−9.18	3.46
11	2.65	23	−9.48	0.28	−8.77	1.82

*Numbers in parentheses indicate the number of data points for the Cooper et al. method (Yeh et al., 1995).

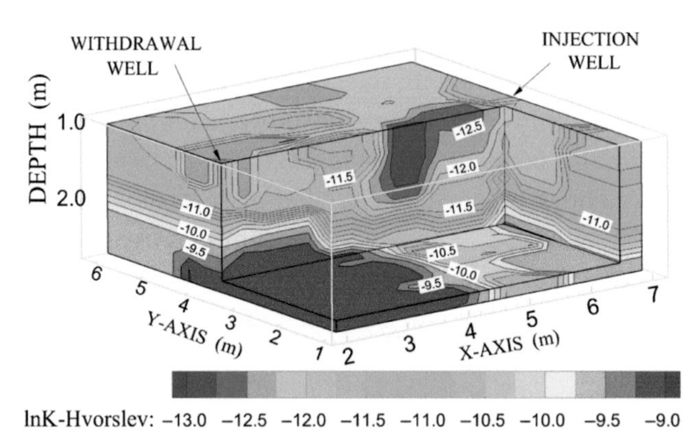

Figure 11.5 The 3-D hydraulic conductivity distribution at the experiment site (Yeh et al., 1995).

at the top and the bottom. These localized heterogeneities explain the difficulty of predicting the BTC at Well B using the calibrated model and parameters for the BTC at Well A in the May 1990 experiment.

Two tracer experiments were conducted at the new site during May 1992 and August–September 1992. Rainfall and changes in regional groundwater hydraulic gradients were minimal in May 1992. During August–September 1992, there were five or six heavy rainfalls, causing the rise of the water table and changes in regional gradient.

The first experiment (May 1992) injected a chloride tracer (KCL) at a concentration of approximately 230 mg/L for 16 hours at IW#2 after a steady-state flow was established under a constant injection and withdrawal rate of 3.78 L/min at IW#2 and WW wells.

During the first 12 days, tracer samples were simultaneously collected from the five depths of the 28 monitoring wells to obtain a three-dimensional spatial distribution or "snapshots" of the chloride plume at various times. They were collected by simultaneously extracting groundwater from all depths of sampling wells, with an approximate rate of 100 mL/min, starting from the five wells closest to the injection well. The complete operation of sampling the entire array of wells took less than 20 min; therefore, each tracer snapshot could be considered instantaneous. Snapshots were taken every 2 hours during the first 30 hours. Then, they were taken at 4- to 5-hour intervals until 100 hours and at longer intervals until the end of the experiment. A continuous breakthrough curve was also recorded at the withdrawal well.

The second injection experiment took place during August–September 1992. This experiment was designed to test NOM chemistry mechanisms (McCarthy et al., 1995). The experiment ran for 43 days with an injection period of 659 hours. A chloride solution was metered into the injection stream with a target final concentration of 100 mg/L. The metering pump's instability resulted in fluctuating chloride concentrations from 50 to 250 mg/L during the injection. Further, several heavy rainstorms occurred during the first 20 days of injection, complicating this experiment. These rainstorms raised the regional water table and limited the injection rate to 3.0 L/m. Because of the high regional water table, a withdrawal of 4.5 L/m was implemented to maintain the same gradient from the injection to withdrawal wells as in the May experiment. Groundwater was also withdrawn from the well (IW#1) to intercept regional groundwater moving toward the withdrawal well. The pumping rate in IW#1 varied with the regional water table, from 5.5 L/m after heavy rain to 0 L/m after a week without rain.

Eighteen snapshots were taken at variable intervals during the experiment's early times and following the tracer injection shutdown. Continuous BTCs were collected at the withdrawal well and at 2.6 m depth in four sampling wells (wells 6, 13, 21, and 28) along the line between the injection and the withdrawal wells.

11.3.1 Analysis

Based on ADE, a variably saturated flow and transport in three-dimension (VSAFT3, formally called MMOC3, Srivastava and Yeh, 1992) was used to simulate the tracer experiments. The simulation domain ($9.0 \times 7.0 \times 1.5$ m) was discretized into small, rectangular elements in either model. This discretization results in a finite element mesh of 27 × 20 × 23 nodes and 10,868 variable-size elements. During the simulation of solute transport, the maximum time step size was set to 3 mins.

Simulation of the May 1992 chloride tracer experiment adopted an HPHM and a layered conceptual model. For both models, no-flow boundary conditions were defined for the simulation domain's top and bottom boundaries. Constant head values were assigned to the lateral boundaries. These head values were estimated using Theis's equation's analytical solution for the well doublet, accounting for the site's average natural hydraulic gradient (0.023 m/m) during the experiment. Nodes representing the injection and withdrawal wells were assigned constant head values throughout the depth. They were adjusted such that the calculated total fluxes at the injection and withdrawal wells matched with measured values.

The aquifer's initial chloride concentration was 11 mg/L, corresponding to the mean background concentration. Top and bottom boundaries were specified as no flux. The boundary nodes on the side close to the withdrawal well were defined as constant concentration nodes with groundwater background concentration. The other three sides of the domain were treated as non-dispersive flux boundary nodes. We assigned a time-variant concentration condition to the nodes corresponding to the injection well to reflect the input concentration variation.

11.3.1.1 Highly Parameterized Heterogeneous Model (HPHM)

HPHM assigns a local-scale K value to each element (local-scale CV). However, the number of elements in the simulation is greater than the number of conductivity values measured. An interpolation/extrapolation scheme was used to overcome this problem. It assigned the K value to the element without measured value using the nearest eight elements' measured K values and their inverse distance as weights. The interpolation's vertical search radius was restricted to one slug test interval as suggested by the geologic logs to preserve the aquifer's layered structure. This interpolation scheme and the K value sets derived from the H and C methods at each slug test location created H conductivity and C conductivity fields for the entire domain.

The average porosity of 0.25 was selected for the entire aquifer, close to the value of 0.3 derived from the analysis of the pore structure of intact cores of aquifer materials from one borehole. This porosity value was assumed constant over the entire simulation domain since porosity variability is much smaller than the conductivity's and has minor impacts on solute transport (Gelhar et al., 1979). The longitudinal dispersivity was assigned a value of 0.05 m, and the transverse dispersivity 0.015 m. These dispersivities represent the effects of heterogeneity at scales smaller than the local-scale CV (i.e., pore-scale and below).

11.3.1.2 Layered Approach

The layered conceptual model defined eleven homogeneous layers of 0.15 m thickness. It adopts an ensemble REV for each layer. The K value of each layer was assigned as the geometric mean of all measured conductivity values (from the H conductivity set) at that depth (Table 11.2). A macrodispersivity concept was adopted to account for the effect of conductivity variability within each layer, neglected by the layer homogeneity assumption. The variability likely encompasses lenses, laminations, cross-bedding, and others much larger than pores. As a result, the term macrodispersivity was used.

Macrodisperivity theorem relating to the variability of local-scale K for a nonuniform flow regime is unavailable. As initial values, longitudinal and transverse macrodispersivities were estimated using the result of Gelhar and Axness (1983), assuming two-dimensional uniform flow within each layer with statistically isotropic properties. According to their formulation, the longitudinal and transverse macrodispersivities (A_{xx} and A_{yy}, respectively) for each layer were calculated by

$$\begin{aligned} A_{xx} &= \sigma^2_{\ln K}\lambda/\gamma^2 \\ A_{yy} &= \frac{\sigma^2_{\ln K}\alpha_L}{8\gamma^2}\left(1 + 3\frac{\alpha_T}{\alpha_L}\right) \end{aligned} \tag{11.3.1}$$

where $\sigma^2_{\ln K}$ is the variance of ln K; λ is the correlation scale of ln K; α_L and α_T are the longitudinal and transverse local dispersivities, respectively; and $\gamma = 1 + \sigma^2_{\ln K}/6$. These values were input to VSAFT3 as the initial estimates of the simulation's longitudinal and transverse dispersivities. Note that the vertical transverse dispersivity was also set equal to the value of A_{yy}. Since the estimates were derived under the uniform flow condition, the initial dispersivity values were subsequently adjusted by model calibration.

11.3.2 Results and Discussions

11.3.2.1 Snapshots

The simulated three-dimensional plumes based on the three approaches (i.e., using H and C conductivities and the layer model) generally agree with those observed. A closer comparison between the observed and simulated plumes is given in Fig.11.6, depicting the behavior of the plumes at different times in a cross-section along a line between the injection and withdrawal wells. Fig. 11.16 illustrates that simulated results using either the H or the C conductivity reasonably reproduced the observed plume's shape. Generally, the conductivity distribution's layered

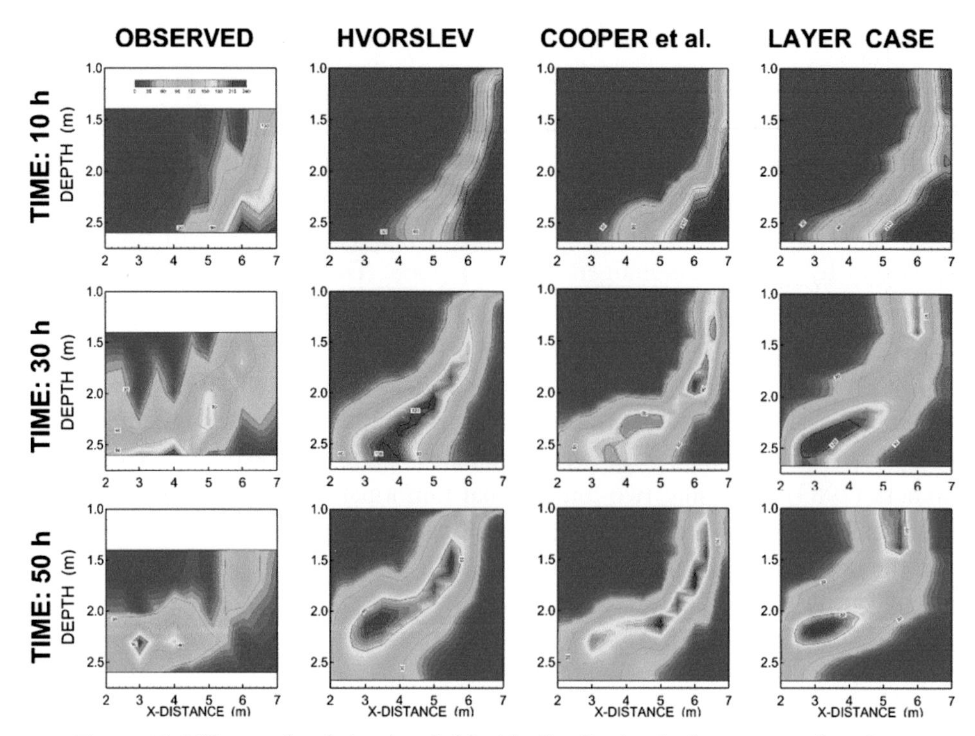

Figure 11.6 Observed and simulated chloride distribution in the cross section along the line connecting the injection and the withdrawal wells (May 1992 test). Concentration contour interval is 30 mg/L (Yeh et al., 1995).

structure dictated the solute's movement. The solute's fast and slow movements reflected the high and low conductivity values at the aquifer's bottom and top, respectively. At late times (50 hours after the chloride injection), this effect of layered heterogeneity became even more profound. The plume split into two portions: One part moved through the lower layers, whereas the other part stayed at the upper layers of the aquifer because of the presence of the low-permeability inclusion in the upper central portion of the aquifer (see Fig. 11.5), which enhanced the lateral migration of the plume (*Mas-Pla,* 1993).

11.3.2.2 Spatial Moments

Spatial moments of the observed three-dimensional plume were calculated using Eq. (10.2.1) to Eq. (10.2.6). As illustrated in Fig. 11.7a, the total masses of the observed and simulated plumes increase first due to continuous 16 hours of injections and then decrease, reflecting the effects of the divergent and convergent dipole flow field (Fig. 11.2) and the limited area coverage of the sampling network. The simulated mass (zeroth moment) using the H conductivity and the layer model agrees well with the observed mass before 40 hours. We also observe

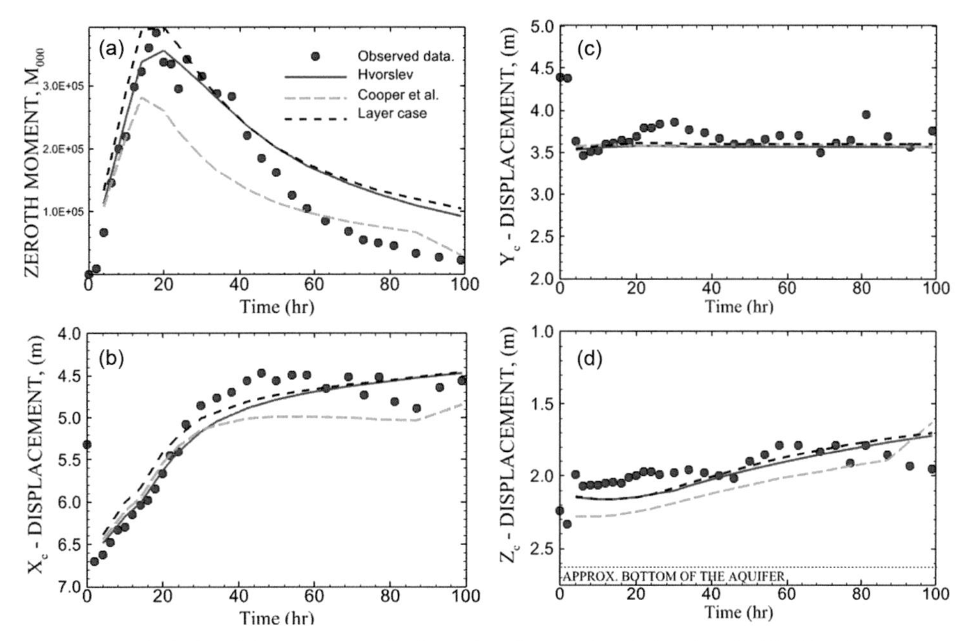

Figure 11.7 Spatial moment analysis of the observed and simulated chloride plumes (May 1992 test). (a) Zeroth moment, or total mass versus time. (b) Location of the center of mass x_c;(c) y_c; and (d) z_c, versus time (Yeh et al., 1995).

that the mass estimated from the observed data is much less than the simulation after $t > 40$ hours. The smaller sampling network than the simulation domain and errors in the extrapolation scheme could explain the difference.

In contrast, the simulation using the C conductivity produced an overall lower mass than the observed. The reason is the larger mean K at the bottom layers of the C conductivity than the H conductivity. It produced a more significant amount of chloride, leaving the simulation domain than the H conductivity.

In Fig. 11.7b, the centers of mass (the first moment) of observed and simulated plumes in the x-direction are plotted as a time function. Note that the origins of x, y, and z are the land surface of the injection well. The z is the depth (or positive downward). Generally, the agreement between the simulated and observed mass centers is satisfactory. At the early time, most of the mass moved through the high-permeability bottom layer toward the withdrawal well, resulting in a rapid change of x_c. Afterward, some portion of the plume remained in the upper zone of low permeability; meanwhile, the other part in the bottom high-permeability zone had already reached the withdrawal well – consequently, a slowdown in the center of mass movement of the plume. At $t > 40$ hours, the plume in the upper portion of the aquifer encountered the low-permeability inclusion around $x = 4.5$ m (Fig. 11.5), and the plume's movement thus became extremely slow.

The center of mass of the observed plume y_c versus time (Fig. 11.17c) indicates some horizontal drift away from the central line connecting the injection and withdrawal wells. Such drift may be due to the interpolation scheme, sampling network, and natural hydraulic gradient variation neglected in the simulation.

From Fig. 11.7d, we observe that the mass center of the observed plume varied slightly along the vertical direction z_c. The mass center remained approximately constant at 2 m below the ground surface, moving slightly upward. This behavior conforms to the upper low permeable zone's effect, delaying the solute's movement in the upper zone.

11.3.2.3 Breakthroughs at Point Sampling Ports

The observed and simulated BTCs at five different depths of wells 6, 13, 21, and 28 are shown in Fig. 11.8. The simulated and the observed BTCs near the injection

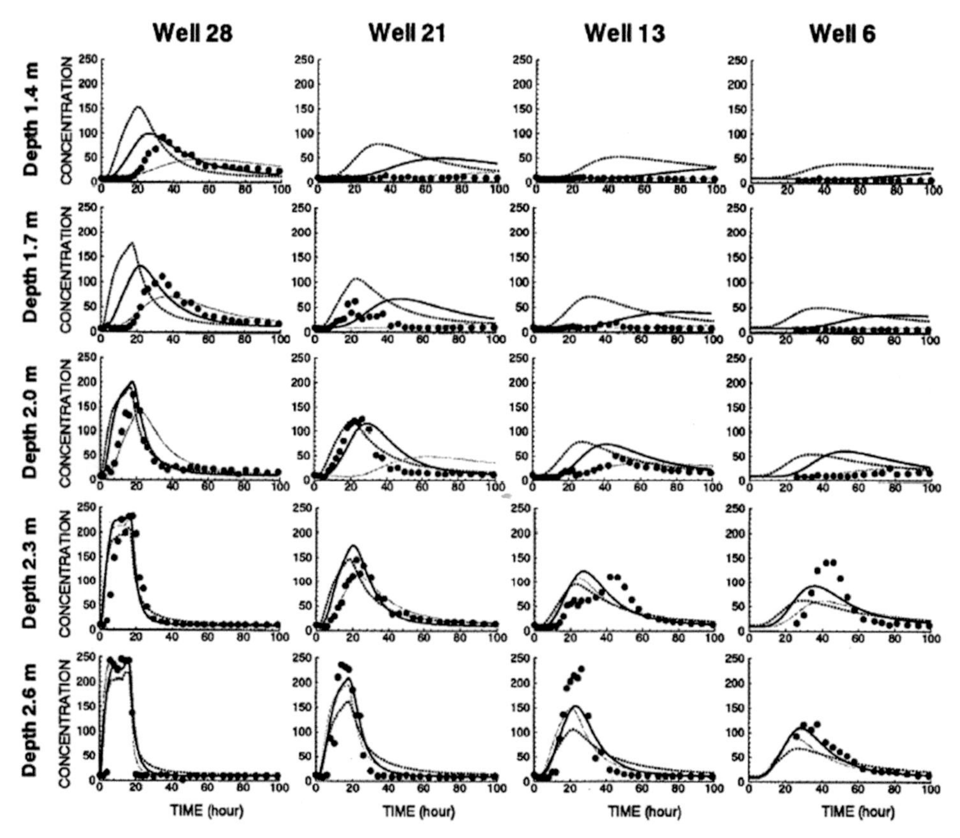

Figure 11.8 Observed and simulated chloride breakthrough curves at wells 6, 13, 21, and 28 (May 1992 test). Solid circles denote observed data; solid line denotes H conductivity value set simulation; dotted line denotes C conductivity value set simulation; and dashed line denotes layer case (Yeh et al., 1995).

well and at the bottom layers of the aquifer, consisting of relatively homogeneous material (variance of $\ln K = 0.28$, see Table 10.2), are in excellent agreement.

The zeroth and first temporal moments of the observed and simulated BTC at all wells were also analyzed. For a discrete BTC, the zeroth moment (the mass of BTC) and the first moment are given by

$$t_0 = \sum_{i=1}^{n} C_i \Delta t; \quad t_1 = \sum_{i=1}^{n} C_i t_i \Delta t \tag{11.3.3}$$

where C_i is the concentration, above background, at location (x, y, z) and at time t_i. Because the observed BTCs were sampled at irregular time intervals, a linear interpolation was employed to estimate the concentration values at an equal interval. The mean arrival time is given by the ratio of the first and the zeroth temporal moments (t_1/t_0) .

Temporal moments of BTCs provide a quantitative comparison of the BTCs in Fig. 11.9. Therefore, temporal moments of the observed and simulated BTCs at the five depths of wells 5, 6, 12, 13, 19, 20, 21, 22, 26, 27, 28, and 29 were calculated. The zeroth moments (area) of BTCs based on the H and the C conductivity and the observed at different depths and wells are plotted in Fig. 11.9a and b. The areas of the BTCs using the H conductivity are greater than those observed at shallow depths (Fig. 11.9a). In contrast, the areas of the BTCs at the shallow depths using the C conductivity are smaller than those observed (Fig. 11.9b). The areas of the simulated BTCs at the bottom of the aquifer (bluc circles) using either the H or the C conductivity agree well with those observed.

Arrival times of the center of mass (the first temporal moment) of the simulated BTCs using the H and the C conductivity are compared against the observed in Fig. 11.9c and d, respectively. The H conductivity reproduced the observed arrival times better than the C. The agreement between the simulated and observed arrival times is excellent for the bottom layers (the layers at 2.0, 2.3, and 2.6 m) (blue circles) and progressively worsens when depth decreases. Unresolved heterogeneity, misrepresentation of the water table condition, and measurement errors in hydraulic conductivity or concentration at these upper layers may be the causes. Furthermore, the first-moment estimates may be inaccurate due to the flat BTCs observed at the sampling ports at the upper layers far from the injection well (see Fig. 11.8). Both BTC plots and moment analysis suggest that the simulation based on the H conductivity agrees with the observed BTCs better than the C conductivity.

These comparisons between the observed and simulated BTCs manifest the difficulties in simulating solute transport behaviors at small scales, manifesting the ensemble mean nature of ADE – it predicts likely behaviors only.

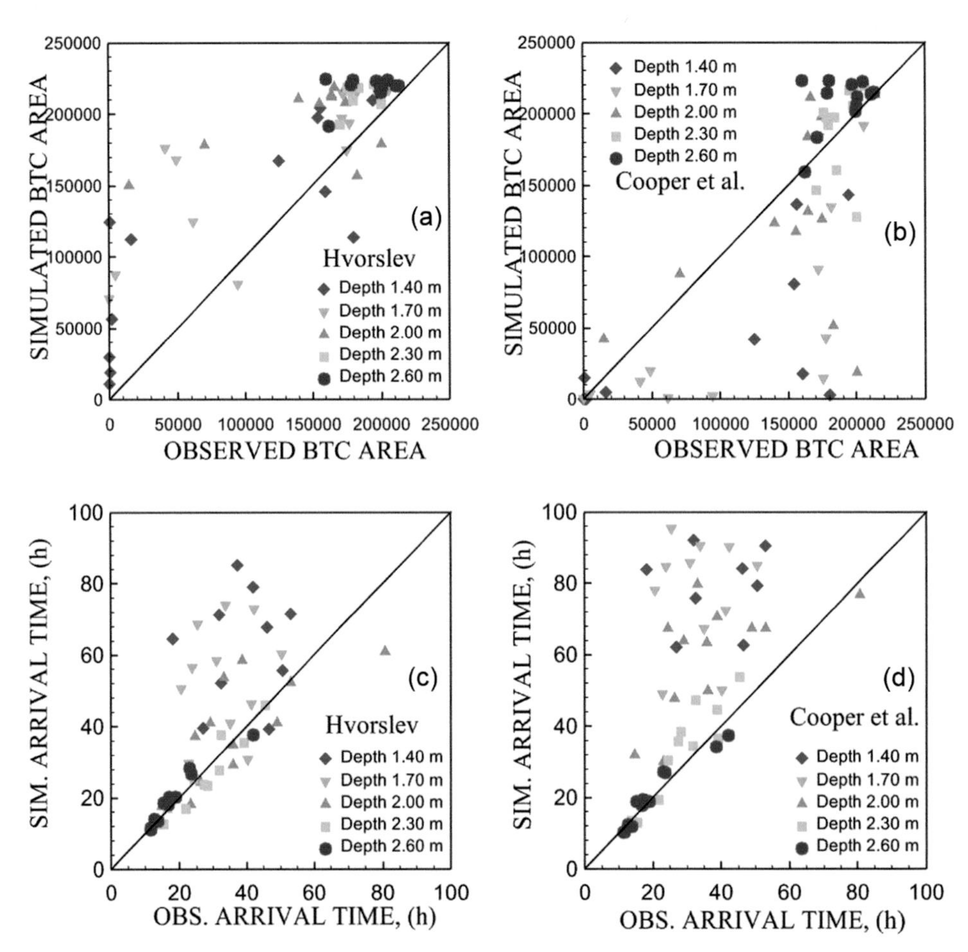

Figure 11.9 Comparison of the observed and simulated areas and centers of mass (mean arrival time) of the breakthrough curves at all sampling ports at wells 5, 6, 12, 13, 19, 20, 21, 22, 26, 27, 28, and 29. Areas are in units of milligrams per liter times minutes (Yeh et al., 1995).

11.3.2.4 Breakthrough Curve at the Withdrawal Well

BTC at the withdrawal well integrates BTCs from all converging streamlines (see Fig. 11.2a). While BTC from each streamline is affected by the streamline velocity, local-scale dispersion, and molecular diffusion, the integrated BTC at the withdrawal well is dictated by the differences in the arrival time of the solute among all streamlines (large-scale velocity variations). Reproducing BTC at the withdrawal well becomes a suitable metric to judge HPHM with H or C conductivity and the layered approach for reproducing the large-scale velocity field.

Simulated BTCs at the withdrawal well are shown in Fig. 11.10 and were obtained by averaging the concentrations at the well nodes using nodal fluxes as

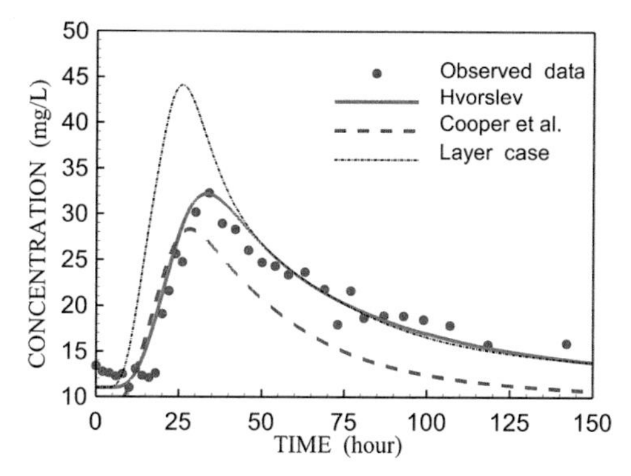

Figure 11.10 Observed and simulated chloride breakthrough curves at the withdrawal well (Yeh et al., 1995).

weights. As demonstrated in the figure, the BTC from the H conductivity successfully reproduced the observed. Conversely, the simulation based on the C conductivity predicted an earlier peak arrival and underestimated the tail of the observed breakthrough. These results corroborate the overestimated *Ks* of the bottom layers of the aquifer by the C method, as remarked before.

As emphasized before, macrodispersivity is designed to account for the local-scale K variation within the purported homogeneous layer. Specifically, these layers may have heterogeneity of scales larger than pores (e.g., gravel aggregates, plant roots, clay lenses). Thus, the layered model used the macrodispersivity values from Eq. (11.3.1) for this purpose. As shown in Fig. 11.10, the layer model predicted a higher peak and earlier arrival of the BTC than the observed.

Theoretically, the BTC at the withdrawal well, which integrates the contributions from many different parts of the aquifer, should reach the ergodic condition and be close to the ensemble BTC. As depicted in Fig. 11.10, the layer model reproduces the observed BTC at the withdrawal well, but it is still not as satisfactory as those based on the HPHM. More importantly, The result of HPHM is free of calibration effort, indicating the importance of detailed site characterization.

The stochastic analysis of macrodispersion (Chapter 9) unveils that dispersion in the ensemble mean sense is a function of mean, variance, correlation scales, and the mean gradient. In this experiment, the simulated divergence and convergence of the flow field in the dipole test captures the large-scale mean gradient, the density slug tests characterized the layering structure and, in turn, the correlation scales, and the variation in local-scale K, making up the variance. The converging flow to the withdrawn well finally facilitates the ensemble average (mixing) to satisfy ergodicity. The agreement between the observed and simulated breakthroughs using HPHM is a testimony of the seminal contribution of the stochastic analysis of solution transport.

11.4 August–September 1992 Injection Experiments (Validation)

As demonstrated in the May 1992 tracer test simulations, HPHM reasonably reproduces the groundwater velocity field without any calibration and is superior to the layered approach. Also, the H conductivity set yielded more satisfactory results than the C conductivity set. For this reason, the simulation of August–September 1992 experiments used HPHM with the H conductivity. Note that this simulation is solely a prediction without calibration – validating the model and parameters from previous experiments.

The observed and simulated chloride concentration distributions along the cross-section between the injection and withdrawal wells at different times are illustrated in Fig. 11.11. The simulation results are not as satisfactory as in the May 1992 experiment (Fig. 11.6). The simulated plume moved faster than the observed, although the patterns were very similar. The discrepancies became significant at 84 hours and after. Similar to the May 1992 experiment simulation, the greatest discrepancies occur in the upper layers. Such differences likely stem from the incorrect representation of the changes in boundary conditions resulting from rainfall events and factors discussed in the simulation of the May 1992 experiment. The simplified quasi-steady-state approach for tackling the effect of pumping at the second withdrawal well may augment these discrepancies. Consequently, the simulation overestimated the flow rate of the entire aquifer.

Nevertheless, Fig. 11.12 shows that the agreement between the simulated and the observed BTCs at wells 28, 21, 13, and 6 at the bottom of the aquifer (depth 2.6 m) is excellent up to approximately 300 hours before the effect of rainfall events. The impacts from the additional pumping became significant in the lower portion of the aquifer. A similar agreement was observed at the withdrawal well (Fig.11.13). However, we emphasize that a slight discrepancy between the simulated and observed peak arrival times corresponds to several hours, which is significant because of the short distance between the injection and withdrawal wells.

The discrepancy between the simulated and the observed BTC affirms the critical role of large-scale flow in solute transport.

11.5 Lessons Learned

- Unlike the natural gradient experiments (Chapter 10), the significant large-scale dipole flow velocity pattern created by the forced gradient experiments dictates the bulk solute movements and distributions. Heterogeneity at different scales then modify detailed solute behaviors at different observation scales (e.g., integrated BTCs at the pumping well, and BTCs at multilevel point sampling ports).

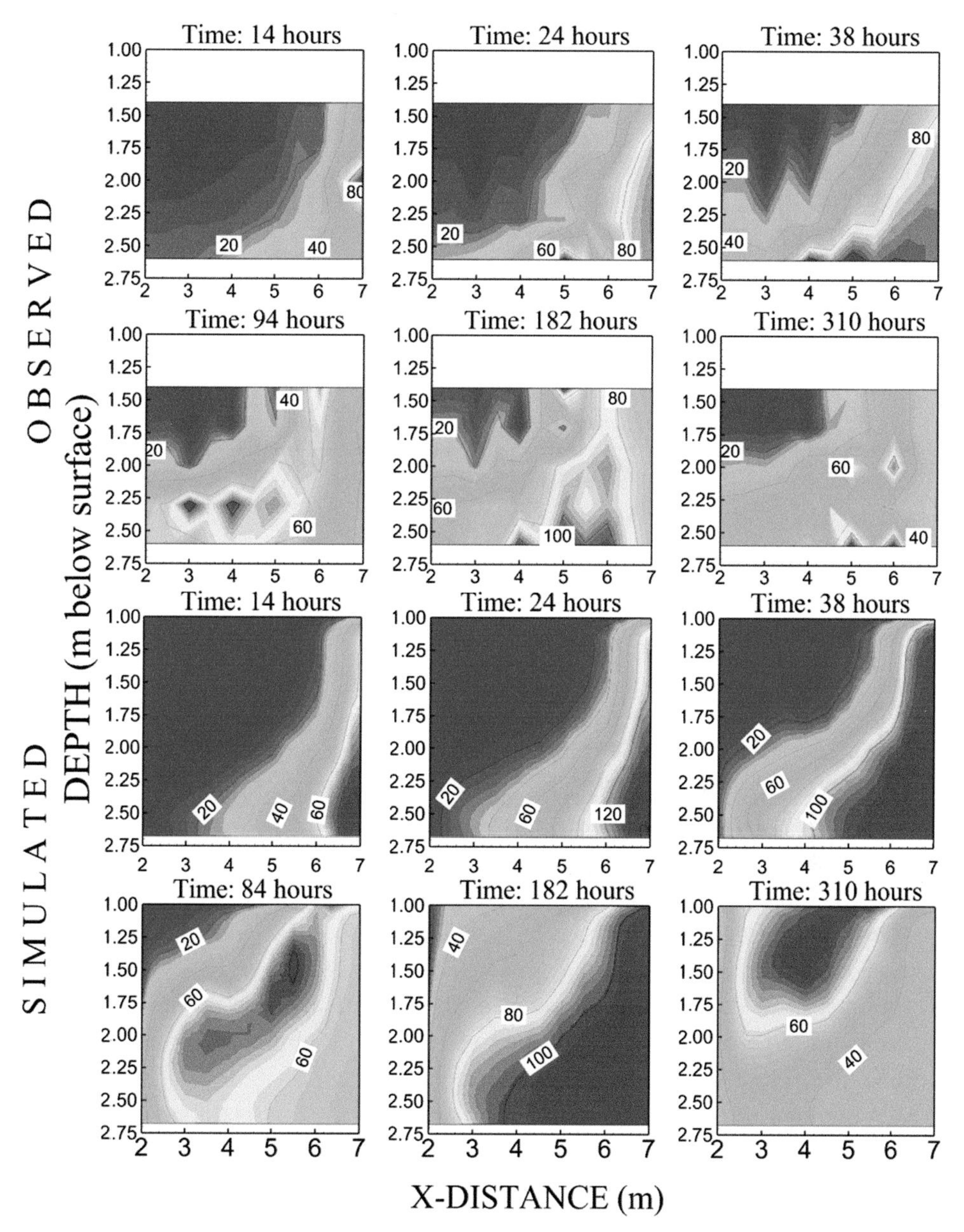

Figure 11.11 Observed and simulated plume distributions along a cross-section in August–September 1992 experiment (Yeh et al., 1995).

- The dominant large-scale heterogeneities, namely, the aquifer stratification and the low-permeability inclusions between the injection and the withdrawal wells, control the secondary modifications of the bulk plume behaviors.

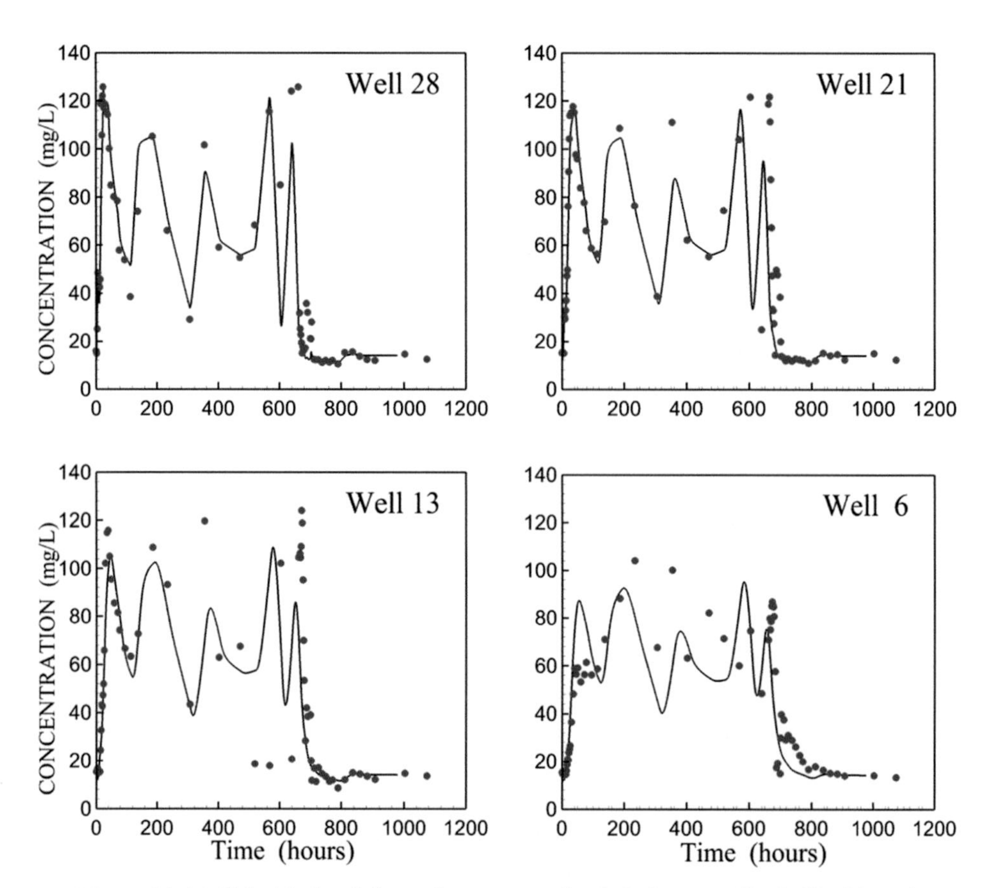

Figure 11.12 Chloride breakthrough curves at depth 2.6 mat wells 6, 13, 21, and 28 (August–September 1992 test). Red dots are observed data, and solid lines are predictions (Yeh et al., 1995).

- The tracer test during August and September 1992 exposed the importance of change in large-scale flow fields due to temporal variability in boundary conditions, overriding the effects of heterogeneity.
- Measurement errors in slug tests, uncertainties in the interpretation of slug test results, errors in the interpolation of conductivity values to finite elements, and difficulties in controlling numerical dispersion make exact predictions of BTCs at small-scale sampling ports in the aquifer intractable.
- Despite our inability to simulate BTCs at every location, HPHM with the ADE predicted the BTCs within the relatively conductive and less variable layer adequately in both May and August–September experiments.
- The forced gradient experiments and analyses unequivocally are a testimony to the effects of multi-scale groundwater velocity variations in time and space on

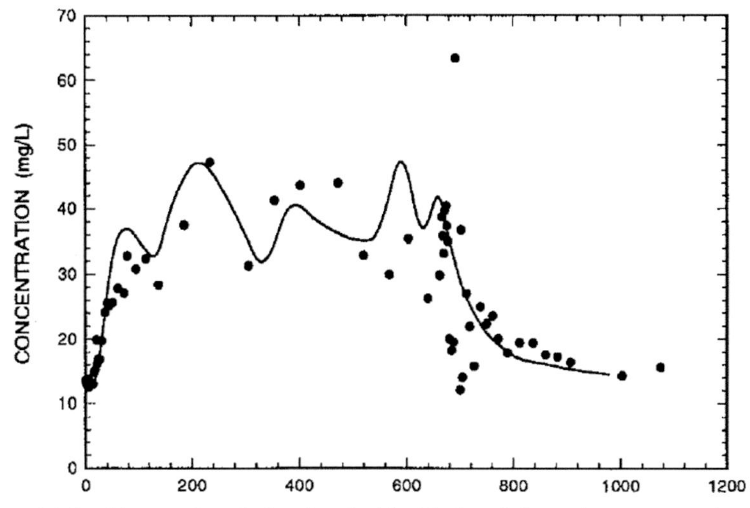

Figure 11.13 Observed and simulated chloride breakthrough curves at the withdrawal well (August–September 1992 test). Black dots are observed and line is the predicted concentration (Yeh et al., 1995).

different observation scales. Velocity variations due to regional flow dominate, and those caused by layered structure and interlayered heterogeneity follow.

- Regardless of the scales of our observation and interests, capturing spatiotemporal variation of multi-scale velocity is the key.
- For these reasons, fine-resolution characterization of temporal and spatial variation of boundary conditions, significant source or sink, and heterogeneity of the aquifers should be the ultimate solution to solute transport analysis.

12

High-Resolution Characterization (Tomographic Surveys)

Hydraulic Tomography and Geophysics ERT Surveys

12.1 The Need for Highly Parameterized Heterogeneous Models

As we explained throughout the book, scientists and engineers embrace the ensemble-mean Fick's law to describe the seemingly random meandering of tracer molecules in a fluid in viewing the scale of our observation, interest, and model. Similarly, our interest in the integrated tracer concentration BTC from a pipe promotes our ignorance of the detailed fluid-dynamic scale velocity variation in the pipe and acceptance of one-dimensional ensemble-mean ADE, built upon Fick's law. For tracer experiments in soil columns, a similar school of thought prevails. We ignore pore-scale velocity and concentration variations in the column and focus on the integrated BTC at the end of the soil column, accepting Fick's law or searching for other ensemble mean models such as a dead-end, mobile/immobile, and fractional derivative ADE. Applying the same idea to study the field-scale solute migration, we often overlook the scales of observation and interest, which are much smaller than the heterogeneity in the field. For practicality, we assume field homogeneity and adopt field-scale ensemble-mean ADEs with macrodispersion, mobile/immobile, or other non-Fickian models to analyze the solute plume migration in aquifers.

These approaches built upon the ensemble mean concept could satisfy our needs only in the ensemble mean sense, rather than at one field site unless the tracer cloud reaches ergodicity or the tracer plume has experienced enough heterogeneity. Such requirements may be met if the heterogeneity remains statistically homogeneous and invariant in space, and the mean flow field remains constant as the solute cloud migrates, which is improbable. Therefore, the applicability of theories in previous chapters to an aquifer – one realization of the ensemble – is often questionable at our interest and observation scales. These scale issues demand a high-resolution delineation of the multi-scale heterogeneity in one realization. In the following sections, we introduce new technology that may advance the understanding and prediction of solute migration in field-scale geologic media.

12.2 Hydraulic Tomography (HT)

HT, narrated in plain language, is a technique that uses the monitoring wells as recorders to capture the pressure signals created by pumping or injection of water (source of excitation) at a given location. As the pressure disturbance travels from the source to recorders, it encounters heterogeneity (low/high permeability layers, zones, or fractures) and transforms. The pressure histories (hydrographs) recorded at the observation wells become a snapshot of the heterogeneity. Repeating the above procedure with different source locations, we have many snapshots. Each snapshot bears some differences due to changes in flow paths and heterogeneity as the pressure propagates from the source location. Interpreting all these snapshots jointly, we obtain a more precise image of the hydraulic heterogeneity than using one snapshot. HT's principle is straightforward – common sense.

Indeed, electrical resistivity, seismic, or acoustic tomography adopt the same principle, but these tomographies use electrical current, seismic shock, or sound generators as excitation sources. They collect responses from the subsurface to image anomalies that modify the signals. To illustrate the HT concept and principle, we examine the following simple numerical experiment as an example.

12.2.1 HT Concepts and Principles

This example considers a pumping test at a well (Pw1 at $x = 0$, $y = 0$, red circle in Fig. 12.1) in a 2-D plane confined aquifer, which has impermeable boundaries on the left and the bottom and constant heads on the right and the top boundaries at considerable distances away from the pumping well. The aquifer is homogeneous with $T = 1$ and $S = 0.001$ (the black background) but has an anomaly (the red rectangle) with $T = 0.01$ and $S = 0.001$. This aquifer's responses to continuous pumping at a rate of 1 at the pumping well are observed at observation wells (from

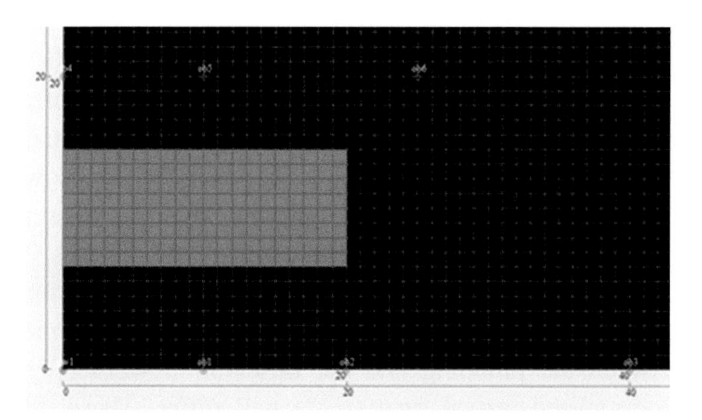

Figure 12.1 Schematic illustration of the aquifer for HT principle.

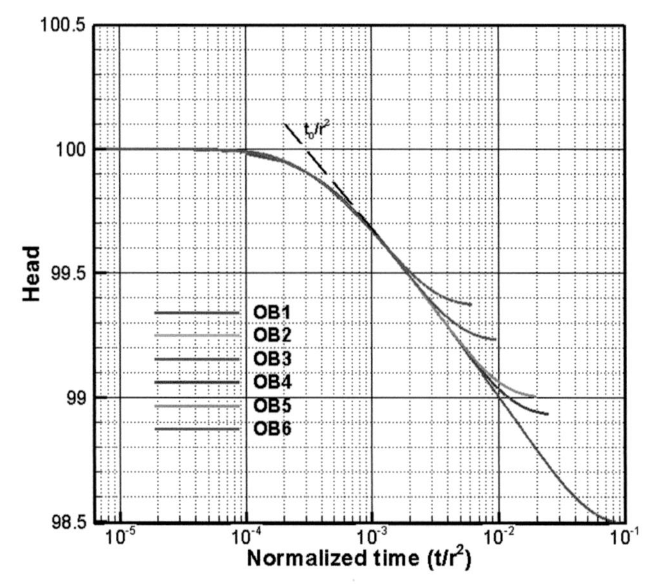

Figure 12.2 Change in pressure head as a function of log (t/r^2) at the six observation wells in the hypothetical aquifer with homogeneous T and S. The stabilizing ports of the curves reflect the effects of the constant head boundaries.

ob1 to ob6, marked blue circles in Fig. 12.1). Note that all these parameters and the pumping rate use any consistent units. The observed drawdown-time data at the observation wells (well hydrographs) will be used to determine the arrival times of drawdowns at the observation wells, facilitating locating the anomaly.

Before any analysis, we first recognize that the arrival time of the drawdown at different observation wells will vary according to the distance between the pumping and the observation wells (r), regardless presence of any heterogeneity or not. As a result, we must remove the distance's effect from the observed drawdown to decipher the arrival time for the anomaly location. This removal is accomplished by normalizing the time axis of the observed drawdown-time curve by the square of the distance (r^2). Suppose that the anomaly does not exist (a homogeneous aquifer) and the aquifer is unbounded. The drawdowns with the normalized time at different observation wells should then overlap and arrive at the same normalized arrival time (t_0/r^2), as demonstrated in Fig. 12.2, where the drawdowns are plotted as a function of log (t/r^2). The variable t_0 represents the time when the drawdown is noticeable in the well hydrography at each well.

In contrast, if the observed drawdowns vs log (t/r^2) plot for various observation wells does not overlap, the drawdowns must have encountered some anomalies as they travel between the pumping and the observation wells (Fig. 12.3). Each hydrograph's normalized arrival time (t_0/r^2) will differ. These different arrival

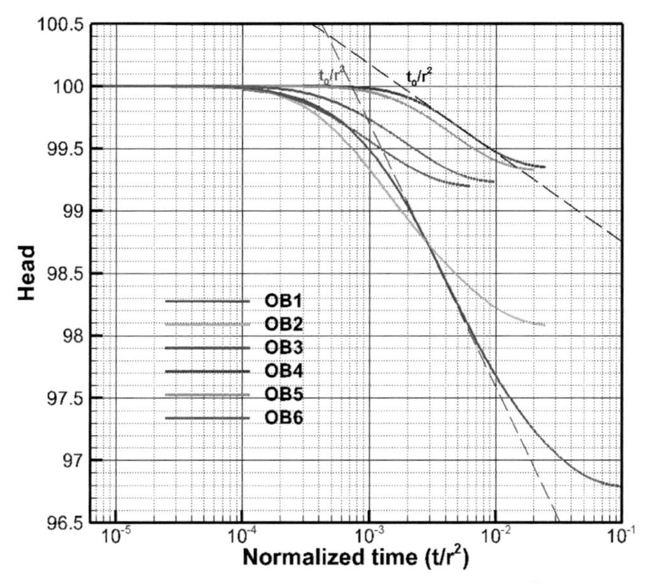

Figure 12.3 Change in pressure head as a function of log (t/r^2) at the six observation wells in the heterogeneous aquifer (Fig. 12.1).

times should inform us of the relative distances between the anomaly and the observation wells, and we can triangulate the anomaly's location.

While this concept is straightforward, quantitative analysis to determine the anomaly's location and property is not trivial. Nevertheless, the Cooper and Jacob straight-line approach (Cooper and Jacob, 1946) for pumping test analysis sheds some light. Following the approach, we draw straight lines tangent to the straight-line portion of the drawdown-log (t/r^2) data and determine the time t_o/r^2 when the drawdown should be zero theoretically. This time represents the first arrival time of the excitation due to the pumping. The approach shows that this arrival time is inversely proportional to $D = T/S$ (see Yeh et al., 2015, or any groundwater textbook):

$$\frac{t_0}{r^2} = \frac{S}{2.25T} = \frac{1}{2.25D}. \tag{12.2.1}$$

Thus, this arrival time determines the D value between the pumping and the observation well. Repeating this procedure leads to the estimates of D values between each pair of the pumping and the observation well. Notice that D is based on the homogeneity assumption and is an average value of Ds covering different proportions of the anomaly. For this reason, we draw a circle at the center of the straight line connecting the pumping and observation wells, using a radius equal to half the distance between the two wells (Fig. 12.4). This circle represents the area

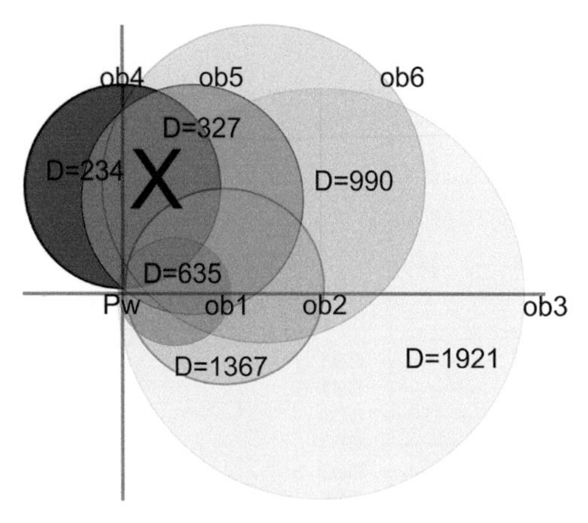

Figure 12.4 The circular areas of D values influence the observed heads at different observation wells most due to pumping at the pumping well. Each circular area has a D value. The overlapped area (X) by circles with D = 234 and 327 is the likely location of the anomaly.

of D value to which the observed head is most sensitive (Sun et al., 2013). A large D value means the hydraulic connectivity between the pumping and the observation is high and vice versa. In this example, there are six observation wells, and six circles are drawn, each with a different D value. We use a light color for a large D value and a deep color for a low D value. With the overlapping colored circles, we could guess that the likely location of the low D areas lies at the overlapping dark-colored area (X in Fig. 12.4). This guess is based on a snapshot of the anomaly.

If the pumping test is repeated at another well and the above analysis is replicated, we have another snapshot from a different perspective. Combining the two snapshots, we have a more certain interpretation of the anomaly's location. Of course, the accuracy increases as the number of observation wells increases or more pumping tests are conducted at different locations. This simple anomaly example elucidates the HT concept without relying on any complex mathematics. However, as the number of anomalies increases and the connectivity network becomes complicated, interpretation of the arrival time becomes a nontrivial task. Indeed, many have developed sophisticated algorithms to decipher the arrival times to map the D distribution (Vasco and Datta-Gupta, 1999; Vasco et al., 1999 and 2000; Brauchler et al., 2003, and many others).

12.2.2 Inverse Model of HT

The above arrival time analysis illuminates the concept of HT. A rigorous analysis of HT requires inverse modeling using observed heads to estimate the parameter distribution of a multidimensional HPHM (Highly Parameterized Heterogeneous

Model). Over the past decades, theories and algorithms for inverse modeling without taking advantage of HT surveys have been developed (see Sun, 1994; McLaughlin and Twonley, 1986). In particular, Kitanidis (1995) and Yeh et al. (1995 and 1996) embraced geostatistic concepts in inverse modeling algorithms. Specifically, they employed the stochastic representation of heterogeneity as prior JPD (Section 9.2) and derived the conditional expectation (the most likely estimates) given some observations (Priestley, 1981). These algorithms rely on Bayesian conditional probability theory and are similar to the Kalman filter in time series analysis (Faragher, 2012). A detailed presentation of this subject is beyond the scope of this book. As such, we will discuss them briefly below.

In effect, HT is not just an inverse modeling analysis. It is also a cost-effective pumping test and intelligent data collection to enhance the estimates of the heterogeneous parameter distribution. Notably, it uses one pumping test to estimate one possible heterogeneous hydraulic property distribution and validates the estimate using a different pumping test. If the validation is not satisfactory, the estimate is adjusted. This process repeats until all data are exhausted.

Many have proposed the HT concept, Yeh and Liu (2002) developed the first steady-state HT inverse algorithm to estimate three-dimensional heterogeneity, and Zhu and Yeh (2005) expanded it to transient HT analysis. These algorithms are built upon the successive linear estimator (SLE, Yeh et al., 1996). In plain language, SLE uses geology in a statistical context (i.e., average hydraulic properties, their standard deviations, average dimensions of layers) as prior (guess) information to yield statistically most-likely spatial distributions of the properties, given available observations of hydraulic responses of the aquifer and point measurements of the properties. Besides, it derives the posterior (or conditional) JPD for quantifying the uncertainty associated with the estimates due to insufficient observations. Details are referred to publications related to HT by Yeh and his colleagues (e.g., Xiang et al., 2009).

12.3 HT Field Examples

This section presents successful field applications of HT to a glaciofluvial aquifer in Canada and a fractured granite formation in Japan.

12.3.1 Glaciofluivial Aquifer

Illman and his colleagues (Alexander et al., 2011; Berg and Illman, 2011, 2013, 2015; Zhao and Illman, 2018) have developed an HT experiment field facility at the North Campus Research Site (NCRS) located on the University of Waterloo campus in Waterloo, Ontario, Canada. The site is situated on a highly

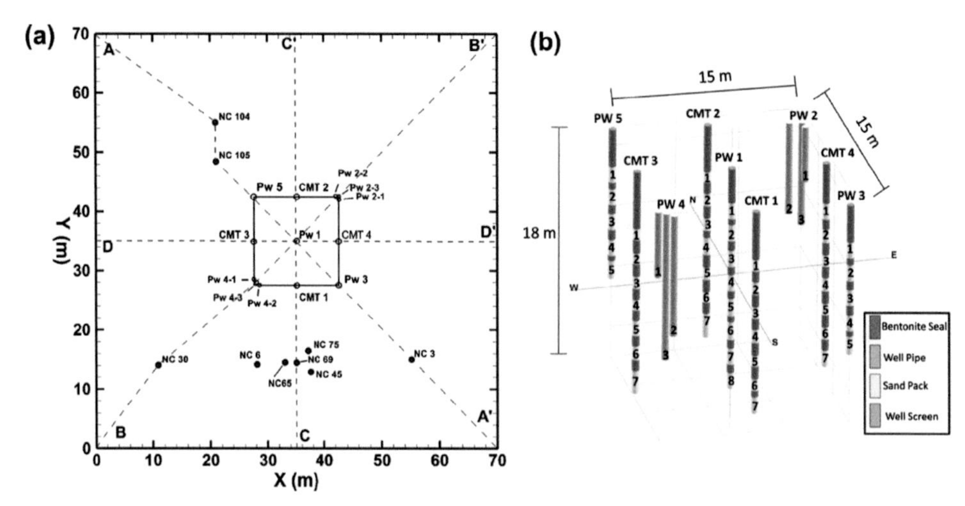

Figure 12.5 (a) Plan view showing well locations used in this study at the North Campus Research Site (NCRS) situated on the University of Waterloo (UW) campus. Solid circles indicate the locations where only geological data are available. Dashed lines indicate four geological cross-sections: A–A', B–B', C–C' and D–D' provided in Fig. 2.(b) Well screen locations shown for wells clustering in the inner 15 by 15 m square area where pumping tests are conducted. From Berg and Illman (2011).

heterogeneous glaciofluvial deposit. The main aquifer zone of the NCRS, composed of high K sand to sandy gravel, exists from 8 to 13 m below the ground surface (mbgs). Core samples indicate that the aquifer consists of two high K units separated by a discontinuous thin aquitard layer, allowing for hydraulic connection between the two units. Aquitard layers composed of low K silts and clays are also presented below and above the main aquifer. The overlying aquitard layer also is discontinue at various locations. Previous pumping tests indicated that the aquifer behaves as a confined to the semi-confined system.

Four continuous multichannel tubing wells (CMTl, CMT2, CMT3, CMT4), three multi-screened wells (PWl, PW3, PW5), and two well clusters (PW2, PW4) were installed in an area of 15 m by 15 m at the NCRS. Fig. 12.5a is a plan view showing well locations. At the same time, Fig. 12.5b provides a three-dimensional perspective view of wells, corresponding pumping, observation intervals, and intervals sealed with bentonite at the site.

Continuous cores were collected to characterize the site geology during the drilling and installation of all CMT and PW wells. After splitting the core into half along its length, soil texture was classified based on the layering observed at the core scale. Then, samples were extracted at 10 or 50 cm intervals for laboratory falling head permeameter tests. The borehole logs of the above nine wells and additional nine wells were compiled to construct the geological model for the

NCRS. Based on the soil types and corresponding depth information, 19 different layers representing seven material types were identified along all boreholes.

A total of 15 pumping/injection tests were conducted to stress the multiple aquifer-aquitard system in a tomographic fashion. Nine pumping tests (PW1–3, PW1–4, PW1–5, PW3–3, PW3–4, PW4–3, PW5–3, PW5–4, and PW5–5) mainly stressed the aquifer layers. No drawdown responses were detected from the bottom ports. Therefore, six additional pumping and injection tests were carried out to stress the aquitard zones at PW1–1 directly, PW1–6, PW1–7, PW2–3, PW3–1, and PW5–1.

During each pumping/injection test, pressure transducers were placed at all observation intervals. Zhao and Illman (2018) utilized 12 pumping/injection test data to perform the HT analysis. Data from eight tests (PW1–1, PW1–4, PW1–6, PW1–7, PW2–3, PW3–3, PW4–3, and PW5–3) were selected for model calibration, while the other four tests (PW1–3, PW1–5, PW5–4, and PW5–5) were reserved for model validation. In total, they selected 522 pressure head data for model calibration and used 348 head data for model validation. The model validation means that the HT estimated K and S fields are used to forecast the flow field under the stresses of different scenarios. Such a validation tests their ultimate objective of aquifer characterization and groundwater forecast.

Zhao and Illman (2018) highlighted the robust performance of the HT through the comparison of different degrees of model parameterization, including (1) the effective parameter approach of EHM; (2) the geological zonation approach relying on borehole logs; and (3) the geostatistical inversion approach considering different prior information (with/without geological data). Results reveal that the simultaneous analysis of eight pumping tests with the geostatistical inverse model yields the best model calibration and validation results. Fig. 12.6 shows the comparison between schematic geologic cross-sections and the K fields estimated by H.

12.3.2 Large-Scale Fractured Granite Site (Mizunami, Japan)

A significant criticism during the development of HT is the myth that the pressure variation initiated by pumping or injection test can travel only for a short distance and is too weak to be useful for aquifer characterization. To bust this myth and elucidate HT's effectiveness and intuitiveness, we briefly discuss an HT field-scale experiment at the Mizunami Underground Research Laboratory (MIU) granite site in Japan (Illman et al., 2009; Zha et al., 2016).

The site consists of a low-permeable fault zone, IF_SB3_02, and fracture lineaments along both sides of the faults (Fig. 12.7a and b). There are nine boreholes at the site (namely, MIZ-1, DH-2, DH-15, MSB-1, MSB-3, 10MI22,

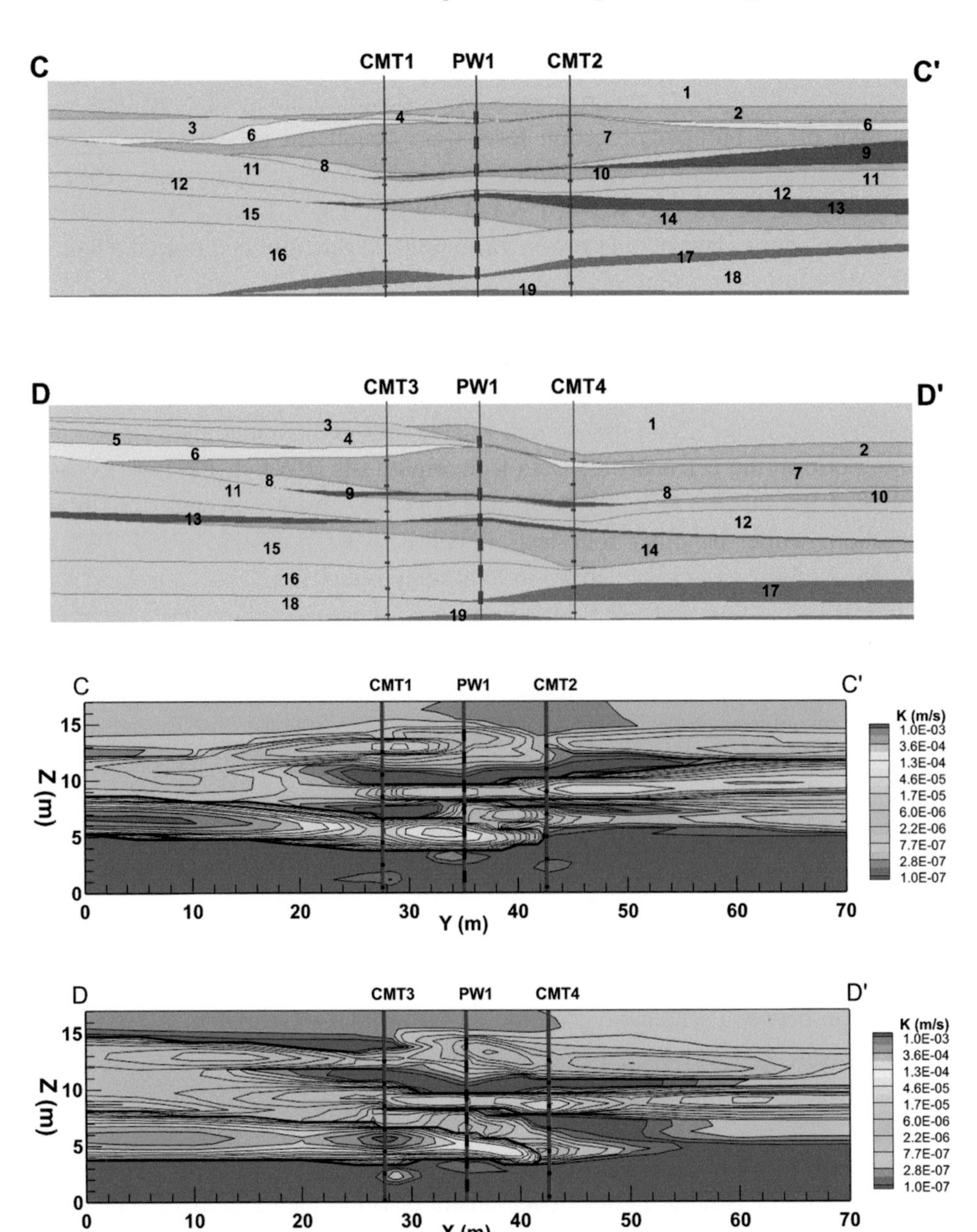

Figure 12.6 Comparison of the schematic geological cross-sections C–C′ and D–D′ in Fig. 12.5 (the upper two diagrams) and the K fields estimated from HT (the lower two). Modified from Zhao and Illman (2018).

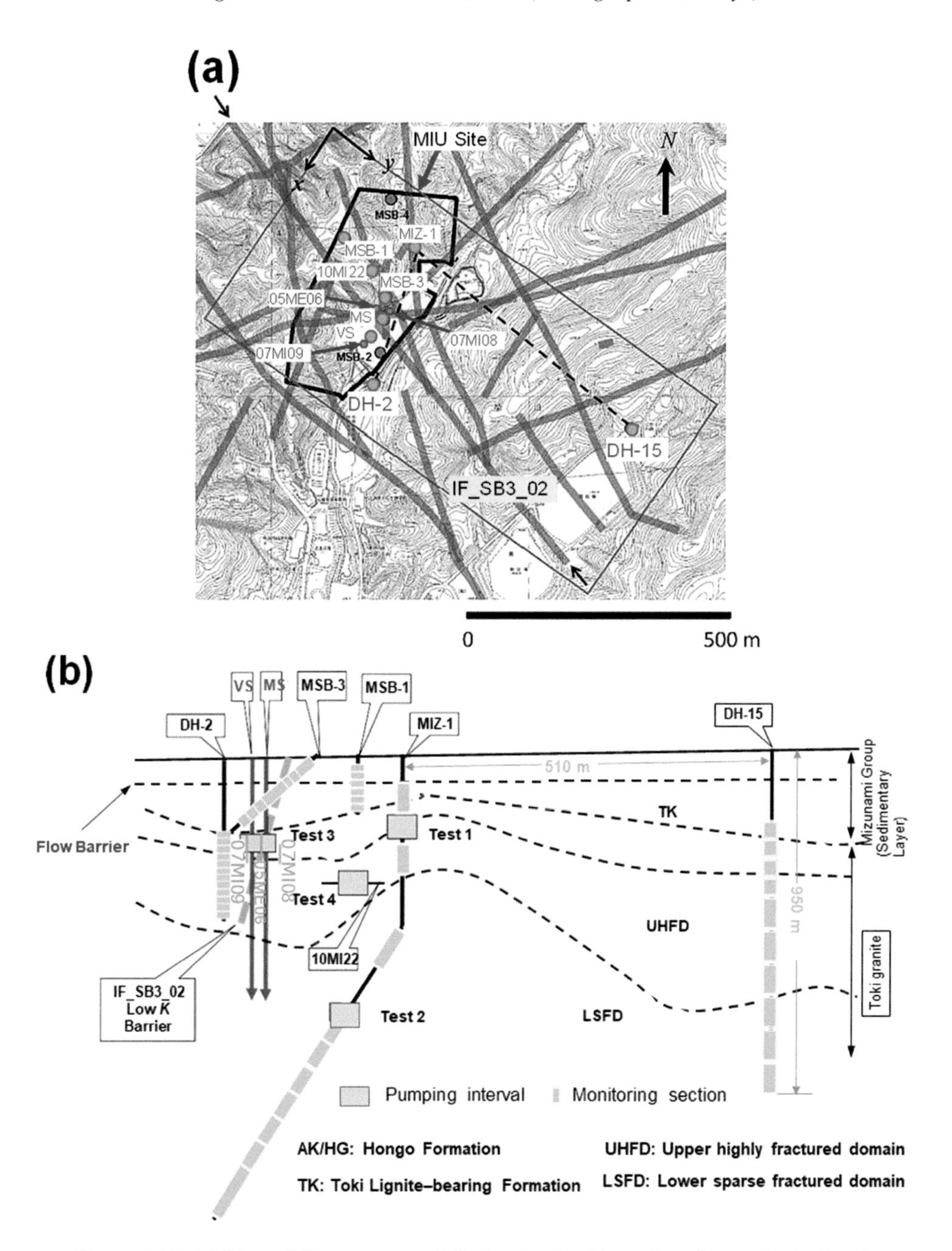

Figure 12.7 (a) Map of lineaments and faults obtained based on the geological and seismic surveys near the MIU site, where borehole locations, and the locations of the main shafts (MS) and ventilation shafts (VS), are shown. (b) A schematic cross-section shows the boreholes, the pumped and observation intervals, and the local geology along the dashed line connecting DH-1, MIZ-1, and DH-15 in Fig. 12.7a. The dashed curves approximately delineate the contact among various geologic units. (Zha et al., 2016).

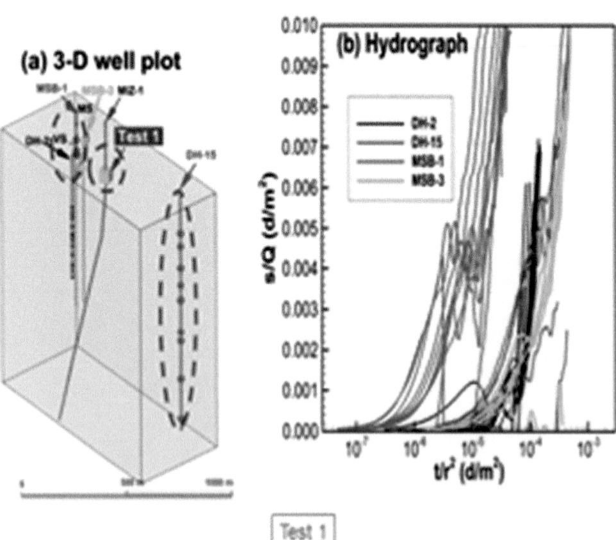

Figure 12.8 three-dimensional illustration of wells (a) and drawdowns (b) during test 1. (Zha et al., 2016).

05ME06, 07MI08, and 07MI09 around the main shaft (MS) and a ventilation shaft (VS) (Fig. 12.7b). MIZ-1 penetrates to a depth of 1300 m, DH-15 is about 1000 m deep, DH-2 is about 500 m deep, and the other boreholes are shallow boreholes with depths between 100 to 200 m. Three pumping tests (tests 1, 2, and 4) were conducted at boreholes MIZ-1 and 10MI22. In addition, during the excavation of the two vertical shafts (MS and VS), groundwater was continuously pumped at the two shafts, and groundwater responses were monitored (test 3).

Fig. 12.8a is a three-dimensional display of the pumping location of test 1 (upper interval of MIZ-1; 191–226 m deep), and monitoring intervals (shown as spheres) at different depths along DH-2, DH-15, MSB-1, and MSB-3. The drawdown-time curves at these intervals are illustrated in Fig. 12.8b. Notice that the time of each hydrography is normalized by the distance square, as in section 12.2.1. These well hydrographs indicate that pressures at most of the intervals of MSB-1 and DH-15 responded first; those of MSB-3 responded next; DH-2 lagged further behind. Notice that DH-15 is located about 700 m horizontally away from the pumped interval, while DH-2 is about 260 m. Both MSB-1 and MSB-3 are about 120 m away. Evidently, monitoring intervals of DH-15, and MSB-1 are well connected with the pumping interval (indicated by red dashed elliptical lines), likely through fractures. Albeit DH-2 is close to the pumping interval, drawdown at its monitoring intervals lagged. Thus, a low K barrier likely exists between DH-2

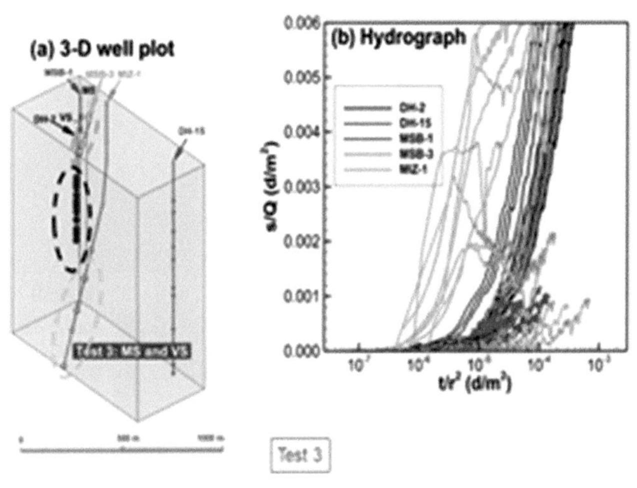

Figure 12.9 3-D illustration of wells (a) and drawdowns (b) during test 3. (Zha et al., 2016).

and the pumping interval. Similar patterns of drawdown-time responses were observed at all the intervals during tests 2 and 4. These responses suggest that the pumped locations of tests 1, 2, and 4 are well connected to DH-15 and MSB-1. At the same time, they are isolated from DH-2, perhaps by the fault (IF_SB3_02), an inferred flow barrier based on various geologic investigations.

The drawdown responses along all boreholes during test 3 (dewatering of MS and VS) is illustrated in Fig. 12.9a like Fig. 12.8a. The color-coded drawdown-time curves during the test are shown in Fig. 12.9b. In contrast to responses during tests 1, 2, and 4, DH-15 and MSB-1 now had the most negligible responses. MSB-3 had rapid and significant responses, and the drawdowns in DH-2 followed. Responses from this test implicate that the two shafts are likely linked with MSB-3, DH- 2, and the lower section of MIZ1 (green, black, and orange ellipses), but separated from DH-15 and MSB-1, perhaps by the low K barrier (IF_SB3_02).

These results again elucidate the commonsense nature of HT. The estimated K field from SLE using data from the tests is shown in Fig. 12.10a with highlighted isosurfaces of $K = 0.06$ m/d (high K zone) and $K = 0.003$ m/d (low K zone). The locations of the high K zones are consistent with those lineaments in Fig. 12.7a. A blowup of the isosurface with $K = 0.003$ m/d (the low K zone) is illustrated in Fig. 12.10b. The location of this low K zone agrees with that of fault IF_SB3_02 (gray color plane) mapped by the geologic investigation, but HT revealed a detailed irregular shape. Note that prior geologic investigation led to the HT design, which placed sources and response monitoring points at different depths

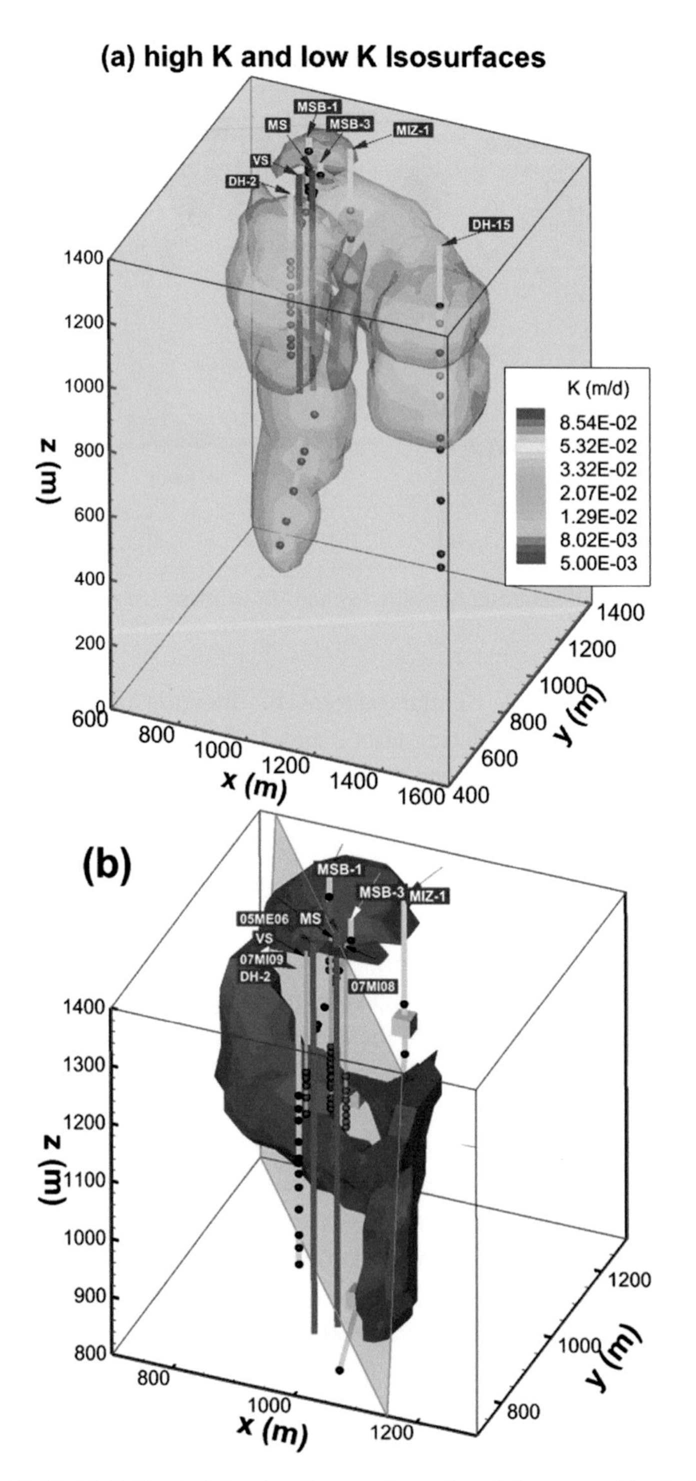

Figure 12.10 (a) Estimated fractured zone (green) and impermeable fault zone (blue) and (b) zooming view of the fault zone. (Zha et al., 2016).

and horizontal locations on both sides of the possible fault. Consequently, complex fractures and fault zones are obtained by HT at the MIU site. Their results confirm the importance of a prior geologic survey.

Tiedeman and Barrash (2020) also reported a similar successful application of HT to map 3-D hydraulic conductivity, fracture network, and connectivity in a small well field at the former Naval Air Warfare Center, West Trenton, New Jersey.

12.4 Electrical Resistivity Tomography (ERT)

HT is a promising high-resolution aquifer characterization technology. Its characterization resolution and implementation costs increase with the number of wells, which could be excessive. Consequently, developing cost-effective pressure sensors without boreholes could alleviate this HT issue. While this new generation sensor remains to be developed, electrical resistivity tomography (ERT) could be a surrogate.

ERT, evolved from the classical dc resistivity method, has been a widely used, inexpensive technique for near-surface investigations over many decades. This classical method estimates the subsurface bulk resistivity using the voltage induced by the current between electrodes at the ground surface. Resistivity data are collected using single or multiple pairs of current and voltage electrodes (dipoles) with known relative positions. They are then interpreted by matching them to theoretical models with varying conductivity subsurface structures. Textbooks such as Keller and Frischknecht (1966) and Koefoed (1979) provide detailed coverages of this method.

Today, the state of the science is to collect extensive current and electrical potential data in a tomography survey fashion – ERT. These data are then interpreted by inversion of multidimensional HPHM, as are the data of HT (e.g., Ellis and Oldenburg, 1994; Li and Oldenburg, 1994; Zhang et al., 1995; Yeh et al., 2006). ERT's operational cost is much less than HT. However, ERT maps the electrical resistivity distribution, which could be related to the hydraulic property (e.g., concentration, moisture content, K, or S_s) distributions. While well-established relationships, e.g., Archie (1942), exist, parameters of the relationship may vary spatially due to physical or geochemical and other environmental factors' variability in geologic media. The time-lapse ERT survey—detecting the difference between successive ERT surveys (e.g., Sawyer et al., 2015) – attempts to overcome this issue. However, the issue remains to be further investigated (e.g., Yeh et al., 2002). Despite this shortcoming, ERT has been accepted as a low-cost and effective monitoring tool for many environmental investigations. Below, we

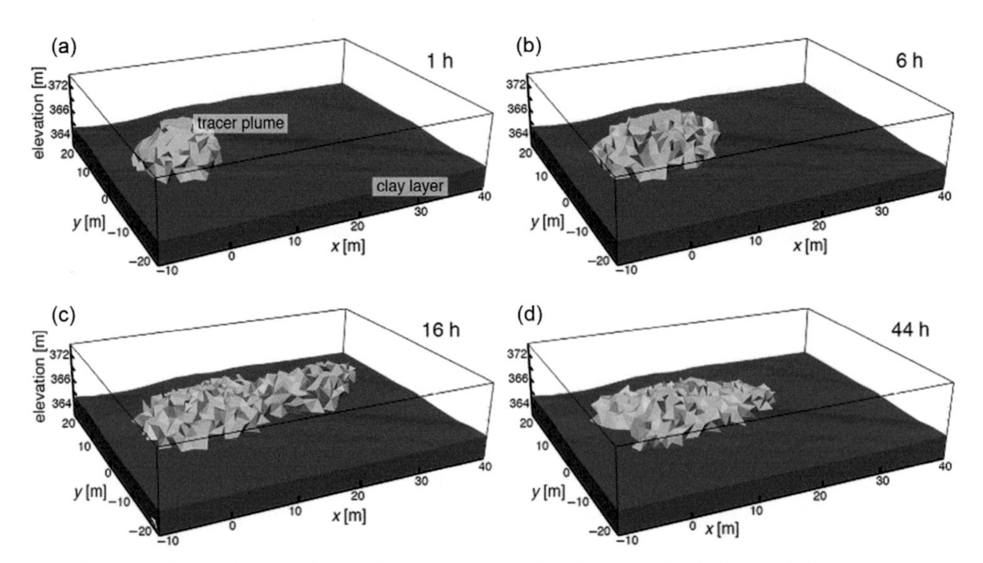

Figure 12.11 Tracer plume defined by a 3% decrease in bulk resistivity with respect to the baseline model (a) 1 hour, (b) 6 hours, (c) 16 hours, and (d) 44 hours after the start of the tracer injection. (Doetsch et al., 2012).

present a recent study of the application of ERT to monitor salt solution migration in a riparian groundwater system.

Doetsch et al. (2012) conducted a 3D ERT time-lapse survey at an alluvial aquifer adjacent to the Thur River in northeastern Switzerland. This experiment involved injecting a salt tracer into a gravel aquifer through an injection well and then monitoring the evolution of the tracer plume using surface ERT and continuous measurements of hydraulic head and water electrical resistivity in observation wells. The measurements were carried out on a gravel bar overlying a highly permeable groundwater system in direct contact with a restored stretch of the Thur River.

The evolutions of the three-dimensional ERT images in Fig. 12.11 reflect the migration of the salt-tracer plume one hour after salt injection began and 6, 16, and 44 hours later. The moment analysis (Chapters 4, 8, 9, and 10) found that the center of mass moves with a velocity of 2×10^{-4} m/s, and its front appears to move more than twice as fast at 5×10^{-4} m/s. After 6 hours, the plume mass center slowed down while its front continued moving at a relatively high velocity. A region of preferential flow along which the plume rapidly propagated was observed at 16 hours. Shortly after, the front of the plume had moved $\sim$35 m and then left the area covered by the electrode array. At times $>$20 h, the plume appeared to shrink, likely attributed to an effect of ERT's resolution and detection limitations. While the actual plume was continuously increasing in size, the remaining tracer mass in the aquifer's well-resolved region decreased as the tracer

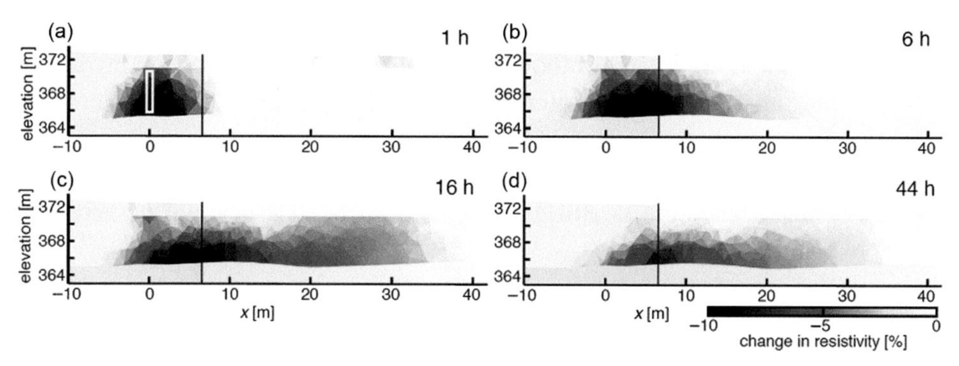

Figure 12.12 Cross-sections of changes in resistivity (a) 1 hour, (b) 6 hours, (c) 16 hours, and (d) 44 hours after the start of the tracer injection. The change in resistivity is calculated with respect to the initial resistivity. The white box in (a) marks the injection interval. The vertical black line marks the position of observation well. (Doetsch et al., 2012).

became more diluted and moved farther downstream into regions where ERT sensitivities were low. This result means that reliable estimates of the tracer plume can only be obtained within $\sim$16 hours after tracer injection. The remains of the plume within the electrode array moved very slowly near the aquifer base at times $>$40 h.

The density effects associated with the salt tracer are apparent in the cross-section views of resistivity change in Fig. 12.12. The plume's center moved downward with time, especially between 1 and 6 h after tracer injection.

Besides the ERT survey, Doetsch et al. (2012) took advantage of the time-lapse 3D ERT images of the plume to follow the tracer plume's center of mass to estimate the average groundwater velocity (Chapter 10) and investigated preferential flow and tailing for this specific experiment.

Lastly, recent rapid advances in instrumentation and interpretational tools in geoelectrical surveys highlight new opportunities. For example, Slater and Binley (2021) proposed long-term resistivity monitoring (LTRM) to improve understanding and characterization of flow and transport processes operating over months to decadal timescales. The LTRM could be a paradigm shift in high-resolution monitoring of hydrologic processes because of its cost-effectiveness. For instance, the costly and labor-intensive concentration collection efforts in the Borden, George Town, MADE experiments (Chapter 10) could have been avoided. Besides, LTRM may unveil more detailed spatiotemporal evolutions of the solute plume than the sampling schemes in previous experiments. Advances in interpretation tools and these big data undeniably could characterize multi-scale heterogeneity more vividly than ever. Such an LTRM concept is a step toward exploiting natural stimuli for basin-scale characterization championed by Yeh et al.

(2008). That is, making full use of the enormous energy of recurrent natural events (i.e., lightning, flood, and earthquake events) as geophysical or hydraulic tomographic surveys of groundwater basins (see Wang et al., 2020 for river stage tomography).

At this end, we emphasize that data fusion or assimilation technology for high-resolution characterization is still evolving. As such, it is beyond the scope of this book.

12.5 Benefits of High-Resolution Characterization

The previous two sections demonstrate the viability of HT for mapping heterogeneous hydraulic property distributions in field-scale porous and fractured geologic media. The benefits of HT for predicting solution transport remain to be demonstrated, particularly in large-scale field HT and tracer experiments. Most large-scale tracer experiments were conducted much earlier than the development of HT. Because of this time gap, these experiments' designs and data collections were unsuitable for HT analysis. On the other hand, performing new large-scale HT and tracer experiments requires substantial time, effort, and financial support. For these reasons, this section presents the numerical simulation by Ni et al. (2009) to illustrate the utility of HT for predicting solute transport in a hypothetical aquifer.

Ni et al., (2009) considered a 60 m $\times$ 20 m vertical cross-section of a confined aquifer, discretized into 1200 elements (60 $\times$ 20) of 1 m $\times$ 1 m in size. A random field generator created the distributions of hydraulic conductivity, porosity, and longitudinal and transverse dispersivities in the aquifer, assuming normal JPDs with given means, variances, and exponential covariance (Chapter 9). Fig. 12.13 shows the generated distributions of the parameter fields. The K field's geometric mean is 1.0 m/day, and that of the porosity (n) field is 0.3. The variance of ln$K$ is 1.0, and that of ln n is 0.01. Likewise, the geometric mean of longitudinal dispersivity α_L is 1.0 m, while that of transverse dispersivity α_T is 0.1 m. The variances of ln α_L and ln α_T are assumed 0.1 since little information about their variability.

In addition, all the generated parameter fields are statically anisotropic (Chapter 9), having the same correlation length (λ) of 40 m for the x-direction and 5 m for the z-direction. Note that all the parameter fields were generated independently, representing the worst scenario for predicting flow and transport in such an aquifer. Specifically, the knowledge of one parameter field does not yield any information about the others. These K, porosity, and longitudinal and transverse dispersivity fields are considered the reference or true fields for their study.

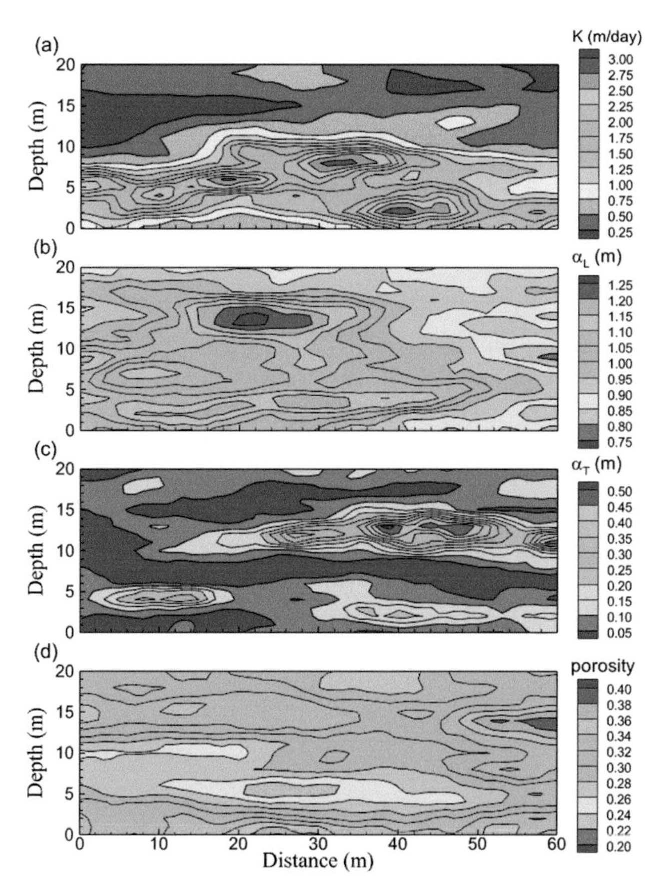

Figure 12.13 (a) The K, (b) the longitudinal dispersivity, (c) the transverse dispersivity, and (d) the porosity distributions of the aquifer. Modified from Ni et al. (2009).

After creating the heterogeneous fields, the aquifer's left and right sides were assigned constant hydraulic heads of 100.6 m and 100 m, respectively. A natural gradient steady-state flow field under the given boundary conditions was simulated. Under this simulated flow field, an instantaneous slug of tracer (100 ppm) was released at the tracer source zone on the left side of the aquifer (Fig. 12.14). The transient migration of the slug of tracer was simulated by solving the HPHM with the local-scale flow equation and ADE (Eqs. 9.3.2 through 9.3.6 in Chapter 9) with the method of characteristics (Chapter 5) (VSAFT2, Yeh et al., 1993). No flux boundaries were on the aquifer's top and bottom, whereas zero dispersive flux boundaries were assumed on the left and right.

Snapshots of the simulated plume at times 500 and 1000 days and associated streamlines and velocity field are displayed in Fig. 12.15a and b, respectively. The snapshots show that the peak of the plume noticeably lags behind the plume front,

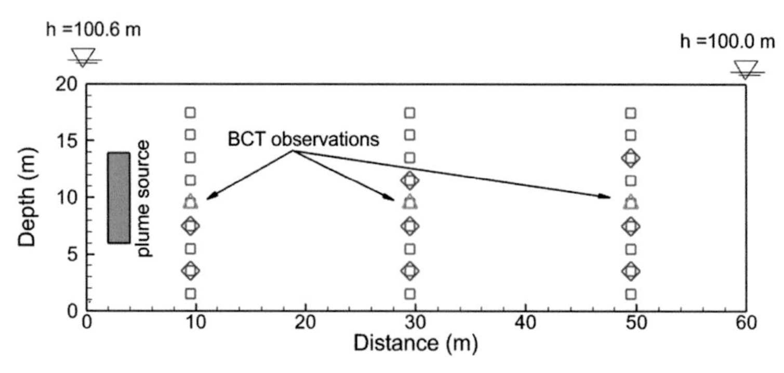

Figure12.14 A view of the synthetic aquifer with the source zone of the tracer, three vertical wells, and boundary conditions. Red diamonds are pumping locations, while blue squares denote the head observation ports and orange triangles the concentration monitoring points. Modified from Ni et al. (2009).

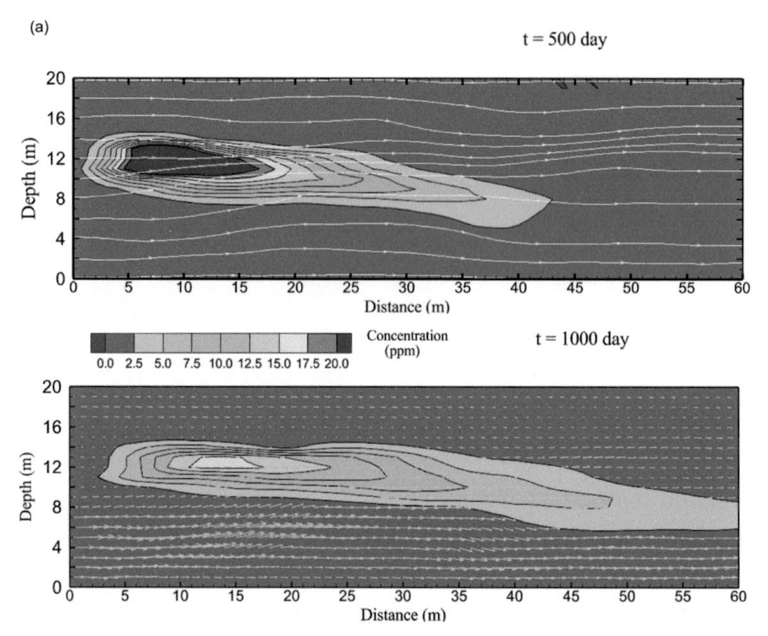

Figure 12.15 (a) The snapshot of simulated plumes at t = 500 days and (b) t = 1000 days after tracer release at the source zone. White continuous lines are the streamlines, and white arrows are the velocities. Modified from Ni et al. (2009).

creating non-Gaussian and skewed spatial distributions similar to the observed in the MADE site. (Fig. 10.10, Chapter 10). Again, we reiterate that these plumes were simulated using the local-scale flow equation and the local-scale (i.e., soil column) classical ADE in HPHM, without using non-Fickian, scale-dependent, mobile/immobile, or fractional derivative transport theories. The cause of this skew distribution is evident if one carefully examines the heterogeneous K field in Fig. 12.13, the solute source location in Fig. 12.14, streamlines, and velocity vectors in Fig. 12.15. Specifically, the source zone resides in a low permeability region, which retards the migration of the majority of the solute plume. In contrast, the high permeability regions below and in front of the low permeability zones move the solute rapidly forward once the solute has arrived. These phenomena resemble those observed in the MADE experiment (Fig. 10.10 in Section 10.3). The effects of variability in other parameters such as porosity and dispersivities seemly play minor roles. The local-scale variability of the local-scale K heterogeneity dominates the shape of the solute distribution. We reiterate that the simulation was carried out using the classical local-scale ADE based on Fick's law.

This example stresses the importance of local-scale K heterogeneity, which partially dictates the local-scale velocity variation in the aquifer. The logical question is: Could the previously discussed HT or other approach improve our ability to predict the migration of the solute migration in this aquifer?

Ni et al. (2009) answered the question. Their work compared the estimated K fields from HT and kriging method (a spatial interpolation and extrapolation method) using some known K measurements to the true K field. Afterward, it compared the simulated head and velocity fields from the two estimated K fields to the true fields. They finally examine the solute transport predictions from the macrodispersion approach in addition to those predicted from the two estimated K fields.

To compare the predictions of solute migrations from the three approaches, they installed three wells at $x = 10$ m, 30 m, and 50 m. Each well fully penetrated the entire depth of the aquifer with nine open ports, which were used as either head or concentration sampling ports or pumping ports for HT survey (See Fig. 12.14). Besides, nine aquifer samples were taken for K measurements at the port locations, resulting 27 K measurements.

HT was conducted, pumping at one of the eight pumping ports at a 10 m^3/day rate to reach a steady state. Then, they monitored the heads at all the 27 ports. The pumping test was repeated at the other seven pumping ports. After completing the HT survey, there were eight snapshots, each consisting of 27 heads at different locations. SLE (Section 12.2.2) was used to interpret the snapshots to derive the K field (Fig. 12.16b).

In addition to HT, a variogram analysis (Chapter 9) was conducted using the 27 direct K measurements and obtained approximated ranges of 20.1 m and 5.1 m for x and z directions, respectively. Kriging was used to interpolate and extrapolate the 27 K values to obtain an estimated K field for this aquifer (Fig. 12.16a).

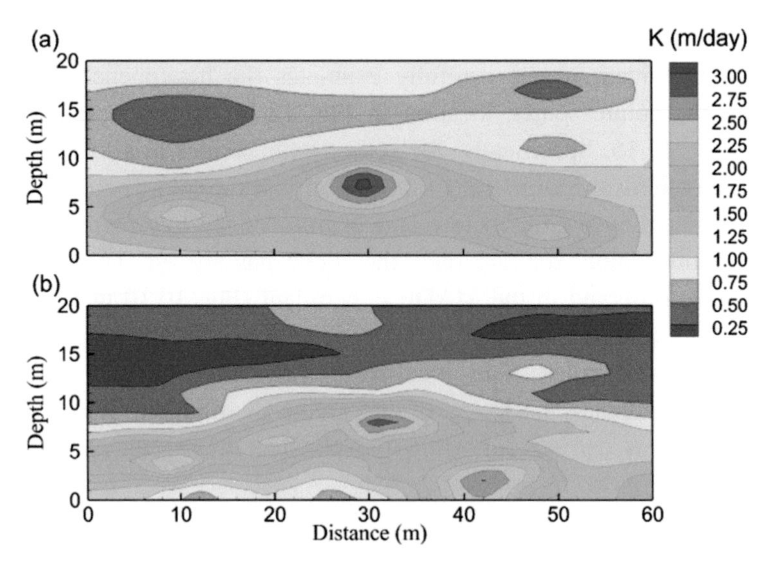

Figure 12.16 (a) Kriged K field and (b) the K field estimated from HT survey. Modified from Ni et al. (2009).

Comparing the HT K fields in Fig. 12.16b with the reference K field in Fig. 12.13a, the HT closely captures the reference field. However, the kriging estimated field (Fig. 12.16a) only gives a relatively smooth K distribution.

Quantitative comparison was accomplished by visually examining scatter plots of the two estimated versus reference K fields (Fig. 12.13a) and evaluating some statistic metrics. First, the Pearson coefficient correlation (typically, correlation) is calculated, looking for a possible relationship between two variables (the estimated and the true).

$$\rho = \sum_{i=1}^{n}(x_i - \overline{x})(y_i - \overline{y}) / \sqrt{\sum_{i=1}^{n}(x_i - \overline{x})^2 (y_i - \overline{y})^2}. \tag{12.5.1}$$

In the equation, $x_{i,}$ and y_i respectively represent the estimated and the true. $\overline{x}$ and $\overline{y}$ are their corresponding means. n is the total number of the data, and i indicates the data number. If the correlation is high, a linear relationship exists. A linear model is then fitted to the pairs of the true and estimated K values. The linear model is

$$\widehat{y}_i = b_0 + b_1 x_i. \tag{12.5.2}$$

The term $\widehat{y}_i$ is the estimate given the true x_i or the true given the estimate. The slope and the intercept of the line are b_1 and b_0, respectively. The coefficient of determination (R^2, R-square, or goodness-of-fit) is then used to determine how close the fitted regression line is to the data.

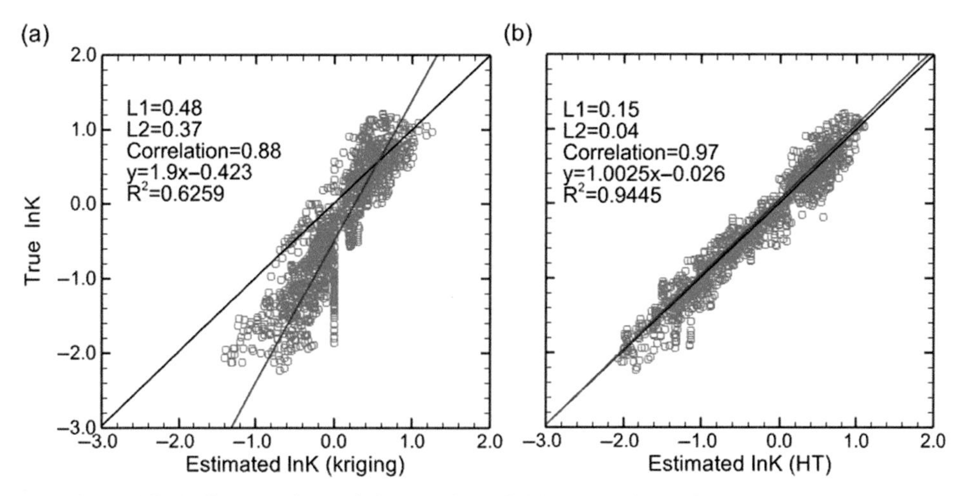

Figure 12.17 Scatter plots of the true lnK field vs. (a) the estimates from kriging, and (b) from HT survey. The red lines are the linear regression lines. Modified from Ni et al. (2009).

$$R^2 = \sum_{i=1}^{N} (y_i - \bar{y})^2 / \sum_{i=1}^{N} (\widehat{y}_i - \bar{y})^2. \tag{12.5.3}$$

A high R^2 value means the regression line closely represents the relationship between the estimates and the true. A regression line with a slope close to one and an intercept close to zero indicates minimal unbiased estimates of the actual field.

In addition, the mean absolute error (L1), the mean square error (L2) between the estimated and true are determined:

$$L1 = \frac{1}{n}\sum_{i=1}^{n} |x_i - y_i| \text{ and } L2 = \frac{1}{n}\sum_{i=1}^{n} (x_i - y_i)^2 \tag{12.5.4}$$

A small L1 means that the mean of the predicted values is close to the true mean, while a small L2 value indicates the overall deviation of the estimated values from measured values is small.

According to Fig. 12.17, the correlation between the true and the kriging estimate is high (0.88), indicating a linear relationship. The regression line has a slope equal to 1.9 and an intercept of −0.423, with $R^2 = 0.626$. Visually, kriging estimates with values from −1.4 to 0.0 are larger than the true, overestimating the K field. This fact also is reflected in the greater slope of the linear regression line than one and its large intercept value. This bias is due to the non-representativeness of the 27 K measurements.

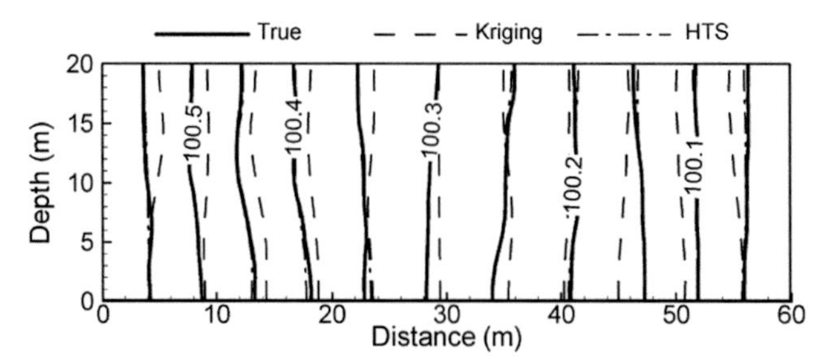

Figure 12.18 The comparison of the reference head field and those simulated based on the kriged K field and those based on HT survey. The true heads and those simulated from HT estimates are almost identical. Modified from Ni et al. (2009).

In contrast, the HT's K estimates are highly correlated with the actual field with a correlation of 0.97. They are slightly scattering around the 1:1 line, indicating that the HT estimates capture the actual field pattern but not the details, as indicated by the scattering. The slope of the regression line is 1.01, and the intercept is 0.05 with $R^2 = 0.9445$. The superiority of HT's estimate to the kriging estimate can be attributed to several facts. HT considers the observed head, the pumping rate, flow, and the governing flow equation, which reveals connectivity between the pumping ports at eight locations and head monitoring ports and average K values. Specifically, one pumping test reveals possible hydraulic connectivity between pumping and observation ports and an average K, and the subsequent pumping test validates and then improves them. As a result, the HT estimate is close to the pattern of the actual K field after the eight pumping tests.

They subsequently simulated the head and velocity distributions using these two estimated K fields and the boundary conditions under a natural gradient flow condition. The comparison between the head fields from the two K fields and the head field based on the reference K field is shown in Fig. 12.18. The head field from HT's K estimate is identical to that from the reference K field. However, the flow field from the kriged K field shows noticeable differences.

Likewise, Fig. 12.19a and b shows the comparison of the velocity components in x-direction velocities, u_x, of the reference field and from the kriging estimates and HT estimates, respectively. The comparisons of the velocity component in z-direction u_z are displayed in Fig. 12.20a and b, respectively. These figures also include the statistic metrics for these plots. According to Fig. 12.19a, u_x values simulated using the kriged K field are biased, and the values in the range of 0–0.01 m/day are larger than the reference velocity values. This velocity component simulated via the HT's K estimates is much closer to the reference one than the kriging K estimates, although with some scattering. The scatterings are the results of unresolved heterogeneity by HT survey.

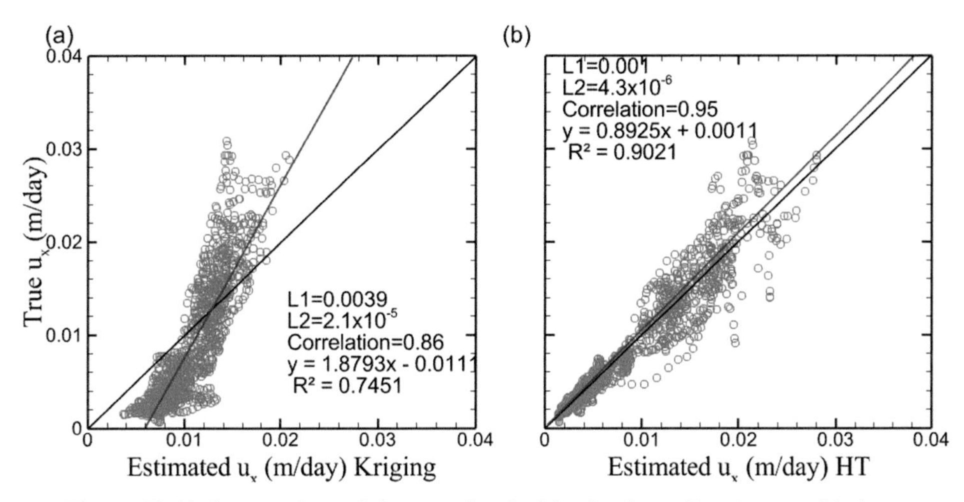

Figure 12.19 Scatter plots of the actual velocities in the x-direction vs. (a) those simulated based on the K estimates from kriging and (b) those simulated based on the *K* field from HT survey. The red lines are the linear regression lines. Modified from Ni et al. (2009).

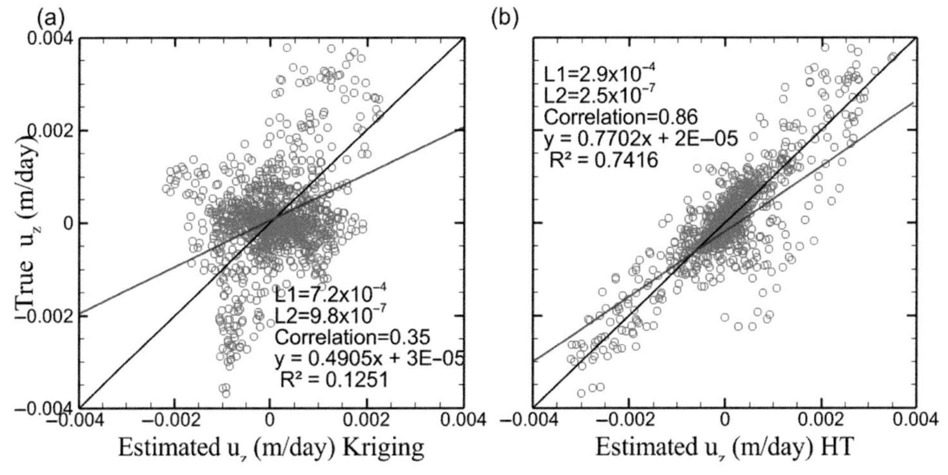

Figure 12.20 Scatter plots of the actual velocities in the z-direction vs. (a) those simulated based on the K estimates from kriging and (b) those simulated based on the K field from HT survey. The red lines are the linear regression lines. Modified from Ni et al. (2009).

It should be emphasized here that, as shown in Figs. 12.18, 12.19, and 12.20, the head is less sensitive to the *K* variability than the velocities. The former is the manifestation of energy propagation, and the latter is the advection of fluid itself. This distinction plays an essential role in understanding solute transport in geologic media.

12.6 Predictions of Transport with Different Degrees of *K* Resolutions

Ni et al. (2009) also compared the snapshots of the reference plume at times 500 and 1000 days to those derived by EHM (Equivalent Homogeneous Model) with macrodispersion and HPHM with estimated *K* fields by kriging and HT. Application of the macrodisperison approach requires some estimates of the macrodispersivities and the effective hydraulic conductivity for EHM as remarked in Chapters 9 and 10. Following the macrodispersion theory (Gelhar and Axness, 1983; Gelhar, 1993) as did Sudicky (1986) in Chapter 10, Ni et al. (2009) obtained estimates of longitudinal and transverse macrodispersivity values as 6.3 m and 0.4 m, respectively. Subsequently, the geometric mean of the local-scale *K* was used to simulate the ensemble mean flow field, and the macrodispersivity and the ensemble mean ADE were employed to derive the ensemble mean concentration distributions (see Chapter 9).

On the other hand, the HPHM with the local-scale flow equation (Chapter 9) was adopted to calculate the flow and velocity fields at the local scale with the kriging and HT estimated *K* fields. Assuming the reference local-scale dispersivity and porosity fields are known, they conducted solute transport simulations using the HPHM with local-scale ADE using the velocity fields from the kriging and HT estimated *K* fields. This assumption allows them to emphasize the effect of local-scale velocity rather than the dispersivity and porosity.

The simulated snapshots of the tracer plumes at times 500 and 1000 days based on macrodispersion, kriging estimated *K* field, and HT are compared with those of the reference in Fig. 12.21. The two plots on the top row are the reference solute distributions, slightly different from those presented in Fig. 12.15a and b because the contour levels are adjusted to the same as those in other simulation results.

The plots in the second row show that using EHM with the macrodispersion concept leads to the Gaussian distributions, as should be when the ensemble-mean ADE is used. The simulated snapshots do not reproduce the retarded plume peaks and fast spreads of the plume front of the reference snapshots and their details. This result is expected since the approach aims to derive the most likely distributions in the ensemble mean sense under a piece of limited information (spatial statistics) about this particular field site (one realization of the ensemble). More importantly, we stress that other ensemble-mean EHM approaches, adopting the scale-dependent, mobile-immobile zones, fractional derivative transport equation, encounter the same difficulties unless the approaches are calibrated with observed snapshots. After the calibration, they still would fail to reproduce the reference plume distribution in detail because they are merely ensemble-mean equations with different joint probability distribution conceptualizations.

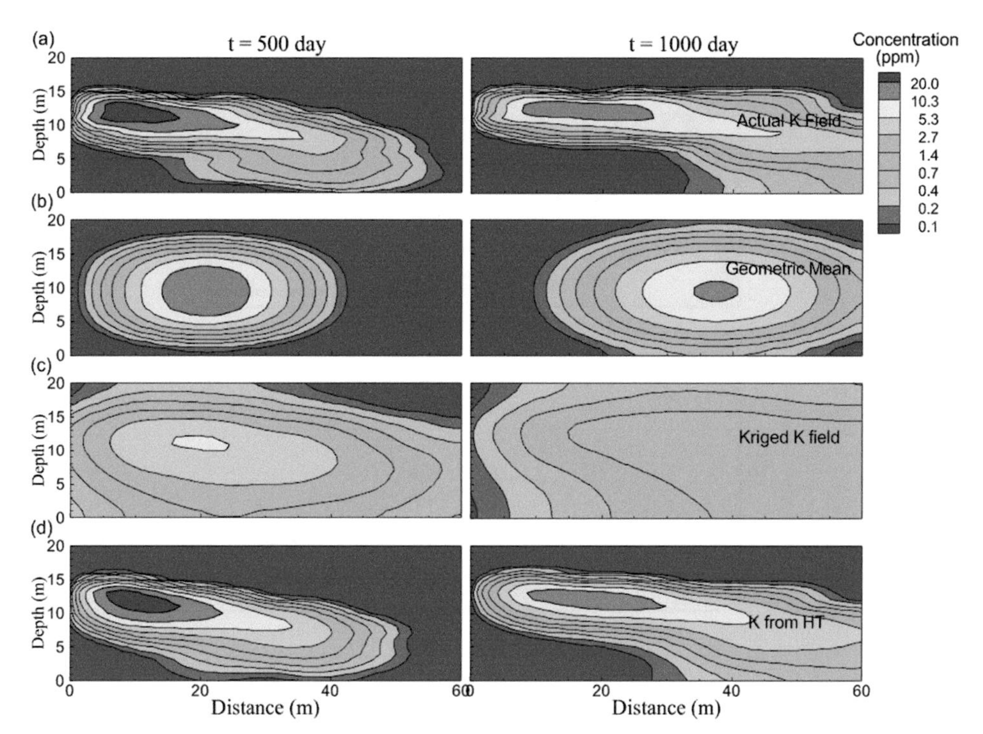

Figure 12.21 Snapshots of the plume at t = 500 days and 1,000 days: (a) the reference, (b) the simulated using the effective K and macrodispersivity, (c) the simulated using the kriged K field, and (d) the simulated based on the K field from HT survey. Modified from Ni et al. (2009).

Simulated plume snapshots based on the kriged *K* field are presented in the third row of the figure. As mentioned early, the kriged *K* field overestimated the reference *K* fields. The overestimation effects are vividly illustrated in the simulated snapshots. The plume rushes and spreads out over a great area since the spread is velocity-dependent even though the dispersivity and porosity fields are the same.

Plots in the last row are the simulated snapshots from the *K* field estimated by HT. They closely resemble those in the actual field, showing the delay of the major part of the trace plume, rapid dispersal of its front, and non-Gaussian concentration distributions. The robustness of HT for predicting solute transport was also demonstrated by its improvements in predicting the BTC at sampling ports in the three wells.

To compare the simulated BTCs based on different approaches, they examined the BTCs at monitoring ports at $z = 10$ m of the three monitoring wells ($x = 10$, 30, and 50 m) (see Fig. 12.22). The figure shows that the HPHM based on the local-scale *K* field estimated by HT results in the best predictions at all three

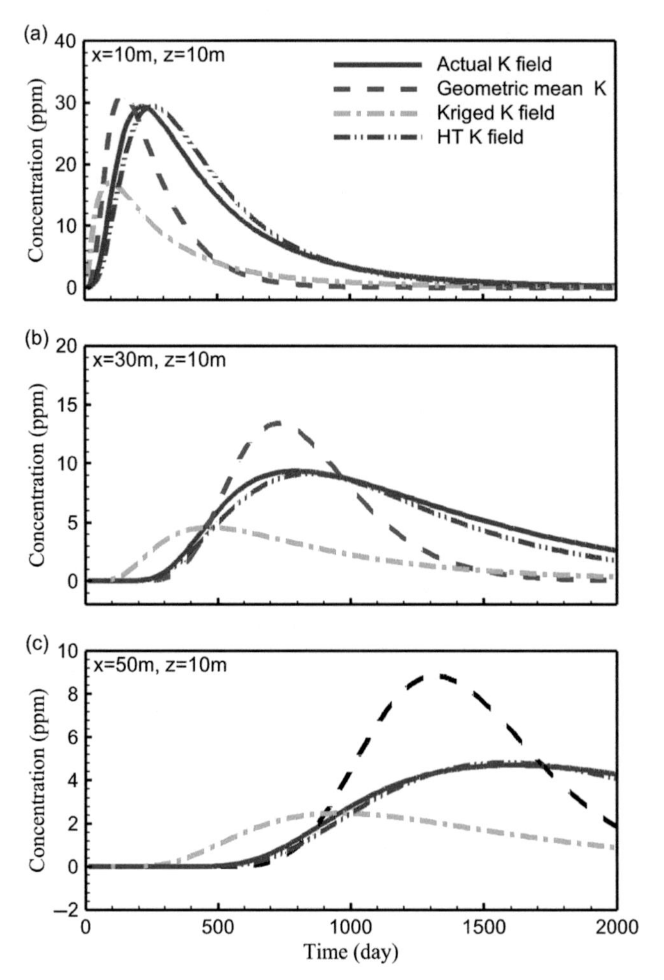

Figure 12.22 Comparisons of the actual BTCs with the simulated using the macrodispersion approach, the kriged K field, and the HT K field at the elevation $z = 10$ m of the three wells ($x = 10$ m 30m, and 50 m). Modified from Ni et al. (2009).

locations compared with the reference BTCs. The estimate from HT yielded accurate BTC peaks and arrival times at different locations. In contrast, the BTCs based on the HPHM with the kriged K field show poor predictions of BTC peaks and underestimations of their arrival times. The macrodispersion approach using the EHM also yields unacceptable BTC predictions, failing to capture the long tailing of the actual BTC, which has been explained as the non-Fickian solute transport based on the ensemble-mean Fickian ADE approach.

These predictions are based solely on the estimated K fields or the estimated spatial statistics (EHM approach); they were not calibrated to match any observed plume characteristics. In addition, the HPHM used the local-scale K field estimated from HT or kriging and the same referenced heterogeneous dispersivity and porosity fields. Because of this, we conclude that the knowledge of the actual local-scale velocity fields, dictated by the local-scale K and head variability, is more critical than the variability of the local-scale dispersivity and porosity values under the given characteristics variability. Specifically, variability of the local-scale K over the entire domain (large-scale heterogeneity) dominates the large-scale behavior of the tracer plume, and the role of dispersivity and porosity representing pore-scale heterogeneity is minor. This conclusion deviates from Feehley et al. (2000).

12.7 Importance of Large-Scale Flow Variations

Chapters 9 and 10 focus on situations where flow is unidirectional mean flow or forced gradient dipole flow fields during the discussions of the mixing or dispersion in soil columns and fields. Real-world field situations often involve complex, large-scale regional flow patterns. For this reason, this section explores the impacts of large-scale flow variations on the solute transport in geologic media at our observation and interest scales using a numerical experiment of the hyporheic mixing process.

The hyporheic zone refers to the region where the surface water from the river enters the streambed, travels along short groundwater flow paths, and returns to the river (Cardenas et al., 2004; Sawyer and Cardenas, 2009; Su et al., 2018). During this process in the hyporheic zone, surface water and groundwater meet and mix at their interface below the streambed. Such a surface water-groundwater mixing zone is called the hyporheic mixing zone. In this zone, sharp gradients in physical, chemical, and biological conditions exist, giving rise to unique processes that do not occur elsewhere in the overlying stream or underlying aquifer (Packman and Salehin, 2003; Sawyer and Cardenas, 2009). For this reason, hyporheic mixing controls magnitudes and rates of physical, biogeochemical, and thermal processes and influences the transport and fate of a wide range of ecologically relevant substances, including nutrients, carbon, and contaminants (Brunke and Gonser, 1997; Cardenas and Wilson, 2007; Packman and Salehin, 2003; Su et al., 2016; Winter et al., 1998).

Su et al. (2020) conducted numerical simulations of flow and solute transport in hypothetic streambeds (Fig. 12.23b) to explore the effects of spatial variance and correlations of the K field, surface water velocity, and upwelling groundwater flux on the mixing zone below the river bed. To avoid the unit's dependence, they normalized the domain by the dune length (the distance between one trough and another, $l = 1$ m). As a result, the streambed was 3.0 long and 1.2 deep with three dunes. The surface water depth was 0.5, and the dune stoss height and length were

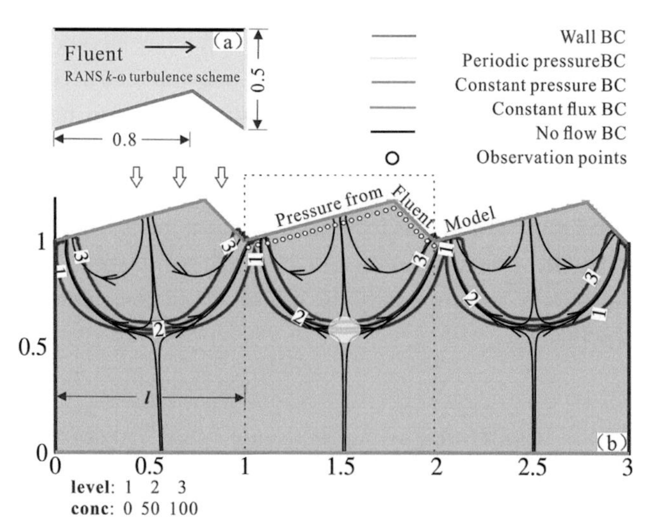

Figure 12.23 The simulation domain and boundaries for (a) the steady-state 2-D turbulent surface water flow model, Fluent and (b) the steady-state 2-D groundwater flow and transient solute transport model (VSAFT2). The upper boundary of the VSAFT model is the pressure distribution resulting from the Fluent model with the surface water velocity = 6, and the bottom boundary is a constant upwelling groundwater flux of 0.005. The other two side boundaries are no flux boundaries. The K field is homogeneous ($K = 0.74$ m/day). Black lines with arrowheads are the streamlines: The arrow indicates the flow direction. The white circle in (b) represents the stagnant zone between the surface water and groundwater circulation regimes. From Su et al. (2020).

0.2 and 0.8, respectively. Square grids with the size 0.01×0.01 discretize the domain for both flow and solute transport simulations.

To synthesize the heterogeneity of the streambed, they considered $F(\mathbf{x})$ (i.e., ln$K(\mathbf{x})$) of the streambed as a random field with specified variance σ_f^2 and correlation scales: λ_x and λ_z. The porosity n of the streambed was uniform, with a value of 0.4. The correlation scales λ_x and λ_z were normalized by l. The variance of flux $\left(\sigma_q^2\right)$ and σ_f^2 were normalized as σ_q^2/μ_K^2 and $\sigma_f^2/\overline{F}^2$, respectively. $\mathbf{q}$ and q_G were normalized by the mean of $K(\mu_K)$ (0.74 m/day). Since surface water velocity, v_s, is often reported as cm/s and the hydraulic conductivity, K, is m/day, we normalize v_s as $v_s/(864\mu_K)$ (i.e., 864 is the conversion factor between m/day and cm/s). The original notation for these variables in the following analysis is retained for convenience, even though they are normalized.

Six scenarios (Table 12.1) were investigated, where uniform local-scale longitudinal and transverse dispersivity values were assigned to be 1×10^{-7} m, ensuring that the simulated solute transport process is advection-dominated. This setup aimed to illustrate that the smearing or mixing of the local-scale concentration distribution $C(\mathbf{x}, t)$ can be attributed to the local-scale flux variation.

Table 12.1 *The parameters representing the six different scenarios (Su et al., 2020).*

Scenario	Reference	A	B	C	D	E	F
$\overline{F}$ *mean of* $\ln K$	−0.3	−0.3	−0.3	−0.3	−0.3	−0.3	−0.3
σ_f^2 *variance of perturbation of* $\ln K$	0	0.2	0.2	0.2	0.2	2	2
I *statistical anisotropy ratio*	0	1.5	7.5	1.5	1.5	1.5	7.5
v_S *surface water velocity*	6	6	6	10	6	6	6
q_G *upwelling groundwater velocity*	0.005	0.005	0.005	0.005	0.03	0.005	0.005

A line source of continuous concentration of 100 g/L was imposed on the dune surface for the simulations, and the bottom boundary concentration was set to zero, representing upwelling groundwater concentration. Starting with zero initial concentration over the domain, they simulated solute transport for 50 days until the local-scale concentration distributions $C(\mathbf{x}, t)$ reach approximately steady states.

12.7.1 Mixing Zone and Influencing Factors in a Single Realization

One single realization of random K fields was used to simulate Scenarios A, B, C, D, E, and F (Table 12.1). These simulations were compared with those in the reference case, where the streambed is homogeneous. The homogeneity, in this case, means that the spatial REV exists, within which pore-scale velocity variation is mild. Thus, its effects on solute transport is modeled by using local-scale dispersivity of 10^{-7}. Results of the six scenarios are illustrated in Fig. 12.24a–f, where the colored contours depict spatial variability of K, and the black lines with arrows indicate streamlines from surface water and groundwater. The arrows indicate the direction of the flow.

The interface between the surface streamlines (originating from the dune surface) and groundwater ones (originating from the bottom boundary) outlines the surface water and groundwater circulation regions as well as the stagnation zones (indicated by a white circle), where the velocity is almost zero. The concentration

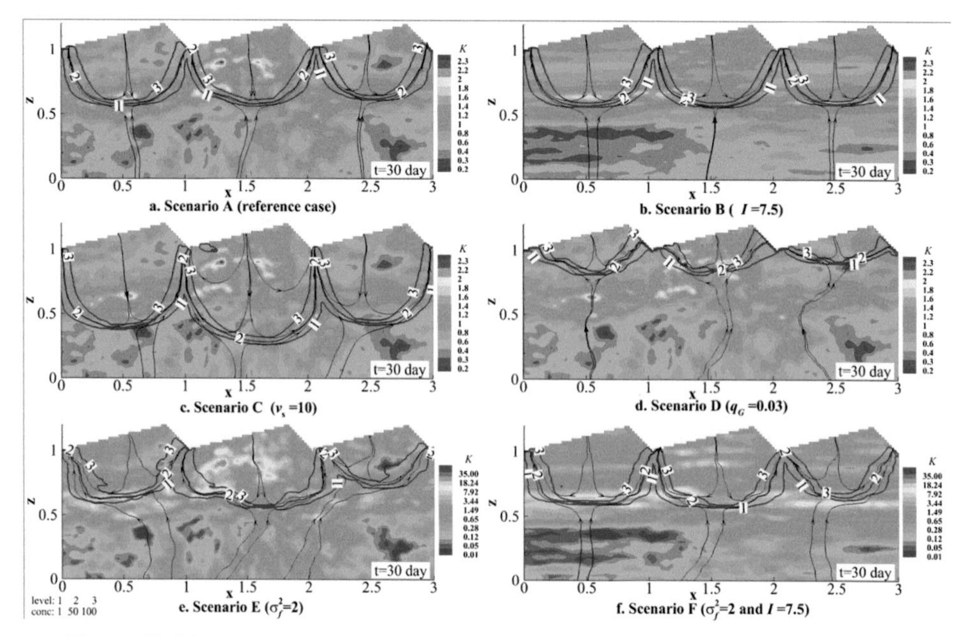

Figure 12.24 The concentration contour lines (red lines) and hyporheic streamlines (black lines) when time = 30 days in a single realization of the random field (a) Scenario A (where σ_f^2, I, v_s and q_G equal 0.2, 1.5, 6.0, and 0.005, respectively); (b) Scenario B, I increased to 7.5; (c) Scenario C, v_s increased to 10; (d) Scenario D, q_G increased to 0.03; (e) Scenario E, σ_f^2 increased to 2. (f) Scenario F, $\sigma_f^2 = 2$ and I = 7.5. The contour maps represent the heterogeneous hydraulic conductivity field in each scenario. From Su et al. (2020).

contours in Figs. 12.23 and 12.24 are the results at 30 days when the solute transport processes reach approximately steady states.

12.7.1.1 Homogeneous Case

The steady-state flow pattern in the reference case (Fig. 12.23b) shows that all the surface water enters the streambed from the dune's surface and diverges due to the pressure gradients along the irregular dune surface. This flow divergence creates two circulation regimes separated near the middle of the dune stoss: one flows toward the trough on the left side of the dune and the other on the right side. Likewise, the groundwater upwelling from the bottom of the domain diverges at the bottom boundary near the center of the dune stoss, producing two flow regimes. A stagnation zone exists at the interface between the four surface and groundwater flow regimes. Away from the stagnation zone, surface and groundwater flow parallel to each other. They then converge to exit the streambed at the troughs on both sides of a dune.

For solution transport simulations, the upper and low boundaries were assigned a constant concentration = 100 and 0, respectively. The red lines represent concentration contour lines when time equals 30 days after the tracer was released. This case is the reference case. The red contour lines (labeled 1, 2, and 3) in Fig. 12.23 denote 0, 50, and 100% of the simulated tracer concentration distribution 30 days after the tracer was released from the dune surface. The area behind the contour line with label 3 represents the area where the water is 100% of surface water. Meanwhile, the area below the contour line labeled 1 denotes where the water contains 0% surface water but 100% groundwater. In other words, the contour lines with labels 1 and 3 outline the dispersion or mixing zone between the surface and groundwater flow regimes.

A mixing zone is not expected in this case because of the very small dispersivity (10^{-7}). Accordingly, the noticeable mixing zone (the area between contours 1 and 3) manifests numerical dispersion (see Chapter 5). Such numerical dispersion always exists unless the grid size is infinitesimally small to a mathematical point.

Even though it is a numerical artifact, numerical dispersion produces results analogous to physical dispersion (small-scale flux variations such as the random motion of molecules or pore-scale velocity variations). In particular, the mixing zone is narrow near the stagnation point, where the flux is nearly zero, and it widens near the exits, where both the surface and groundwater converge to the dune troughs and where the flux is high. The results of the homogeneous case illustrate three key points: (1) the advective flow field beneath the dunes involves convergent and divergent flows and a stagnation zone, (2) purely advective flow (without considering small-scale velocity variation) should not produce any mixing effect in theory, and (3) the appearance of mixing zones in the simulated results is owing to numerical dispersion, manifesting the effects of spatial averaging of concentrations over the finite grid size or a CV (Chapter 1).

12.7.1.2 Effects of Heterogeneity and Boundary Flux

We observe the following by comparing the surface water and groundwater streamlines and their circulation regimes of the six scenarios (Fig. 12.24) with reference cases (Fig. 12.23). An increase in σ_f^2 from 0 (Fig. 12.23b), to 0.2 (Fig. 12.24a) and to 2 (Fig. 12.24e) leads to an increase in the streamline tortuosity, change in flow regime shape, and enlargement in the mixing zone near troughs. On the other hand, as the horizontal to vertical scale ratio increases from $I = 1.5$ (Fig. 12.24a and e) to $I = 7.5$ (Fig. 12.24b and f), the surface-water-flow circulation regimes becomes flat and shallow. In contrast, the upper and low boundary fluxes can significantly change the flow regimes and mixing zones, as exhibited in Fig. 12.24c and d. An increase in v_s (surface water velocity) drastically deepens and enlarges the

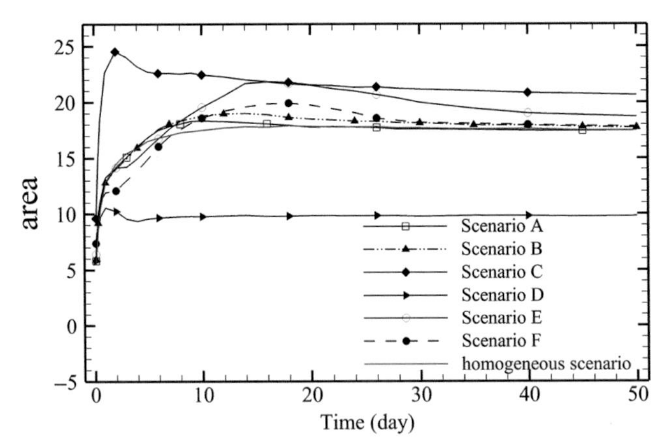

Figure 12.25 The dynamic variation of mixing area in scenarios A, B, C, D, E, F, and the homogeneous scenario (the reference case). From Su et al. (2020).

surface water circulation regime (Fig. 12.24c), while an increase in q_G (upwelling groundwater flux) noticeably reduces its depth and size (Fig. 12.24d).

Fig. 12.25 shows the change in the total area of the mixing zone from 0 to 50 days in each scenario. The total area sums up the areas of cells with a concentration between 0 and 100. (i.e., the concentration has been diluted or mixed). Overall, the area of all cases increases and then stabilizes at some values. This behavior manifests the time progression of the tracer's expansion, dispersal, and mixing after entering the dune surface. Then, the tracer is impeded at the interface and the stagnation zones due to the counter-directional upwelling groundwater. The groundwater upwelling forces the tracer to return to the stream at the troughs of the dune surface. The stabilization of the mixing area in Fig. 12.25 reflects the balance of the dispersive and the advective flux from the surface circulation field. In other words, there is no further dispersion from the surface water circulation regimes to the groundwater circulation regime. Therefore, the solute transport has reached a steady-state condition, and steady dispersion or mixing exists only at the interface.

The mixing zone area-time behavior is, however, different in each scenario. Specifically, the area in the reference case (homogeneous field) increases smoothly and gradually over time and is maintained at a 17.5 value. Meanwhile, the areas in Scenarios A, B, C, D, E, and F increase to peak values early, then decrease and asymptotically approach different constant values at large times. The timing and magnitude of the area's peak are different and related to the degree of heterogeneity, surface water flux, and upwell groundwater flux in each scenario. In particular, the areas of the mixing zone in Scenarios E and F with $\sigma_f^2 = 2$ first increase to values larger than Scenarios A and B with $\sigma_f^2 = 0.2$. They then decrease and reach asymptotical values slightly greater than those of Scenarios

A and B. Notice that the areas of Scenarios A, B, E, and F over the entire duration are more extensive than that of the homogeneous case. The increase of the variability of the local-scale K increases the variation of q, and the local-scale dispersion coefficient d (according to Eq. (9.3.5) and (9.3.6) in Chapter 9) (albeit the local scale dispersivity was uniform and set to 10^{-7}) and, in turn, the mixing zone at both early and large times.

However, the effects of the increase in surface water velocity and upwelling groundwater flux surpass these effects of heterogeneity. Specifically, the area for Scenario C (i.e., large surface water velocity) rises to the peak (above 24) at about two days – a much higher peak value and earlier time than those of Scenario A, which has the same heterogeneity as Scenario C. It then drops and stabilizes at a value of 21 larger than those of other scenarios. This result indicates that the large downward hydraulic gradient enhances the effect of the small heterogeneity ($\sigma_f^2 = 0.2$) due to the increased surface water velocity. On the other hand, the mixing area in Scenario D has the smallest peak on day 1. It stabilizes at 10 (the smallest mixing area in all these scenarios and the homogeneous case). Since this scenario has the same degree of heterogeneity as Scenario A but a more significant upwelling groundwater flux, the flux effect offsets the effects of heterogeneity.

12.7.2 Hydraulic Gradient and Counter Flow

In Fig. 12.26, Su et al. (2020) presented contour maps of the difference in flux between each scenario and the homogeneous case to describe further the impacts of flux variance on the mixing zone due to the above factors. According to the distribution of reddish and greenish contours in this figure, the overall flux differences increase significantly with an increase in v_s (Scenario C), q_G (Scenario D), and σ_f^2 (Scenario E) in comparison with those in Scenarios A and B. Such significant increases in flux differences in Scenario C and D (Fig. 12.26c and d) further reveal that the differences are driven by the local-scale K heterogeneity and the large-scale hydraulic gradient over the domain, creating large v_s or q_G. Nevertheless, the high upward groundwater flux in the opposite direction to the flux direction of the surface water impedes the area and depth of the surface water circulation regimes. Consequently, the mixing area decreases significantly in Scenario D. We emphasize that the local-scale K at each finite element cell in the above experiments is known. The local-scale dispersion (the ensemble-mean parameter of the local-scale CV) represents the effects of unknown heterogeneity below the CV. Specifically, Fig. 12.26 shows that the velocity variation within the CV that defines the local-scale K is overwhelmed by the velocity variations due to the spatial variability of the local-scale Ks in the domain. Further, the large-scale hydraulic gradient (or regional flow) that induces

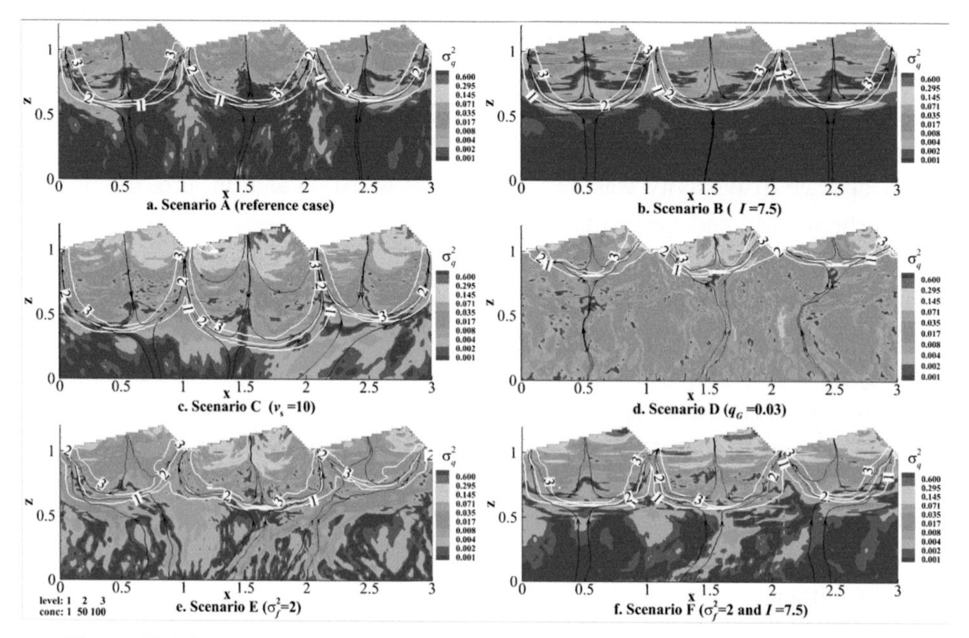

Figure 12.26 The contour map of the difference between the flux in the homogeneous case and that in (a) Scenario A, (b) Scenario B, (c) Scenario C, (e) Scenario E. (f) Scenario F. The white line represents the contour line of the mixing area. From Su et al. (2020).

groundwater upwelling or stream downward flow dictates the effects of local-scale K heterogeneity and, in turn, the mixing.

12.8 Illustrative MC Simulation

The results above are based on one realization of the random local-scale K field, which is fully specified. In real-world situations, this heterogeneity exists at multiple scales and is unknown. Hence, the EHM approach or MC simulation (MCS) is most appropriate (Sections 9.7 and 9.8, Chapter 9). Employing the EHM approach, nevertheless, requires effective hydraulic conductivity (not the mean K) and the macrodispersivity associated with the macrodispersion coefficient D_{eff} of the ensemble REV for the equivalent homogeneous streambed under the counter flow conditions. Because theoretical formulas for this flow situation are not readily available, they resolved to the MCS built on the HPHM local-scale equations (Section 9.5, Chapter 9).

Albeit numerical setups for the MCS are identical to that of Scenario E, the simulation domain in this MCS is limited to one dune, and the MCS used 20 realizations of the random field. Fig. 12.27a shows the simulated local-scale

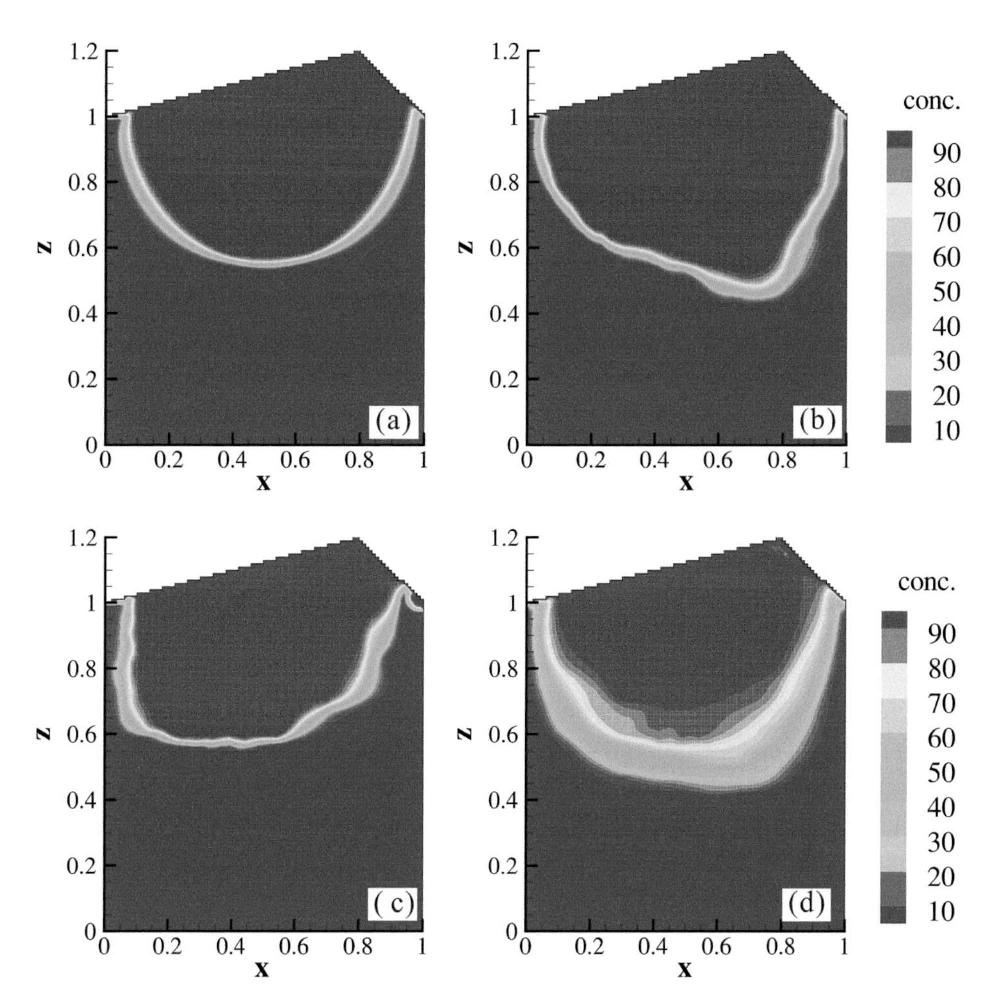

Figure 12.27 Concentration contour maps (a) homogeneous case; (b) and (c) two different realizations of 20 heterogeneous realizations of Scenario E; d) the ensemble average of the 20 realizations. From Su et al. (2020).

concentration distribution in the reference case (homogeneous case) at $t = 30$ days, while Fig. 12.27b and c depict the distributions from two selected realizations from the MC simulation. The ensemble-mean concentration (an average over the 20 realizations), $\overline{C}(\mathbf{x}, t)$, is displayed in Fig. 12.27d. Notice that this ensemble-mean concentration $\overline{C}(\mathbf{x}, t)$ is conceptually the same as the concentration defined over a large-scaled spatial or ensemble REV (see Eq. 9.8.4, Chapter 9).

A comparison of Fig. 12.27a–c with Fig. 12.27d reveals that the ensemble-average concentrations field, $\overline{C}(\mathbf{x}, t)$, (Fig. 12.27d), has a greater mixing (dispersion) zone than the $C(\mathbf{x}, t)$ of the homogeneous and those of the two prescribed heterogeneous K fields (Fig. 12.27b and c). The greater mixing zone arises from the fact that $\overline{C}(\mathbf{x}, t)$ is an average of 20 realizations of the local-scale $C(\mathbf{x}, t)$. The larger mixing zone signifies the effects of the unknown local-scale K distribution in this large CV and its effective hydraulic conductivity on solute transport (Section 9.8, Chapter 9). EHM approach homogenizes an entire heterogenous streambed using the ensemble REV. As such, its mixing or dispersion zone is a by-product of the average of all possible concentration distributions, resulting from flux variations among all possible local-scale K distributions in the streambed with the given spatial statistics. Specifically, this mixing zone reflects the local-scale flux and $C(\mathbf{x}, t)$ uncertainty.

Since the local-scale K of the homogeneous case (Fig. 12.27a) and the two heterogeneous realizations (Fig. 12.27b and c) are prescribed precisely, the mixing zones in these single realizations denote only the effects of unknown flux and concentration perturbations at scales below this local-scale CV. Specifically, had the local-scale dispersion (i.e., d in Eq. (9.3.5), Chapter 9), molecular diffusion, and numerical dispersion been eradicated (pure advective transport) and had the concentration not been the average over the finite element (eliminating numerical dispersion), no mixing zone would have appeared in Fig. 12.27a–c.

According to the above reason, the appearance of the mixing zone in the ensemble mean concentration (Fig. 12.27d) is a testament to our uncertainty about solute advection fronts due to the homogenization of multi-scale heterogeneous streambed. Precisely, diffusion, dispersion, and mixing are a way to express our uncertainty about the velocities at the solute front at molecular, fluid-dynamic, pore- and local-scale scales.

Notice that different numbers of realizations may yield different sizes of mixing zone because the mean and variation in K may not be the same for each realization unless the streambed or the solute plume is sufficiently large such that the ergodicity is met. In this example, increasing the number of realizations in the MC simulation of a limited domain size is necessary to reach ergodicity and stabilize the mixing zone area. Even after the ergodicity is reached, the resulting ensemble mixing zone is still wilder than the mixing zone in each realization, reflecting the unknown local-scale K variability among different realizations. As more local-scale Ks at different locations are depicted and fixed, the uncertainty in the K distribution in the streambed diminishes, and this mixing zone size narrows.

This simplified MC simulation analysis highlights the discussion of the scale issue, the ensemble-mean flow and transport equation, the ensemble-mean concentration, and the variability and uncertainty of fluxes on the mixing in the hyporheic zone or dispersion in aquifers. Interestingly, this simulation shows that different realizations of local-scale K heterogeneity lead to different concentration distributions with different mixing zones, different from that in the homogeneous case even though the transport is advection dominant. Such different outcomes manifest that local-scale flux variation due to the difference in the local-scale K spatial distribution affects the local-scale dispersion (perhaps numerical dispersion).

12.9 Summaries of the Theme of the Book

- Well-mixed models in Chapters 2 and 3 are well suited for spatially integrated behaviors of an aquifer. Their accuracy, nevertheless, rests upon high-resolution spatiotemporal data of inflow and outflow and initial conditions. For this reason, spatially distributed models are highly desirable.
- Chapter 4 introduces the spatially distributed models at the fundamental fluid-dynamic scale; it presents the molecular diffusion concept and Fick's law, which overcomes our inability and lack of interest in prescribing each molecule's movement. This chapter then presents the advection-diffusion equation (ADE), where the advection describes the movement of most molecules, and diffusion accounts for the effect of the seemingly random movement of individual molecules. Analytical and numerical solutions to the ADE are developed in Chapter 5, and the numerical dispersion phenomenon associated with our numerical solution scheme is brought forth.
- Chapter 6 elucidates the theory that extends Fick's law for random motion of molecules to quantify solute spread (dispersion) due to fluid-dynamic-scale velocity variation induced by shear flow in a pipe. Most importantly, it explicates the conditions required for Fick's law to be valid. Extending the shear flow dispersion theory to the soil column is articulated in Chapter 7, where the dispersion is employed to account for pore-scale velocity variation deviating from Darcy-scale velocity induced by pore-scale heterogeneity. Non-Fickian dispersion phenomena are explored. Chapter 8 presents various methods for estimating advection velocity and dispersion parameters for soil column experiments. As of now, it becomes clear that diffusion and dispersion are parsimonious concepts. They are stochastic descriptions of uncertainty in the velocity variations below the scale of our advection models (i.e., fluid dynamics scale velocity, average velocity over the cross-section of a pipe, or Darian velocity in soil columns).
- As our interest moves beyond the soil column to the aquifer scale, a pragmatic equivalent homogenous model (EHM) with macrodispersion for aquifers is presented in Chapter 9. It derives the ensemble-mean concentration of the solute plume, which could resemble point observation if the solute spreading reaches Fickian's regime. Chapter 10 examines the application of the approach to the

natural gradient tracer experiment in the Borden site aquifer of mild local-scale K heterogeneity and explains its success. It also discusses its failure in the tracer experiment at MADE site, Mississippi, where significant multi-scale heterogeneity prevails such that the tracer plume could not reach the ergodic condition. Debates about the mobile-immobile zone and fractional derivative models are presented.

- Chapter 11 reviews field tracer experiments under force-gradient conditions. It explores the benefits of a highly parameterized heterogeneity model(HPHM) with densely measured local-scale K for simulating solute transport. It also unveils the importance of temporal variability of regional flow fields on the tracer migration besides the spatial heterogeneity.
- Previous chapters elucidate the following key points:
- Diffusion, dispersion, and macrodisperion are ramifications of our attempt to simplify the description of velocity variations below our CV scales (i.e., the molecule-, fluid-dynamic-, pore-, zonation-, or basin-scales). The validity of Fickian law depends on the scale of velocity variation, our observation, and interest scales.
- Scale-dependent dispersion and non-Fickian behaviors in geologic media are scientific issues arising from our attempt to simplify multi-scale heterogeneous aquifers by adopting the EHM and macrodispersion approach.
- EHM with macrodispersion theory is a valuable and practical tool to predict the statistically most likely solute migration under situations where complex velocity fields are inaccessible. This approach could yield satisfactory predictions if the solute plume reaches the ergodic condition or Fickian regime.
- EHMs with scale-dependent dispersivity, mobile-immobile, non-Fickian, or fractional derivative models are other ensemble mean models. They are more versatile than EHM with the macrodispersion concept, but model calibration to the observations is often necessary. Calibrating any solute transport model requires multidimensional, detailed snapshots of the solute plume. Besides, a solute plume migration may behave differently in various geologic media (e.g., Borden, Georgetown, MADE sites). Further, the plume's behavior may vary with the source's size, location, and heterogeneity encountered in the same aquifer. As a result, the model must be recalibrated to available solute plume snapshots and loses the independent predictive capability.
- For this reason, the development of statistical theories relating spatial variability of the local-scale hydraulic properties to parameters of these models (e.g., the fraction of dead-end zone and mass transfer coefficient of the mobile/immobile model) may facilitate predicting the solute transport in aquifers without relying on model calibration. The statistics should therefore address the uncertainty associated with their predictions. These models invoke the ensemble REV as

does EHM with macrodispersion. Their predictions are equally uncertain at the scale of our observations and interest. Besides, large-scale velocity spatiotemporal variations could alter the behavior of solute migration.

- Ultimately, the demand for high-resolution predictions, which reduces uncertainty to satisfy our fine-scale interest and observation, champions the highly parameterized heterogeneous model (HPHM). Adopting HPHM as the objective, Chapter 12 elucidates the viability and cost-effectiveness of HT (hydraulic tomography) for characterizing spatially varying hydraulic properties of aquifers over large areas. Many other tools (such as machining learning, hydrogeophysical joint inversion of related hydrologic and geophysical information) would likely improve our ability to characterize the local-scale velocity variations.
- LTRM (long-term resistivity monitoring) and exploiting natural stimuli as large-scale tomography could be a viable, cost-effective tool for monitoring solute transport in aquifers.
- Multi-scale heterogeneity and unpredictable spatiotemporal variations in multi-scale flow fields dictate solute migration in environments. Furthermore, our inability to characterize the variability below the scale of our model, observation, and interest CV exacerbates the difficulty in forecasting solute migration. For this reason, molecular-scale diffusion, Fick's law and soil column-scale dispersion concepts, and classical ADE remain invaluable even if the large-scale variability can be identified. As a result, the stochastic conceptualization of the problem is necessary.

12.10 Homework

1. Repeat the HT example (Section 12.2.1) using more pumping data to increase the resolution of HT image.

References

Adams, E. E., & Gelhar, L. W. (1992). Field study of dispersion in a heterogeneous aquifer: 2. Spatial moments analysis. *Water Resources Research*, *28*(12), 3293–3307. https://doi.org/10.1029/92WR01757

Alexander, M., Berg, S. J., & Illman, W. A. (2011). Field study of hydrogeologic characterization methods in a heterogeneous aquifer. *Groundwater*, *49*(3), 365–382. https://doi.org/10.1111/j.1745-6584.2010.00729.x

Amirbahman, A., & Olson, T. M. (1993). Transport of humic matter-coated hematite in packed beds. *Environmental Science & Technology*, *27*(13), 2807–2813. https://doi.org/10.1021/es00049a021

Anderson, M. P. (1979). Using models to simulate the movement of contaminants through groundwater flow systems. *Critical Reviews in Environmental Science and Technology*, *9*(2), 97–156. https://doi.org/10.1080/10643387909381669

Anderson, M. P., & Bowser, C. J. (1986). The role of groundwater in delaying lake acidification. *Water Resources Research*, *22*(7), 1101–1108. https://doi.org/10.1029/WR022i007p01101

Archie, G. E. (1942). The electrical resistivity log as an aid in determining some reservoir characteristics. *Transactions of the AIME*, *146*(01), 54–62. https://doi.org/10.2118/942054-G

Aris, R. (1956). On the dispersion of a solute in a fluid flowing through a tube. *Proceedings of the Royal Society of London. Series A. Mathematical and Physical Sciences*, *235*(1200), 67–77. https://doi.org/10.1098/rspa.1956.0065

Aris, R. (1958). On the dispersion of linear kinematic waves. *Proceedings of the Royal Society of London. Series A. Mathematical and Physical Sciences*, *245*(1241), 268–277. https://doi.org/10.1098/rspa.1958.0082

Barlebo, H. C., Hill, M. C., & Rosbjerg, D. (2004). Investigating the Macrodispersion Experiment (MADE) site in Columbus, Mississippi, using a three-dimensional inverse flow and transport model. *Water Resources Research*, *40*(4). https://doi.org/10.1029/2002WR001935

Bear, J. (1972). *Dynamics of fluids in porous media*, American Elsevier, New York, 764 pp.

Bear, J. (1979). *Hydraulics of groundwater*. McGraw-Hill Inc, 569 pp.

Benson, D. A., Schumer, R., Meerschaert, M. M., & Wheatcraft, S. W. (2001). Fractional dispersion, Lévy motion, and the MADE tracer tests. *Transport in Porous Media*, *42*(1), 211–240.

Berg, S. J., & Illman, W. A. (2011). Three-dimensional transient hydraulic tomography in a highly heterogeneous glaciofluvial aquifer-aquitard system. *Water Resources Research*, *47*(10). https://doi.org/10.1029/2011WR010616

Berg, S. J., & Illman, W. A. (2013). Field study of subsurface heterogeneity with steady-state hydraulic tomography. *Groundwater*, *51*(1), 29–40. https://doi.org/10.1111/j.1745-6584.2012.00914.x

Berg, S. J., & Illman, W. A. (2015). Comparison of hydraulic tomography with traditional methods at a highly heterogeneous site. *Groundwater*, *53*(1), 71–89. https://doi.org/10.1111/gwat.12159

Boggs, J. M., & Adams, E. E. (1992). Field study of dispersion in a heterogeneous aquifer: 4. Investigation of adsorption and sampling bias. *Water Resources Research*, *28*(12), 3325–3336. https://doi.org/10.1029/92WR01759

Boggs, J. M., Beard, L. M., Long, S. E., McGee, M. P., MacIntyre, W. G, Antworth, C. P., & Stauffer, T. B. (1993). *Database for the second macrodispersion experiment (MADE-2)*. Technical Report TR-102072, Electric Power Research Institute, Palo Alto, CA.

Boggs, J. M., Schroeder, J. A., & Young S. C. (1995). Data to support model development for natural attenuation study. Report No. WR28–2-520-197. TVA Engineering Laboratory, Tennessee Valley Authority, Norris, TN.

Boggs, J. M., Young, S. C., Beard, L. M., Gelhar, L. W., Rehfeldt, K. R., & Adams, E. E. (1992). Field study of dispersion in a heterogeneous aquifer: 1. Overview and site description. *Water Resources Research*, *28*(12), 3281–3291. https://doi.org/10.1029/92WR01756.

Boggs, J. M., Young, S. C., Benton, D. J., & Chung, Y. C. (1990). Hydrogeologic characterization of the MADE Site. Interim Report EN-6915, Electric Power Research Institute, Palo Alto, CA.

Boulton, N. S. (1954). Unsteady radial flow to a pumped well allowing for delayed yield from storage. International Association of Scientific Hydrology *Publication*, *2*, 472–477.

Brauchler, R., Liedl, R., & Dietrich, P. (2003). A travel time based hydraulic tomographic approach. *Water Resources Research*, *39*(12). https://doi.org/10.1029/2003WR002262

Brauner, J. S., & Widdowson, M. A. (2001). Numerical simulation of a natural attenuation experiment with a petroleum hydrocarbon NAPL source. *Ground Water* 39(6), 939–952.

Bredehoeft, J. D., & Pinder, G. F. (1973). Mass transport in flowing groundwater. *Water Resources Research*, *9*(1), 194–210. https://doi:10.1029/wr009i001p00194

Brunke, M., & Gonser, T. O. M. (1997). The ecological significance of exchange processes between rivers and groundwater. *Freshwater Biology*, *37*(1), 1–33. https://doi.org/10.1046/j.1365-427.1997.00143.x

Cardenas, M. B., & Wilson, J. L. (2007). Dunes, turbulent eddies, and interfacial exchange with permeable sediments. *Water Resources Research*, *43*(8). https://doi.org/10.1029/2006WR005787

Cardenas, M. B., Wilson, J. L., & Zlotnik, V. A. (2004). Impact of heterogeneity, bed forms, and stream curvature on subchannel hyporheic exchange. *Water Resources Research*, *40*(8). https://doi.org/10.1029/2004WR003008

Carslaw, H. S., & Jaeger, J. C. (1959). *Conduction of Heat in Solids*. Oxford: Clarendon Press.

Carslaw, H. S., & Jaeger, J. C. (1988). *Conduction of Heat in Solids*. Oxford Science Publications. New York: The Clarendon Press, Oxford University Press.

Chatwin, P. C. (1970). The approach to normality of the concentration distribution of a solute in a solvent flowing along a straight pipe. *Journal of Fluid Mechanics*, *43*(2), 321–352.

Cheng, R. T., Casulli, V., & Miford, S. N. (1984). Eulerian–Lagrangian solution of the convection-dispersion equation in natural coordinates. *Water Resources Research*, *20* (7), 944–952.

Coats, K. H., & Smith, B. D. (1964). Dead-end pore volume and dispersion in porous media. *Society of Petroleum Engineers Journal*, *4*(01), 73–84. https://doi.org/10.2118/647-PA

Cooper Jr, H. H., & Jacob, C. E. (1946). A generalized graphical method for evaluating formation constants and summarizing well-field history. *Eos, Transactions American Geophysical Union*, *27*(4), 526–534. https://doi.org/10.1029/TR027i004p00526

Cooper Jr, H. H., Bredehoeft, J. D., & Papadopulos, I. S. (1967). Response of a finite-diameter well to an instantaneous charge of water. *Water Resources Research*, *3*(1), 263, 269. https://doi.org/10.1029/WR003i001p00263

Crank, J. (1956). *The Mathematics of Diffusion*. Oxford: Clarendon Press.

Csanady, G. T. (1973). *Turbulent Diffusion in the Environment*. D. Reidel Publishing Company, Dordrecht-Holland.

Dagan, G. (1982). Stochastic modeling of groundwater flow by unconditional and conditional probabilities: 1. Conditional simulation and the direct problem. *Water Resources Research*, *18*(4), 813–833. https://doi.org/10.1029/WR018i004p00813.

Dagan, G. (1982). Stochastic modeling of groundwater flow by unconditional and conditional probabilities: 2. The solute transport. *Water Resources Research*, *18*(4), 835–848. https://doi.org/10.1029/WR018i004p00835

Dagan, G. (1984). Solute transport in heterogeneous porous formations, *Journal of Fluid Mechanics.*, *145*, 151–177.

Dagan, G. (1987). Theory of solute transport by groundwater. *Annual Review of Fluid Mechanics*, *19*(1), 183–213. https://doi.org/10.1146/annurev.fl.19.010187.001151

DeSmedt, F., & Wierenga, P. J. (1979). A generalized solution for solute flow in soils with mobile andimmobile water. *Water Resources Research*, *15*(5), 1137–1141. https://doi.org/10.1029/WR015i005p01137.

Doetsch, J., Linde, N., Vogt, T., Binley, A., & Green, A. G. (2012). Imaging and quantifying salt-tracer transport in a riparian groundwater system by means of 3D ERT monitoring. *Geophysics*, *77*(5), B207–B218. https://doi.org/10.1190/geo2012-0046.1

Drost, W., Klotz D., Koch, A., Moser, H., Neumaier, F., & Rauert, W. (1968). Point dilution methods of investigating groundwater flow by means of radioisotopes. *Water Resources Research*, *4*(1), 125–146. https://doi.org/10.1029/WR004i001p00125

Dunnivant, F. M., Jardine, P. M., Taylor, D. L., & McCarthy, J. F. (1992). Cotransport of cadmium and hexachlorobiphenyl by dissolved organic carbon through columns containing aquifer material. *Environmental Science & Technology*, *26*(2), 360–368. https://doi.org/10.1021/es00026a018

Einstein, Albert, E. (1905). "Über die von der molekularkinetischen Theorie der Wärme geforderte Bewegung von in ruhenden Flüssigkeiten suspendierten Teilchen" [On the Movement of Small Particles Suspended in Stationary Liquids Required by the Molecular-Kinetic Theory of Heat]. *Annalen der Physik*, *322*(8), 549–560.

Ellis, R. G., & Oldenburg, D. W. (1994). The pole–pole 3-D Dc-resistivity inverse problem: A conjugategradient approach. *Geophysical Journal International, 119* (1), 187–194. https://doi.org/10.1111/j.1365-246X.1994.tb00921.x

Fahim, M. A., & Wakao, N. (1982). Parameter estimation from tracer response measurements. *The Chemical Engineering Journal, 25*(1), 1–8. https://doi.org/10.1016/0300-9467(82)85016-8

Faragher, R. (2012). Understanding the basis of the kalman filter via a simple and intuitive derivation [lecture notes]. *IEEE Signal Processing Magazine, 29*(5), 128–132. https://doi.org/10.1109/MSP.2012.2203621

Feehley, C. E., Zheng, C., & Molz, F. J. (2000). A dual-domain mass transfer approach for modeling solute transport in heterogeneous aquifers: Application to the Macrodispersion Experiment (MADE) site. *Water Resources Research, 36*(9), 2501–2515. https://doi.org/10.1029/2000WR900148

Fick, Adolf, F. (1855). Ueber Diffusion. *Annalen der Physik und Chemie, 170*(1), 59–86. https://doi.org/10.1002/andp.18551700105

Fischer, H. B., List, J. E., Koh, C. R., Imberger, J., & Brooks, N. H. (1979). *Mixing in Inland and Coastal Waters*. New York: Academic Press.

Fletcher, C. A. J. (1988). *Computational Techniques for Fluid Dynamics 1: Fundamental and General Techniques*. Berlin and New York: Springer-Verlag.

Freyberg, D. L. (1986). A natural gradient experiment on solute transport in a sand aquifer: 2. Spatial moments and the advection and dispersion of nonreactive tracers. *Water Resources Research, 22*(13), 2031–2046. https://doi.org/10.1029/WR022i013p02031

Garabedian, S. P., LeBlanc, D. R., Gelhar, L. W., & Celia, M. A. (1991). Large-scale natural gradientracer test in sand and gravel, Cape Cod, Massachusetts, 2, Analysis of spatial moments for a nonreactive tracer, *Water Resources Research, 27*(5), 911–924, 1991.

Gelhar, L. W. (1993). Stochastic Subsurface Hydrology. Prentice-Hall, Inc. 390 pp.

Gelhar, L. W., & Axness, C. L. (1983). Three-dimensional stochastic analysis of macrodispersion in aquifers. *Water Resources Research, 19*(1), 161–180. https://doi.org/10.1029/WR019i001p00161.

Gelhar, L. W., Gutjahr, A. L., & Naff, R. L. (1979). Stochastic analysis of macrodispersion in a stratified aquifer. *Water Resources Research, 15*(6), 1387–1397. https://doi.org/10.1029/WR015i006p01387

Gelhar, L. W., & Wilson, J. L. (1974). Ground-water quality modeling, *Ground Water, 12* (6), 339–408.

Goode, D. J. (1990). Particle velocity interpolation in block-centered finite difference groundwater flow models. *Water Resources Research, 26*(5), 925–940. https://doi.org/10.1029/WR026i005p00925

Guan, J., Molz, F. J., Zhou, Q., Liu, H. H., & Zheng, C. (2008). Behavior of the mass transfer coefficient during the MADE-2 experiment: New insights. *Water Resources Research, 44*(2). https://doi.org/10.1029/2007WR006120

Harvey, C., & Gorelick, S. M. (2000). Rate-limited mass transfer or macrodispersion: Which dominates plume evolution at the Macrodispersion Experiment (MADE) site? *Water Resources Research*, 36, 637–650. https://doi.org/10.1029/1999Wr900247

Herr, M., Schäifer, G., & Spitz, K. (1989). Experimental studies of mass transport in porous media with local heterogeneities. *Journal of Contaminant Hydrology*, 4, 127–137.

Hill, M. C., Barlebo, H. C., & Rosbjerg, D. (2006). Reply to comment by F. Molz et al. on "Investigating the Macrodispersion Experiment (MADE) site in Columbus, Mississippi, using a three-dimensional inverse flow and transport model". *Water Resources Research, 42*(6), 1–4. https://doi:10.1029/2005WR004624

Huyakorn, P. S., Jones, B. G., & Andersen, P. F. (1986). Finite element algorithms for simulating three-dimensional groundwater flow and solute transport in multilayer systems. *Water Resources Research, 22*(3), 361–374. https://doi.org/10.1029/WR022i003p00361

Huyakorn, P. S., P. F. Andersen, O. Giiven, & F. J. Molz. (1986). A curvilinear finite element model for simulating two-well tracer tests and transport in stratified aquifers. *Water Resources Research 22*(5), 663–678.

Hvorslev, M. J. (1951). *Time Lag and Soil Permeability in Ground-Water Observations.* Waterways Experiment Station, Corps of Engineers, US Army.

Illman, W. A., Liu, X., Takeuchi, S., Yeh, T. C. J., Ando, K., & Saegusa, H. (2009). Hydraulic tomography in fractured granite: Mizunami Underground Research site, Japan. *Water Resources Research, 45*(1). https://doi.org/10.1029/2007WR006715

Istok, J. (1989). *Groundwater Modeling by the Finite Element Method.* Washington: American Geophysical Union.

Julian, H. E., Boggs, J. M., Zheng, C., & Feehley, C. E. (2001). Numerical simulation of a natural gradient tracer experiment for the Natural Attenuation Study: Flow and physical transport. *Ground Water 39*(4), 534–545.

Keller, G. V., & Frischknecht, F. C. (1966). *Electrical Methods in Geophysical Prospecting*. Oxford: Pergamon Press.

Killey, R. W. D., & Moltyaner, G. L. (1988). Twin Lake tracer tests: Setting, methodology, and hydraulic conductivity distribution, *Water Resources Research, 24*(10), 1585–1612, 1988.

Kitanidis, P. K. (1995). Quasi-linear geostatistical theory for inversing. *Water Resources Research, 31*(10), 2411–2419. https://doi.org/10.1029/95WR01945

Koefoed, O., & Principles, G. (1979). *Geosounding Principles, I: Resistivity Sounding Measurements*. New York: Elsevier Scientific Publishing Company.

Kreft, A., & Zuber, A. (1978). On the physical meaning of the dispersion equation and its solutions for different initial and boundary conditions. *Chemical Engineering Science, 33*(11), 1471–1480. https://doi.org/10.1016/0009-2509(78)85196-3

Kurotori, T., Zahasky, C., Hejazi, S. A. H., Shah, S. M., Benson, S. M., & Pini, R. (2019). Measuring, imaging and modelling solute transport in a microporous limestone, *Chemical Engineering Science*, *196*, 366–383. https://doi.org/10.1016/j.ces.2018.11.001

LeBlanc, D. R., Garabedian, S. P., Hess, K. M., Gelhar, L. W., Quadri, R. D., Stollenwerk, K. G., & Wood, W. W. (1991). Large-scale natural gradient tracer test in sand and gravel, Cape Cod, Massachusetts: 1. Experimental design and observed tracer movement. *Water Resources Research, 27*(5), 895–910. https://doi.org/10.1029/91WR00241

Li, Y., & Oldenburg, D. W. (1994). Inversion of 3-D DC resistivity data using an approximate inverse mapping. *Geophysical Journal International, 116*(3), 527–537. https://doi.org/10.1111/j.1365-246X.1994.tb03277.x

Libelo, E. L., Stauffer, T. B., Geer, M. A., MacIntyre, W. G., & Boggs, J. M. (1997). A field study to elucidate processes involved in natural attenuation, In *4th International In Situ and On-Site Bioremediation Symposium.* New Orleans, Louisiana.

Llopis-Albert, C., & J. E. Capilla. (2009). Gradual conditioning of non-Gaussian transmissivity fields to flow and mass transport data: 3. Application to the Macrodispersion Experiment (MADE-2) site, on Columbus Air Force Base in Mississippi (USA). *Journal of Hydrology 371*(1–4), 75–84. https://doi.org/10.1016/j.jhydrol.2009.03.016.

MacIntyre, W. G., Boggs, J. M., Antworth, C. P., & Stauffer, T. B. (1993). Degradation kinetics of aromatic organic solutes introduced into a heterogeneous aquifer. *Water Resources Research 29*(12), 4045–4051.

Mackay, D. M., Freyberg, D. L., Roberts, P. V., & Cherry, J. A. (1986). A natural gradient experiment on solute transport in a sand aquifer: 1. Approach and overview of plume movement. *Water Resources Research*, *22*(13), 2017–2029. https://doi.org/10.1029/WR022i013p02017

Magee, B. R., Lion, L. W., & Lemley, A. T. (1991). Transport of dissolved organic macromolecules and their effect on the transport of phenanthrene in porous media. *Environmental Science & Technology*, *25*(2), 323–331. https://doi.org/10.1021/es00014a017

Mao, D., Wan, L., Yeh, T. C. J., Lee, C. H., Hsu, K. C., Wen, J. C., & Lu, W. (2011). A revisit of drawdown behavior during pumping in unconfined aquifers. *Water Resources Research*, *47*(5). https://doi.org/10.1029/2010WR009326

Marinov, I., & Marinov, A. M. (2014). The influence of a municipal solid waste landfill on groundwater quality: A modeling case study for raureni–ramnicu valcea (romania). *International Journal of Computational Methods and Experimental Measurements*, *2* (2), 184–201. https://doi.org/10.2495/CMEM-V2-N2-184-201

Mas-Pla, J. (1993). *Modeling the transport of natural organic matter in heterogeneous porous media: Analysis of a field-scale experiment at the Georgetown site, South Carolina* (Doctoral dissertation, The University of Arizona).

Mas-Pla, J., Yeh, T. C. J., McCarthy, J. F., & Williams, T. M. (1992). A forced gradient tracer experiment in a Coastal Sandy Aquifer, Georgetown Site, South Carolina. *Ground Water*. *30*(6), 958–964.

Matheron, G., & De Marsily, G. (1980). Is transport in porous media always diffusive? A counterexample. *Water Resources Research*, *16*(5), 901–917. https://doi.org/10.1029/WR016i005p00901

McCarthy, J. F., Gu, B., Liang, L., Mas-Pla, J., Williams, T. M., & Yeh, T. C. (1996). Field tracer tests on the mobility of natural organic matter in a sandy aquifer. *Water Resources Research*, *32*(5), 1223–1238. https://doi.org/10.1029/96WR00285

McCarthy, J. F., Marsh, J. D., & Tipping, E. (1995). Mobilization of actinides from disposal trenches by natural organic matter. In *209th ACS National Meeting*.

McCarthy, J. F., Williams, T. M., Liang, L., Jardine, P. M., Jolley, L. W., Taylor, D. L., Palumbo, A. V., & Cooper,L. W. (1993). Mobility of natural organic matter in a study aquifer. *Environmental Science & Technology*, *27*(4), 667–676. https://doi.org/10.1021/es00041a010

McLaughlin, D., & Townley, L. R. (1996). A reassessment of the groundwater inverse problem. *Water Resources Research*, *32*(5), 1131–1161. https://doi.org/10.1029/96wr00160

Molz, F. J., Zheng, C., Gorelick, S. M., & Harvey, C. F. (2006). Comment on "Investigating the Macrodispersion Experiment (MADE) site in Columbus, Mississippi, using a three-dimensional inverse flow and transport model" by Heidi Christiansen Barlebo, Mary C. Hill, and Dan Rosbjerg. *Water Resources Research*, *42*(6). https://doi:10.1029/2005WR004265.

Murphy, E. M., Zachara, J. M., Smith, S. C., Phillips, J. L., & Wietsma, T. W. (1994). Interaction of hydrophobic organic compounds with mineral-bound humic substances. *Environmental Science & Technology*, *28*(7), 1291–1299. https://doi.org/10.1021/es00056a017

Naff, R. L., Yeh, T. C. J., & Kemblowski, M. W. (1988). A note on the recent natural gradient tracer test at the Borden site. *Water Resources Research*, *24*(12), 2099–2103. https://doi.org/10.1029/WR024i012p02099

Neuman, S. P. (1972). Theory of flow in unconfined aquifers considering delayed response of the water table. *Water Resources Research*, *8*(4), 1031–1045, https://doi:10.1029/WR008i004p01031

Ni, C. F., Yeh, T. C. J., & Chen, J. S. (2009). Cost-effective hydraulic tomography surveys for predicting flow and transport in heterogeneous aquifers. *Environmental Science & Technology*, *43*(10), 3720–3727. https://doi.org/10.1021/es8024098

O'Connor, D. J., & Mueller, J. A. (1970). A water quality model of chlorides in Great Lakes. *Journal of the Sanitary Engineering Division*, *96*(4), 955–975. https://doi.org/10.1061/JSEDAI.0001160

Ogata, A., & Banks, R. B. (1961). A solution of the differential equation of longitudinal dispersion in porous media, Geological Survey Professional Paper 411-A. https://doi.org/10.3133/pp411A

Packman, A. I., & Salehin, M. (2003). Relative roles of stream flow and sedimentary conditions in controlling hyporheic exchange. *Hydrobiologia*, *494*(1), 291–297. https://doi.org/10.1023/A:1025403424063

Padilla, I. Y., Yeh, T. C. J., & Conklin, M. H. (1999). The effect of water content on solute transport in unsaturated porous media. *Water Resources Research*, *35*, 3303–3313. https://doi.org/10.1029/1999Wr900171

Pang, L., Goltz, M., & Close, M. (2003). Application of the method of temporal moments to interpret solute transport with sorption and degradation. *Journal of Contaminant Hydrology*, *60*(1–2), 123–134. https://doi.org/10.1016/S0169-7722(02)00061-X

Parker, J. C., & Van Genuchten, M. T. (1984). Flux-averaged and volume-averaged concentrations in continuum approaches to solute transport. *Water Resources Research*, *20*(7), 866–872. https://doi.org/10.1029/WR020i007p00866

Pickens, J. F., & Grisak, G. E. (1981). Scale-dependent dispersion in a stratified granular aquifer. *Water Resources Research*, *17*(4), 1191–1211. https://doi.org/10.1029/WR017i004p01191

Pinder, G. F., & Gray, W. G. (1977). *Finite Element Simulation in Surface and Subsurface Hydrology*. New York: Academic Press.

Priestley, M. B. (1981). *Spectral Analysis and Time Series*, Academic Press, 890 pp.

Rehfeldt, K. R., Boggs, J. M., & Gelhar, L. W. (1992). Field study of dispersion in a heterogeneous aquifer: 3. Geostatistical analysis of hydraulic conductivity. *Water Resources Research*, *28*(12), 3309–3324. https://doi.org/10.1029/92WR01758

Ritzi, R. W., & Yeh, T.-C. J. (1988). Comment on "The role of groundwater in delaying lake acidification" by M. P. Anderson and C. J. Bowser. *Water Resources Research*, *24*(5), 787–790. https://doi.org/10.1029/WR024i005p00787

Roache, P. J. (1976), *Computational Fluid Dynamics*. New Mexico: Hermosa Publishers.

Roberts, P. V., Goltz, M. N., & Mackay, D. M. (1986). A natural gradient experiment on solute transport in a sand aquifer: 3. Retardation estimates and mass balances for organic solutes. *Water Resources Research*, *22*(13), 2047–2058. https://doi.org/10.1029/WR022i013p02047

Ryan, J. N., & Gschwend, P. M. (1990). Colloid mobilization in two Atlantic Coastal Plain aquifers: Field studies. *Water Resources Research*, *26*(2), 307–322. https://doi.org/10.1029/WR026i002p00307

Ryan, P. J., Harleman, D. R. F., & Stolzenbach, K. D. (1974). Surface heat loss from cooling ponds. *Water Resources Research*, *10*(5), 930–938. https://doi:10.1029/wr010i005p00930

Saffman, P. G. (1959). A theory of dispersion in a porous medium. *Journal of Fluid Mechanics*, *6*(3), 321–349. https://doi.org/10.1017/S0022112059000672
Sawyer, A. H., & Cardenas, M. B. (2009). Hyporheic flow and residence time distributions in heterogeneous cross-bedded sediment. *Water Resources Research*, *45*(8). https://doi.org/10.1029/2008WR007632
Sawyer, A. H., Zhu, J., Currens, J. C., Atcher, C., & Binley, A. (2015). Time-lapse electrical resistivity imaging of solute transport in a karst conduit. *Hydrological Processes*, *29*(23), 4968–4976. https://doi.org/10.1002/hyp.10622
Slater, L., & Binley, A. (2021). Advancing hydrological process understanding from long-term resistivity monitoring systems. *Wiley Interdisciplinary Reviews: Water*, *8*(3), e1513. https://doi.org/10.1002/wat2.1513
Srivastava, R., & Yeh, T. C. J. (1992). A three-dimensional numerical model for water flow and transport of chemically reactive solute through porous media under variably saturated conditions. *Advances in Water Resources*, *15*(5), 275–287. https://doi.org/10.1016/0309-1708(92)90014-S
Stapleton, R. D., Sayler, G. S., & Boggs J. M. (2000). Changes in subsurface catabolic gene frequencies during natural attenuation of petroleum hydrocarbons. *Environmental Science and Technology 34*(10), 1991–1999.
Su, X., Shu, L., & Lu, C. (2018). Impact of a low-permeability lens on dune-induced hyporheic exchange. *Hydrological Sciences Journal*, *63*(5), 818–835. https://doi.org/10.1080/02626667.2018.1453611
Su, X., Shu, L., Chen, X., Lu, C., & Wen, Z. (2016). Interpreting the cross-sectional flow field in a river bank based on a genetic-algorithm two-dimensional heat-transport method (GA-VS2DH). *Hydrogeology Journal*, *24*(8), 2035–2047. https://doi.org/10.1007/s10040-016-1459-y
Su, X., Yeh, T. C. J., Shu, L., Li, K., Brusseau, M. L., Wang, W., ... & Lu, C. (2020). Scale issues and the effects of heterogeneity on the dune-induced hyporheic mixing. *Journal of Hydrology*, *590*, 125429. https://doi.org/10.1016/j.jhydrol.2020.125429
Sudicky, E. A. (1986). A natural gradient experiment on solute transport in a sand aquifer: Spatial variability of hydraulic conductivity and its role in the dispersion process. *Water Resources Research*, *22*(13), 2069–2082. https://doi.org/10.1029/WR022i013p02069
Sun, N.-Z. (1994). *Inverse Problems in Groundwater Modeling*, Kluwer Acad., Norwell, Mass.
Sun, R., T.-C. J. Yeh, D. Mao, M. Jin, W. Lu, & Y. Hao (2013). A temporal sampling strategy for hydraulic tomography analysis, *Water Resources Research*, *49*. https://doi.org/10.1002/wrcr.20337
Suzuki, M., & Smith, J. M. (1971). Kinetic studies by chromatography. *Chemical Engineering Science*, *26*(2), 221–235. https://doi.org/10.1016/0009-2509(71)80006-4
Taylor, G. I. (1921). Diffusion by continuous movements. *Proceedings of the London Mathematical Society*, *2*(1), 196–212. https://doi.org/10.1112/plms/s2-20.1.196
Taylor, G. I. (1953). Dispersion of soluble matter in solvent flowing slowly through a tube. *Proceedings of the Royal Society of London. Series A. Mathematical and Physical Sciences*, *219*(1137), 186–203. https://doi.org/10.1098/rspa.1953.0139
Tiedeman, C. R., & Barrash, W. (2020). Hydraulic tomography: 3D hydraulic conductivity, fracture network, and connectivity in mudstone. *Groundwater*, *58*(2), 238–257. https://doi.org/10.1111/gwat.12915

Turner, G. A. (1972). *Heat and Concentration Waves*, Academic, New York: Elsevier. University Pres, 343 pp.
Updegraff, C. D. (1977). *Parameter Estimation for a Lumped-parameter Gound-water Model of the Mesilla Valley*. New Mexico: New Mexico Water Resources Research Institute.
Valocchi, A. J. (1985). Validity of the local equilibrium assumption for modeling sorbing solute transport through homogeneous soils. *Water Resources Research*, *21*(6), 808–820. https://doi.org/10.1029/WR021i006p00808
Vasco, D. W., & Datta-Gupta, A. (1999). Asymptotic solutions for solute transport: A formalism for tracer tomography. *Water Resources Research*, *35*(1), 1–16. https://doi.org/10.1029/98WR02742
Vasco, D. W., Keers, H., & Karasaki, K. (2000). Estimation of reservoir properties using transient pressure data: An asymptotic approach. *Water Resources Research*, *36*(12), 3447–3465. https://doi.org/10.1029/2000WR900179
Vasco, D. W., Seongsik, Y., & Datta-Gupta, A. (1999). Integrating dynamic data into high-resolution reservoir models using streamline-based analytic sensitivity coefficients. *Spe Journal*, *4*(04), 389–399. https://doi.org/10.2118/59253-PA
Wang, Y.-L., Yeh, T.-C. J., Wen, J.-C., Gao, X., Zhang, Z., & Huang, S.-Y. (2019). Resolution and ergodicity issues of river stage tomography with different excitations. *Water Resources Research*, *55*. https://doi.org/10.1029/2018WR023204
Wang, Y.-L., Yeh, T.-C. J., Wen, J.-C., Huang, S.-Y., Zha, Y., Tsai, J.-P., et al. (2017). Characterizing subsurface hydraulic heterogeneity of alluvial fan using riverstage fluctuations. *Journal of Hydrology*, *547*, 650–663. https://doi.org/10.1016/j.jhydrol.2017.02.032
Williams, T. M., & McCarthy, J. F. (1991). Field scale tests of colloid-facilitated transport. *In National Research and Development Conference on Control of Hazardous Materials. Hazardous Mat. Control Inst Greenbelt, MD*, 179–184.
Winter, T., Harvey, J., Franke, O., & Alley, W. (1998). *Ground Water and Surface Water: A Single Resource*. US Geological Survey.
Woodbury, A. D., & Sudicky, E. A. (1991). The geostatistical characteristics of the Borden aquifer. *Water Resources Research*, *27*(4), 533–546. https://doi.org/10.1029/90WR02545
Xiang, J., Yeh, T. C. J., Lee, C. H., Hsu, K. C., & Wen, J. C. (2009). A simultaneous successive linear estimator and a guide for hydraulic tomography analysis. *Water Resources Research*, *45*(2). https://doi.org/10.1029/2008WR007180
Ye, M., Khaleel, R., & Yeh, T. C. J. (2005). Stochastic analysis of moisture plume dynamics of a field injection experiment. *Water Resources Research*, *41*(3). https://doi.org/10.1029/2004WR003735
Yeh, T. C. J., & Liu, S. (2000). Hydraulic tomography: Development of a new aquifer test method. *Water Resources Research*, *36*(8), 2095–2105. https://doi.org/10.1029/2000WR900114
Yeh, T. C. J., Gutjahr, A. L., & Jin, M. (1995). An iterative cokriging-like technique for ground-water flow modeling. *Groundwater*, *33*(1), 33–41. https://doi.org/10.1111/j.1745-6584.1995.tb00260.x
Yeh, T. C. J., Khaleel, R., & Carroll, K. C. (2015). *Flow Through Heterogeneous Geologic Media*. Cambridge University Press, 343 pp.
Yeh, T. C. J., Lee, C. H., Hsu, K. C., Illman, W. A., Barrash, W., Cai, X., . . . & Winter, C. L. (2008). A view toward the future of subsurface characterization: CAT scanning groundwater basins. *Water Resources Research*, *44*(3). https://doi.org/10.1029/2007WR006375

Yeh, T. C. J., Liu, S., Glass, R. J., Baker, K., Brainard, J. R., Alumbaugh, D., & LaBrecque, D. (2002). A geostatistically based inverse model for electrical resistivity surveys and its applications to vadose zone hydrology. *Water Resources Research, 38* (12), https://doi.org/10.1029/2001WR001204.

Yeh, T. C. J., Mas-Pla, J., Williams, T. M., & McCarthy, J. F. (1995). Observation and three-dimensional simulation of chloride plumes in a sandy aquifer under forced-gradient conditions. *Water Resources Research, 31*(9), 2141–2157. https://doi.org/10.1029/95WR01947

Yeh, T. C. J., Srivastava, R., Guzman, A., & Harter, T. (1993). A numerical model for water flow and chemical transport in variably saturated porous media. *Groundwater, 31*(4), 634–644. https://doi.org/10.1111/j.1745-6584.1993.tb00597.x

Yeh, T.-C. J., Khaleel, R., & Carroll, K. C. (2015). *Flow Through Heterogeneous Geologic Media*. Cambridge University Press. New York, USA.

Yeh, T.-C. J., Jin, M., & Hanna, S., (1996). An iterative stochastic inverse method: Conditional effective transmissivity and hydraulic head fields. *Water Resources Research, 32*(1), 85e92. http://dx.doi.org/10.1029/95WR0286

Yeh, T. C. J., Zhu, J., Englert, A., Guzman, A., & Flaherty, S. (2006). A successive linear estimator for electrical resistivity tomography. *Applied hydrogeophysics. (p45-p74). Springer, Dordrecht, the Netherland.*

Zha, Y., Yeh, T. C. J., Illman, W. A., Tanaka, T., Bruines, P., Onoe, H., ... & Wen, J. C. (2016). An application of hydraulic tomography to a large-scale fractured granite site, Mizunami, Japan. *Groundwater, 54*(6), 793–804. https://doi.org/10.1111/gwat.12421

Zhang, J., Mackie, R. L., & Madden, T. R. (1995). 3-D resistivity forward modeling and inversion using conjugate gradients. *Geophysics, 60*(5), 1313–1325. https://doi.org/10.1190/1.1443868

Zhang, M., & Zhang, Y. (2015). Multiscale solute transport upscaling for a three-dimensional hierarchical porous medium. *Water Resources Research, 51*(3), 1688–1709. https://doi:10.1002/2014WR016202

Zhang, Y., & Gable, C. W. (2008). Two-scale modeling of solute transport in an experimental stratigraphy. *Journal of Hydrology, 348*(3–4), 395–411. https://doi.org/10.1016/j.jhydrol.2007.10.017

Zhao, Z., & Illman, W. A. (2018). Three-dimensional imaging of aquifer and aquitard heterogeneity via transient hydraulic tomography at a highly heterogeneous field site. *Journal of Hydrology, 559*, 392–410. https://doi.org/10.1016/j.jhydrol.2018.02.024

Zheng, C., Bianchi, M., & Gorelick, S. M. (2011). Lessons learned from 25 years of research at the MADE site. *Groundwater, 49*(5), 649–662. https://doi.org/10.1111/j.1745-6584.2010.00753.x

Zhu, J., & Yeh, T. C. J. (2005). Characterization of aquifer heterogeneity using transient hydraulic tomography. *Water Resources Research, 41*(7). https://doi.org/10.1029/2004WR003790

Zsolnay, A. (1992). Effect of an organic fertilizer on the transport of the herbicide atrazine in soil. *Chemosphere, 24*(5), 663–669. https://doi.org/10.1016/0045-6535(92)90220-L

Zsolnay, A. (1993). The relationship between dissolved organic carbon and basal metabolism in soil. Mitt. d. *Österr. Bodenk. Ges, 47*, 83–95.

Index

Printed in the United States
by Baker & Taylor Publisher Services